W0107097

MODERN DAIRY TECHNOLOGY

Volume 1

Advances in Milk Processing
Second Edition

Edited by

R. K. Robinson

Department of Food Science and Technology,
University of Reading, UK

SPRINGER-SCIENCE+BUSINESS MEDIA, B.V.

First edition 1986

Second edition 1994

© 1994 Springer Science+Business Media Dordrecht
Originally published by Chapman & Hall in 1994
Softcover reprint of the hardcover 2nd edition 1994

ISBN 978-1-4613-5853-4   ISBN 978-1-4615-2057-3 (eBook)
DOI 10.1007/978-1-4615-2057-3

A catalogue record for this book is available from the British Library

Library of Congress Cataloging-in-Publication data available

Printed on acid-free text paper, manufactured in accordance with
ANSI/NISO Z39.48-1992 (Permanence of Paper).

MODERN DAIRY TECHNOLOGY

Volume 1

Advances in Milk Processing
Second Edition

Preface to the First Edition

The dairy industry is, in many countries, a major contributor to the manufacturing capacity of the food sector, and as more components of milk are utilised in processed foods, so this importance is likely to grow. Already dairy operations range from the straightforward handling of liquid milk through to the production of highly sophisticated consumer items, and it is of note that all this activity is based on a raw material that is readily perishable at ambient temperatures.

This competitive, commercial position, together with the fact that the general public has a high regard for dairy products, is an indication of the extent to which milk producers and processors have combined to ensure that retail products are both nutritious and hygienically acceptable. Achievement of these aims, and at reasonable cost, has depended in large measure on the advances that have been made in the handling of large volumes of milk. Thus, factories designed to handle millions of litres of milk per week are now commonplace, and it is the plant and equipment involved that provides the factual background for this two-volume book.

In some instances, the increased capacity has arisen simply from an expansion of a traditional method, but in others a totally new approach has had to be adopted either for the manufacturing process per se, or for utilisation of the end products. Success has also depended on the derivation of accurate process controls, both through automation, and through improved procedures for quality control. Together, these diverse facets make up the modern dairy industry, and it is to be hoped that these two volumes will do justice to the innovative genius of those who have been involved in its evolution.

R. K. ROBINSON

Preface to the Second Edition

Developments in process technology tend to be driven by the aims of improving the efficiency of an existing process, or increasing market share through the manufacture of a new product range. In either situation, the essentials of maintaining product quality and safety remain of paramount importance, and hence the tendency is for processes to evolve rather than emerge de novo.

Nevertheless, the changes so introduced can result in significant improvements in, for example, the economics of a particular plant or process, and few manufacturers can afford to ignore trends emerging elsewhere within their sector of the food/dairy industry. Similarly, students of dairy science and technology need to be aware of recent developments, hence the aim of this second edition is to highlight the way in which manufacturing procedures have been modified over the past ten years. Inevitably, some sections of the text have required little modification whereas others have needed extensive alteration but, whichever approach has been employed, this new edition seeks to present a totally up-to-date account of the manufacture of dairy products.

R. K. ROBINSON

Contents

Preface to the First Edition v

Preface to the Second Edition vii

1. Heat Treatment of Milk 1
   M. J. Lewis
2. Developments in Cream Separation and Processing . . . 61
   C. Towler
3. Production of Butter and Dairy Based Spreads 107
   R. A. Wilbey
4. Drying of Milk and Milk Products 159
   The late M. E. Knipschildt and G. G. Andersen
5. Protection against Fire and Explosion in Spray Dryers . . 255
   O. Skov
6. Membrane Processing of Milk 273
   A. S. Grandison and F. A. Glover
7. Utilisation of Milk Components: Whey 313
   J. G. Zadow
8. Utilisation of Milk Components: Casein 375
   C. R. Southward
9. Automation in the Dairy 433
   W. M. Kirkland

Index 473

List of Contributors

G. G. Andersen
APV Anhydro AS, 7 Østramarken, DK-2860 Søborg-Copenhagen, Denmark.

F. A. Glover
Formerly National Institute for Research in Dairying, Shinfield, Reading RG2 9AT. Present address: 39b St Peter's Avenue, Caversham, Reading, Berkshire, UK.

A. S. Grandison
Department of Food Science and Technology, University of Reading, Whiteknights, Reading RG6 2AP, UK.

W. M. Kirkland
APV Baker Ltd., PO Box 4, Gatwick Road, Crawley RH10 2QB, West Sussex, UK.

M. E. Knipschildt
APV Anhydro AS, 7 Østramarken, DK-2860 Søborg-Copenhagen, Denmark.

M. J. Lewis
Department of Food Science and Technology, University of Reading, Whiteknights, Reading RG6 2AP, UK.

O. Skov
APV Anhydro AS, 7 Østramarken, DK-2860 Søborg-Copenhagen, Denmark.

C. R. Southward
New Zealand Dairy Research Institute, Private Bag, Palmerston North, New Zealand.

C. Towler

New Zealand Dairy Research Institute, Private Bag, Palmerston North, New Zealand.

R. A. Wilbey

Department of Food Science and Technology, University of Reading, Whiteknights, Reading RG6 2AP, UK.

J. G. Zadow

Formerly Division of Food Research, CSIRO, Dairy Research Laboratory, PO Box 20, Highett, Victoria 3190, Australia. Present address: Z. G. Zadow & Associates, Mordialloc, Victoria, Australia

Chapter 1

Heat Treatment of Milk

M. J. Lewis

Department of Food Science and Technology, University of Reading, UK

In this chapter, milk will refer to bovine milk, either as full cream milk, skim-milk obtained by centrifugal separation, or standardised milk made by combining skim-milk with cream. Currently UK milk is not standardised, although both skim-milk and particularly semi-skim milk containing between 1 and 1·5 per cent fat have become more popular. Sales of semi-skim milk have increased from 4·74 million tonne in 1985 to almost 17.7 million tonne in 1990 (a 373 per cent increase). However, this increase has not occurred in other EC countries: France, 10 per cent; Germany, 5·4 per cent; and Netherlands 17·4 per cent. Other types of milk that require heat treatment are flavoured milk, reconstituted milk, filled milk, evaporated milk, milk modified in composition by demineralisation or lactose hydrolysis and protein-enriched milk produced by ultrafiltration. Milk from other species, such as goats, sheep and buffalo, may also be very important in some countries.

Milk can be regarded as a complete food, containing protein, fat, lactose, vitamins and minerals, together with natural enzymes and those derived from microorganisms within the milk. It has a high nutritional value, but is an excellent medium for microbial growth. Milk is extremely variable in its composition. There are variations between individual cows in a breed, between breeds and between seasons. Variations between species are also very considerable (Jenness, 1982; Walstra and Jenness, 1984).

Milk is heated for a variety of reasons. The main reasons are: to remove pathogenic organisms; to increase shelf-life up to a period of six months; to help subsequent processing, e.g. forewarming before

1

separation and homogenisation; or as an essential treatment before cheesemaking, yoghurt manufacture and the production of evaporated and dried milk. When milk is heated, many changes take place, Fox (1982) has summarised the changes that may lead to protein coagulation:

decrease in pH;
precipitation of calcium phosphate;
denaturation of whey proteins and interaction with casein;
Maillard browning;
modification of casein: dephosphorylation, hydrolysis of κ-casein and general hydrolysis;
changes in micellar structure: zeta potential, hydration changes; association-dissociation.

The casein fraction of milk is very heat stable, whereas the whey protein fraction is heat labile and almost completely denatured at 100 °C. The denatured whey protein complexes with the casein and does not usually precipitate. But when cheese whey is heated, the proteins start to denature, coagulate and precipitate between 75 and 80 °C. This illustrates the protective effect of casein toward coagulation. A solution of sodium caseinate can withstand heating at 140 °C for longer than 60 min at pH 6·7.

Other important changes take place during heating and subsequent storage. They affect the nutritional value and sensory characteristics. Broadly speaking, the two major types of heat treatment are pasteurisation and sterilisation. The major concerns about the resulting products are safety and quality. Heat-treated milk should not be a public health risk. It should have a good keeping quality; provide a good balance of nutrients; and be of desirable sensory characteristics, i.e appearance, colour, flavour and mouthfeel. When milk is heated at a constant temperature, all its constituents and components will be affected, but by different amounts. Increasing the temperature will accelerate reaction rates. But different reactions will be affected to different extents. Physical, chemical, enzymatic and microbial changes will depend principally upon the time–temperature conditions, but will also be influenced by other factors, such as composition, pH, oxygen content and fat globule size. The wide range of reactions taking place when milk is heated will influence the safety and quality of that milk. A brief review is provided of the main kinetic parameters used. The subject is considered in more detail by Kessler (1989) and van Boekel and Walstra (1989).

Effects of Heat on Micro-organisms

Experimentally, it is most often observed that the destruction of vegetative organisms and spores follows first order reaction kinetics. When the log of the population (N) is plotted against time, a straight line relationship results (at constant temperature).

$$-\frac{dN}{dt} = kN$$

therefore

$$-\frac{dN}{N} = k\,dt$$

If this is integrated between the final number (N_F) and the initial number (N_0)

$$\ln \frac{N_0}{N_F} = kt$$

or

$$2 \cdot 303 \log_{10} \frac{N_0}{N_F} = kt$$

where $\log_{10} \dfrac{N_0}{N_F}$ = number decimal reductions

k = reaction velocity constant (s^{-1}) or (min^{-1})
t = heating time (s) or (min).

An alternative way of expressing the heat resistance is using the decimal reduction time (D_T). This is defined as the time required for the population to be reduced by 90 per cent (see Fig. 1) or one log cycle. The number of decimal reductions is given by

$$\log_{10} \frac{N_0}{N_F} = \frac{\text{heating time}}{D_T} = \frac{t}{D_T}$$

combining these equations

$$D_T = \frac{2 \cdot 303}{k}$$

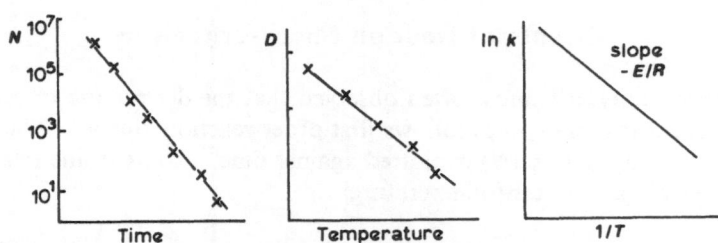

Fig. 1. Representation of kinetic data for the inactivation of microorganisms.

Thus for pasteurisation, decimal reduction times between 60 and 80 °C are important, whereas for sterilisation, values from 110 °C to upwards of 140 °C are relevant. In practice, several types of deviation from first order reaction kinetics may be observed. For a fuller discussion of these, together with the mechanisms for microbial inactivation, the reader is referred to the article by Gould (1989). Deviations are more likely to occur in the inactivation of vegetative organisms, than for spore damage.

Values of D_T are determined experimentally. They depend upon a number of factors, such as pH, media composition and water activity. Consequently different investigators often obtain different D_T values for the same sample. In pasteurisation, an achievement of 6 to 8 decimal reductions for the major pathogens is usually achieved, whereas for sterilisation of low-acid foods, 12 decimal reductions for *Clostridium botulinum* is regarded as a minimum requirement. Inactivation of enzymes can also be represented by a D_T value.

VARIATION OF HEAT RESISTANCE WITH TEMPERATURE

Most thermal processes do not take place at constant temperature. Therefore it is important to be able to evaluate the effects of the heating and cooling periods to the total lethality of the process. There are three ways of representing the variation of reaction rate with temperature.

Activation Energy (E)

When the log of reaction velocity constant (log k) is plotted against the reciprocal of the absolute temperature ($1/T$), a straight line relationship is often observed (Fig. 1). This is described by the Arrhenius equation

$$k = Ae^{-E/RT}$$

or

$$\ln k = \ln A - \frac{E}{RT}$$

where E = activation energy

A = collision frequency.

Z Value

Over a limited temperature range, a straight line relationship also results when the log of decimal reduction time ($\log D_T$) is plotted against temperature. This is used to define the Z value for that reaction. The Z value is the temperature difference which results in a ten-fold change in the D_T value (see Fig. 1). Burton (1983, 1988) reviews Z value data for heat-resistant spores found in milk, and concludes that $Z = 10\,°C$ reasonably describes the data for mesophilic and thermophilic spores. The equation relating D values at different temperatures is

$$\frac{\log D_1 - \log D_2}{T_1 - T_2} = -\frac{1}{Z}$$

Chemical reaction rates (low Z, high E) are much less sensitive to changes in temperature than reaction rates for the inactivation of vegetative organisms and spores. The relationship between E and Z (in °C) is given by

$$E = \frac{2 \cdot 303 R T T_1}{Z}$$

where E = activation energy (J mol^{-1})

T = temperature (K)

T_1 = reference temperature (K)

R = gas constant (J K^{-1} mol^{-1}).

A reference temperature is quoted because most experimental data are determined about a reference temperature, but the equation should only be used if the actual temperature is within one Z value of the reference temperature. Values of Z for spores tend to be higher than those for vegetative organisms.

Q_{10} Value

The Q_{10} value provides a quick check of how sensitive reaction rate is to temperature. It is defined as follows

$$Q_{10} = \frac{\text{Rate of reaction at } (T+10)}{\text{Rate of reaction at } T}$$

where T = temperature (°C)

The relationship between the Q value and the Z value is

$$Z = \frac{10}{\log_{10} Q}$$

For most chemical reactions, Q lies between 2 and 4, whereas values for the inactivation of microorganisms are usually between 10 and 30. Values for vegetative organisms tend to be higher than for spores. Kinetic parameters for a wide range of microbial and chemical reactions occuring in foods are summarised by Lund (1975) and Burton (1988). Some examples for reactions occuring in milk are given in Table I. Other references are provided by Von Boekel and Walstra (1989), Kessler (1989) and Reuter (1989).

In both pasteurisation and sterilisation, the most important parameter is the level of inactivation achieved. For processes where heating and cooling times are relatively short, the simplest procedure is to base the calculation on the holding time and temperature. The high temperature–short time (HTST) process and the ultra-high temperature (UHT) process have relatively short heating and cooling times. For safety reasons, the minimum holding time should be used (see residence times). However, this will give an underestimate of the true level of microbial inactivation. There are a number of procedures to evaluate the contribution of the heating and cooling periods to the total microbial inactivation. Similar procedures have also been used for assessing the amount of chemical damage taking place during sterilisation processes. One procedure relies on integrating the following type of expression.

$$2 \cdot 303 \log_{10} \frac{N_0}{N_F} = \int k \, dt = \int A e^{-E/RT} \, dt$$

If the relationship between temperature and time is known, the expression can be integrated directly. Otherwise it can be evaluated by plotting $A^{-E/RT}$ against time and determining the area under the curve.

TABLE I
Reaction kinetic parameters for some microorganisms, enzymes
and chemical reactions

	$D_{121}(s)$	$Z(°C)$	Q_{10}
B. stearothermophilus NCDO 1096, milk	181	9·43	11·5
B. stearothermophilus FS 1518, conc. milk	117	9·35	11·8
B. stearothermophilus FS 1518, milk	324	6·7	31·1
B. stearothermophilus NCDO 1096, milk	372	9·3	12
B. subtilis 786, milk	20	6·66	31·7
B. coagulans 604, milk	60	5·98	57·6
B. cereus, milk	3·8	35·9	1·9
Cl. sporogenes PA 3679, conc. milk	43	11·3	7·7
Cl. botulinum NCTC 7272	3·2	36·1	1·89
Cl. botulinum (canning data)	13	10·0	10
Proteases inactivation	0·5 to 27 min at 150 °C	32·5 to 28·5	2·0 to 2·3
Lipases inactivation	0·5 to 1·7 min at 150 °C	42 to 25	1·7 to 2·5
Browning		28·2, 21·3	2·26, 2·95
Total whey protein denaturation, 130–150°C		30	2·2
Available lysine		30·1	2·15
Thiamin (B1) loss		31·4 to 29·4	2·08 to 2·19
Lactulose formation		27·7 to 21·0	2·30 to 3·00

Taken from Burton (1988).

An alternative approach is to select a reference temperature and an appropriate Z value, then use the appropriate equations to convert the temperature–time profile of the process to an equivalent processing time at the reference temperature; the following reference temperatures are used.

pasteurisation 72 °C $Z = $ 8 deg (C)
cooking value 100 °C $= $ 20 to 40

F_0 values, sterilisation 121·1 °C   = 10
UHT processing, B^* and C^* 135 °C   = 10·5 (B^*), 31·4 (C^*)

The equation describing two sets of conditions along the thermal death line is

$$\frac{\log t_1 - \log t_2}{\theta_1 - \theta_2} = -\frac{1}{Z}$$

or

$$\log\left(\frac{t_1}{t_2}\right) = \frac{\theta_2 - \theta_1}{Z}$$

The lethality (L) at any experimental temperature (θ_2) is defined as the number of minutes at the reference temperature (θ_1) that would have the same microbial effect as 1 min at the experimental temperature, i.e. $t_1 = L$ and $t_2 = 1$. Therefore

$$\log L = \frac{\theta_2 - \theta_1}{Z}$$

The temperature–time heating profile can be converted to a lethality–time profile. The area under the curve will give the total lethality for the process, expressed as an equivalent time at the reference temperature. Specific examples will be given for pasteurisation and sterilisation later in the text.

The Effects of Raw Milk Quality on Heat Stability

It is well known that good quality milk can withstand high processing temperatures without coagulating. A typical milk is stable for several hours at 100 °C and for about 20 min at 140 °C. The methods for assessing heat stability have been reviewed by Fox (1982) and Walstra and Jenness (1984). The latter have pointed out that never has a process been so extensively researched still to remain so poorly understood. Conditions for coagulating good quality milk are much more stringent than those used for processing most milk. Coagulation rarely occurs, but it is well known that heat exchangers are susceptible to the deposition of milk solids and associated fouling problems. It must be concluded, therefore, that some of the milk being processed is not of the highest quality. The next section reviews those properties of raw milk pertinent to its behaviour during heating. It should also be noted that there are considerable differences in heat stability between milk from different animals. Furthermore, the microbial quality of raw milk has a pronounced effect on the quality of the final product (ICMSF, 1988).

RAW MILK QUALITY

Raw milk from healthy cows (sampled from the bulk tank and handled hygienically) should give rise to a low total colony count, and be free of serious pathogenic organisms. Bramley and McKinnon (1990) provide a comprehensive review on the microbiology of raw milk. They state that initial counts range between less than 10^3 ml^{-1}, indicating minimal contamination, to greater than 10^6 ml^{-1}. Counts of greater than 10^5 ml^{-1} are indicative of serious faults in production hygiene, whereas consistent counts less than 10^4 ml^{-1} reflect good hygiene.

Contamination comes from only three possible sources: from within the udder, from the exterior of the teats and udder, and from milking and storage equipment. Tuberculosis and brucellosis are the two most serious human diseases spread by the consumption of contaminated raw milk. They are most serious because the microorganisms can be excreted from infected animals. Udder disease (mastitis) may result in the presence of *Staphylococcus* species, *E. coli* and other pathogens. During storage some species of *Staphylococcus* produce a heat-resistant enterotoxin, not always completely inactivated by processing. Very occasionally other more pathogenic organisms may affect the udder. *Salmonella* and *Campylobacter* species may also be present from faecal contamination of raw milk after secretion. Occasionally human carriers have been implicated.

Pasteurisation inactivates all of these pathogens, but not the spore-forming organisms, *Clostridium perfringens* and *Bacillus cereus*. Adams *et al.* (1976) showed that growth of Gram-negative psychrotrophs in raw milk led to detectable proteolysis, particularly of κ- and β-casein, even after two days, and it is now well established that several species produce highly heat-resistant enzymes, particularly proteases and lipases. These proteases may survive thermal processing and cause further proteolysis in the stored product, leading eventually to gelation.

If the psychrotrophic count is above 10^6 ml^{-1}, significant levels of the proteases may survive (Van den Berg, 1981). Law *et al.* (1977) found that milk containing more than 8×10^6 ml^{-1} of *Pseudomonas fluorescens* AR11, subjected to UHT treatment, gelled between 10 and 12 days after production. It is useful to monitor psychrotrophic organisms in raw milk, as they are usually predominant among the microorganisms found. Several methods for controlling psychrotrophs have been described by Muir (1990). These include deep cooling to 2 °C, inhibition based on the natural milk enzyme peroxidase, injection with carbon dioxide and thermisation. Evidence of their presence will only become apparent if milk samples stored at 4 °C and examined microscopically for counts at one or two successive intervals of 24 h, show a rapid increase in count. Raw milk also contains acid-producing streptococci and bacilli, and excessive activity will result in a fall in pH, and a decrease in the heat stability of the milk.

Badings and Nester (1978) suggest that fresh milk (raw or low pasteurised) has a bland but characteristic flavour. Many compounds, mostly in sub-threshold concentrations, contribute to its aroma. Off-flavours may come from animal feeds or arise from weed taints. They are also caused by oxidation and light-induced reactions (Badings, 1984). Advocates for raw milk consumption claim that even a mild form of heat treatment, such as pasteurisation, significantly changes the delicate characteristics of its flavour. Raw (untreated) milk sold in the UK must be labelled as 'raw unpasteurised milk'. It must also be labelled that it has not been heat treated and may therefore contain organisms harmful to health. In some countries, there are quality payment schemes based upon total bacterial counts in raw milk, and there is evidence that these schemes are improving the microbial quality of raw milk delivered for heat treatment.

In situations where it is known that raw milk may be held under chilled conditions for some time before use, it may be treated by *thermisation*. In the UK, there is no legal definition of thermisation. Muir (1990) states that 65 to 70 °C for 15 s followed by cooling to below 6 °C would be typical. It is usually followed by a more severe heat treatment.

PASTEURISATION

The following definition of pasteurisation has been adopted by the International Dairy Federation (IDF). 'Pasteurisation is a process applied to a product with the object of minimising possible health hazards arising from pathogenic microorganisms associated with milk by heat treatment, which is consistent with minimal chemical, physical and organoleptic changes in the product.'

Processing Conditions

The UK Milk Special Designation Regulations (1989) state that the following time–temperature combinations should be used for pasteurised milk. It should be:

(a) retained at a temperature of not less than 62·8 °C and not more than 65·6 °C for at least 30 min, and be immediately cooled to a temperature of not more than 10 °C;

(b) retained at a temperature of not less than 71·7 °C (and not more than 78·1 °C in Scotland) for at least 15 s and be immediately cooled to a temperature of not more than 10 °C (or 6 °C in Scotland); or

(c) retained at such a temperature for such a period as may be specified by the licensing authority, with the approval of the minister.

A number of other points are detailed. The milk should not be subject to atmospheric contamination. If the processing temperature is greater than 65·6 °C, there should be provision for diverting milk which is under-processed. Suitable indicating and recording thermometers should be installed, and temperature records should be kept for at least one month. The pasteurised milk should also satisfy the phosphatase test, and the milk should be properly sealed and labelled. The regulations for pasteurising semi-skim and skim-milk (SI 1973/1064) are similar, but these milks must satisfy a methylene blue test. The hygiene and heat treatment regulations (SI 1983/1508) for milk-based drinks were introduced in 1983.

A milk-based drink is defined as a liquid drink (other than a fermented milk) comprising a minimum of 85 per cent milk and other permitted ingredients, a list being supplied. The pasteurisation regulations are the same as for milk, although a coliform test may be necessary if the drink has a colour which interferes with the phosphatase test. The milk regulations, applying at the end of 1986, are summarised by Jukes (1987).

It is interesting to note that the UK regulations permit the use of other time–temperature combinations, and the processing times and temperatures can be plotted on semilog graph paper. The plot can be used to give an indication of the times required at temperatures between 62·8 °C and 71·7 °C, or extrapolated to give processing times at temperatures greater than 71·7 °C. Pasteurisation requiring very short times is known as 'flash pasteurisation'. For example, at 77 °C the extrapolated time would be about one second. All these interpolated time–temperature combinations would be sufficient to inactivate *Mycobacterium tuberculosis* (Society of Dairy Technology, 1983). If data for the disappearance of the cream line are considered, these pasteurisation conditions will not lead to a loss of cream line. This is very important with bottled milk, but less so for milk in nontransparent packaging.

More recently, microbiological standards have been introduced for pasteurised milk. A sample will satisfy the plate count test, if the plate count is below 30,000 per ml, and the coliform test if the coliform count is less than 1 per ml. In 1990, pasteurised milk accounted for about 92 per cent of liquid milk consumed in England and Wales (UK Dairy Facts and Figures, 1991). Statistics for the sales of liquid milk over the period 1980 to 1990, classified according to their heat treatment, for the EC countries, are shown in Table II.

Pasteurisation conditions are fairly uniform worldwide, although slight variations may occur in time–temperature combinations, cooling temperatures and testing procedures. In the USA, the following time–temperature combinations apply to Grade A milk (Busse, 1981).

63 °C	30 min	94 °C	0·1 s
77 °C	15 s	96 °C	0·05 s
89 °C	1 s	100 °C	0·01 s
90 °C	0·5 s		

It is interesting to note that such short residence times were used, because there are difficulties in accurately controlling such short times. This milk will have a shelf-life of 18 days or longer at a maximum of 7 °C. The statutory regulations relating to pasteurised milk in selected countries is reviewed by Staal (1986). In Central Europe, most dairies use 74 °C for between 30 and 40 s, but such harsh conditions are likely to result in a cooked flavour. There is little evidence that harsher processing conditions will increase the shelf-life, as this is more likely to be influenced by storage temperature and the extent of post-processing contamination (Busse, 1981). In fact, in the absence of post-processing contamination, Kessler and Horak (1984) found that more stringent conditions, i.e. 85 °C/15 s and 40 s and 78 °C/40 s reduced the shelf-life slightly. This loss was attributed to the germination of spores stimulated by the harsher processing conditions.

Pasteurisation is a relatively mild form of heat treatment, so most consumers would probably find it difficult to distinguish between raw and pasteurised milks. Pasteurised milk has no readily apparent cooked flavour; no active sulphydryl groups are found. Whey protein denatura-

TABLE II
Sales of liquid milk by type of treatment[a], thousand tonnes

Country	Year	Pasteurised	Sterilised	UHT	Total
Germany[b]	1980	1832	43	1471	3,346
	1990	1923	24	2147	4,094
France[b]	1980	1416	540	1454	3,427[c]
	1990	692	372	3150	4,217[c]
Italy[b]	1981	1392	978	590	3,022[d]
	1987	1415	409	894	2,718[d]
Netherlands	1980	777	–	185 –	962
	1990	841	95	–	936
Belgium[b]	1980	34	521	157	712
	1990	31	383	362	832[e]
Luxembourg[b]	1980	19	2	4	25
	1989	17	...	8	25
United Kingdom[f]	1980/81	5795	413	46	6,254
	1990	4828	241	173	5,242
Irish Republic	1980	465	–	–	465
	1990	507	–	–	507
Denmark	1980	551	...	...	551[g]
	1990	510	...	...	510[g]
Greece	1980	200	–	20[h]	220
	1987	247	–	22	269
Spain	1990	323	582	1525	2,430

a Whole, semi-skimmed and skimmed milk sold by dairies to the domestic market.
b Data relate to the manufacture of liquid milk for consumption.
c Including untreated milk: 1980 – about 17,000 tonnes; 1990 – 2530 tonnes and other unspecified.
d Excluding skimmed milk. Including untreated milk: 1981 – 62,000 tonnes.
e Including 56,000 tonnes of unspecified milk.
f England and Wales: sales to households.
g Including a small volume of sterilised and UHT.
h Mainly imported milk.

Taken from EC Dairy Facts and Figures (1991) (with permission).

tion is low (between 5 and 15 per cent) and there is relatively little loss of the heat-sensitive nutrients. The rennet coagulation properties are not affected. Aboshama and Hansen (1977) observed a slight loss of amino acids, approximately 6 per cent. Scott *et al.* (1984a, b) have surveyed the vitamin contents of milk as delivered to dairies, bulk pasteurised milk, and packaged milk as delivered to the home and after storage in the refrigerator. Pasteurisation was found to have hardly any effect on vitamins, but considerable changes occurred during subsequent transportation and storage. For example, an average value for vitamin C in pasteurised milk was 194 μg ml^{-1}, whereas the range for bottles sampled as delivered was between 0·1 and 18·4 μg ml^{-1}. Some of these reactions may well be more seriously affected if the temperature and residence times are not properly controlled.

Alkaline phosphatase has been used for many years as an index for adequate pasteurisation of milk. This is because conditions required to denature the enzyme are slightly more severe than those required for inactivation of *Mycobacterium tuberculosis*. Analytical equipment is now available to allow its measurement with much more precision. Griffiths (1986) has undertaken a detailed review and further experimental studies on the inactivation of some of the other minor enzymes found in milk. Griffiths studied enzymes such as acid phosphatase, amylases, catalase, lactoperoxidase, lysozyme, ribonuclease and xanthine oxidase, at temperatures used for thermisation and pasteurisation, with a view to using them as indices of adequate heat treatment or even overheat treatment. Lactoperoxidase was virtually completely inactivated by HTST conditions of 78 °C for 15 s, suggesting that it would be a useful indicator of overheat treatment.

Andrews *et al.* (1987) have performed a similar exercise, also covering heat treatments slightly more severe than pasteurisation conditions, on a wider range of enzymes. They have suggested that α-fucosidase would be useful for studying temperatures between 55 and 65 °C; *N*-acetyl-β-glucosaminidase, between 65 and 75 °C; γ-glutamyl transpeptidase, between 70 and 80 °C, and either α-mannosidase or xanthine oxidase, between 80 and 90 °C. Enzymes more resistant to heat treatment are not normally a problem with pasteurised milk for consumption, but could cause problems during the storage of cheese or milk powders made from such milk (Muir, 1990).

Excessive agitation and turbulence may cause damage to the fat globule membranes and stimulate the naturally occurring lipoprotein lipase in raw milk, so causing hydrolysis and a slight impairment of the flavour (Solberg, 1981). Homogenisation downstream of the holding-tube usually eliminates any problems of lipase activity. Homogenisation will also give it a creamier mouthfeel. One of the more recent faults with pasteurised milk is 'sweet curdling' caused by bacterial proteinases, mainly due to *Bacillus cereus*; this is best controlled by close supervision of cleaning and disinfection procedures, and the use of nitric acid which has been found to be the best material for eliminating these spores. There is evidence that these spore counts are seasonal.

Post-processing contamination (PPC) has a decisive influence on the keeping quality of pasteurised milk. It may arise from balance tanks, pipelines, filling and capping machines, and containers downstream of the pasteuriser, and may reduce its keeping quality from two weeks down to only a few days (Ashton and Romney, 1981). Schroeder *et al.* (1982) and Schroeder (1984) have reviewed the effects of psychrotrophic PPC on the keeping quality of milk, and the origins and levels of PPC in commercial dairies in the UK.

Pasteurised milk should be stored at as low a temperature as possible, and kept in the cold chain at all points during its distribution and sale. Under these conditions, psychrotrophic Gram-negative bacteria proliferate and control shelf-life, but as products are stored at warmer temperatures, thermoduric organisms begin to assume dominance. Solberg (1981) makes the novel suggestion of incubating raw milk with certain lactic starter cultures with the aim of lowering the oxidation–reduction potential and inhibiting the growth of psychrotrophic organisms. These starters would be inactivated during subsequent pasteurisation.

The predominant, surviving thermodurics are *Streptococcus salivarius* sub-species, *thermophilus*, certain micrococci, *Arthrobacter* and *Microbacterium*. Spores of bacilli and clostridia are not inactivated by pasteurisation, but clostridia will not develop in milk due to its high oxygen content. Aerobic spore-formers will develop, and may well cause spoilage even in refrigerated milk. Spoilage by thermodurics is most likely to occur when the temperature of the milk rises. There is evidence that some emetic strains of *Bacillus* species occasionally isolated from raw milk, may be a public health risk (Muir, 1990).

Microorganisms arising from post-pasteurisation contamination may initially be present in small numbers. Ashton and Romney (1981) suggest that bacterial counts on the product, together with a repasteurised sample of the product, be undertaken, or that use be made of enrichment techniques for psychrotrophic organisms. Close scrutiny of cleaning procedures will help to reduce PPC. Solberg (1981) suggests that long runs on pasteurisers may give rise to serious bacteriological problems. To avoid long runs, plant should be efficiently cleaned every eight hours. PPC can also be avoided by HTST pasteurisation followed by hot filling into plastic or laminate cartons between 70 and 80 °C in an aseptic atmosphere of protective gas, then fairly rapid cooling to 50 °C followed by slower cooling to either ambient or refrigerated temperature. Many of these chemical, biochemical and micobial aspects have been reviewed in more detail by the IDF (1986).

Batch and Continuous Processing

The batch heating process (Holder process) involves heating the milk to a temperature between 62·8 and 65·6 °C, holding it at that temperature for 30 min, then rapidly cooling it to below 10 °C. It is generally favoured by processors with low throughputs. The equipment is relatively cheap and simple, and can be readily adapted for different volumes and other fluids. However, it is more labour intensive than HTST processes, and its energy costs are higher as it is not easy to incorporate regeneration.

The heating time in a jacketted, heating vessel is given by

$$t = \frac{mc}{AU} \ln \frac{\theta_h - \theta_l}{\theta_h - \theta_F}$$

where t = heating time (s)
$\quad\quad m$ = mass (kg)
$\quad\quad c$ = specific heat (J kg^{-1} K^{-1})
$\quad\quad \theta$ = temperature
$\quad\quad A$ = heat transfer area (m^2)
$\quad\quad U$ = overall heat transfer coefficient (W m^{-2} K^{-1}).

Subscripts h, I and F refer to the heating medium, the initial temperature of the milk and its final temperature respectively. Therefore, the time required to heat 3000 kg of milk from 4 °C to 65 °C using steam (100 °C) in a vessel ($A = 10$ m^2; $U = 730$ W m^{-2} K^{-1}) is given by

$$t = \frac{3000}{10} \times \frac{3900}{730} \ln \frac{100 - 4}{100 - 65}$$

and equals approximately 27·0 min (note the heat transfer coefficient value is taken from Kessler (1981)). Further details about batch pasteurisers are given by the Society of Dairy Technology (1983) and IDF (1986). For higher throughputs, the high temperature–short time continuous process (HTST) is favoured. HTST pasteurisers are available in a range of sizes from 400 to 50,000 l h^{-1}. They can be extremely efficient in terms of energy utilisation by incorporating a regeneration section into the heat exchanger. The layout of a typical HTST pasteuriser is shown schematically in Fig. 2.

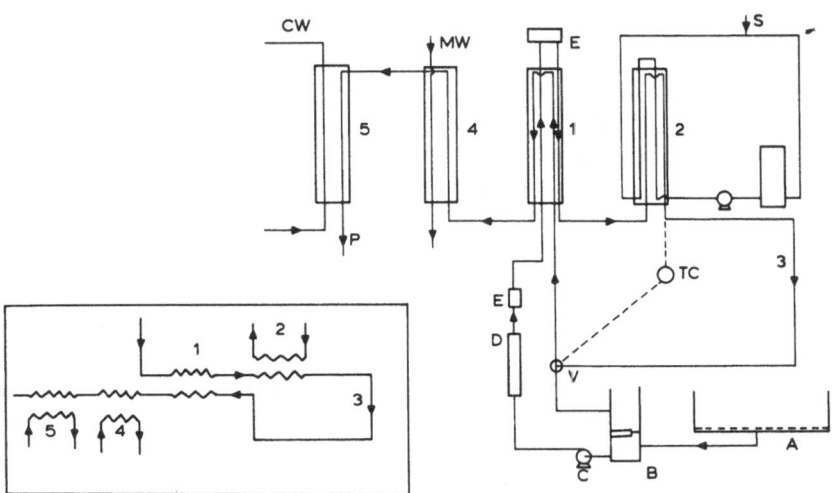

Fig. 2. Layout of HTST pasteuriser (the insert shows a schematic diagram of the heat exchange sections). (A) feed tank (B) balance tank (C) feed pump (D) flow controller (E) filter (P) product (S) steam injection (hot water section) (V) flow diversion valve (MW) mains water cooling (CW) chilled water (TC) temperature controller (1) regeneration (2) hot water section (3) holding tube (4) mains cooling water (5) chilled water cooling.

Raw milk enters the plant through the regeneration section, where it is heated by the hot milk, which in turn cools. After regeneration, it enters a final heating section, where hot water or electricity is used to bring the milk up to the desired temperature. Next, it passes through a holding tube into the regeneration section, then into the cooling sections cooled by mains water and chilled water. The regeneration efficiency (RE) is defined as

$$RE = \frac{\text{Amount of heat supplied by regeneration}}{\text{Total heat load, assuming no regeneration}} \times 100$$

$$= \frac{mc\,(\theta_2 - \theta_1)}{mc\,(\theta_3 - \theta_1)} \times 100$$

where m = mass flow rate $(kg\ s^{-1})$
        c = specific heat
        θ_1 = inlet temperature (°C)
        θ_2 = temperature leaving regenerator
        θ_3 = pasteurisation temperature.

Manufacturers are now claiming regeneration efficiencies in excess of 90 per cent. Therefore, if milk enters at 10°C, it will leave the regeneration section at 68·6 °C. If it is assumed that there is no heat loss over the regeneration section, then

$$(\theta_2 - \theta_1) = (\theta_4 - \theta_3)$$

i.e. the cooled milk will leave the regeneration section at a temperature of 13·1 °C. In this case, the mains water cooling section would be dispensed with, as it is highly unlikely that it would be below this temperature. The product is then cooled from 13·1 °C, to below 10 °C in the chilled water (refrigerated) section. Consequently, regeneration saves energy in terms of both heating and refrigeration costs. Regeneration efficiencies are increased by the use of larger heat exchanger surface areas. But capital costs are higher and the fluid has a longer residence time within the heat exchanger. This latter point is also extremely important in connection with the performance of UHT plants.

Following the layout of Fig. 2, the milk is fed from the main feed tank (A), which contains a coarse filter, to the balance tank (B), which keeps a constant head. It is then pumped by either a centrifugal or positive displacement pump through a flow controller (D) and additional filter (E), into the regeneration section. A fine cloth filter may also be included in the regeneration section. From regeneration the milk goes to the heating section, where energy is supplied by a hot water 'set'. Steam is injected into circulating hot water in a closed system until the temperature reaches the set-point temperature. Steam consumption can be monitored by measuring the amount of steam which condenses. The hot water, set-point temperature is fixed, so that the milk leaves this section at the correct pasteurisation temperature. The milk then enters the holding section, which is normally tubular, and designed to ensure that all the milk resides in the tube for at least 15 s. A temperature probe is located in the holding tube, and is attached to a recorder and a flow diversion valve. If the temperature falls below the desired temperature, the flow diversion valve will operate and direct the under-processed milk back to the balance tank. Under normal processing conditions, the milk will pass into the regeneration section, then through the mains water and chilled water sections before being stored in tanks or packaged into bottles and cartons.

Most HTST pasteurisers are of the plate heat exchanger type, described in more detail by Scott (1970), Kessler (1981) and Society of Dairy Technology (1983). They offer the advantage of providing a large surface area for heat transfer in a compact space. The gaps between the plates are narrow, typically between 3 and 4 mm, to induce turbulence. This turbulence and the pressure drop are borne in mind when the plates are designed. The system is extremely suitable for fluids of low viscosity which are reasonably sensitive to heat, but it cannot cope with particulate matter or fluids of high viscosity, as pressure drops would be too high. It is also unsuitable for liquids which will cause substantial fouling, but milk is not normally susceptible to this problem under pasteurisation conditions. Although steam and hot water are the main heating fluids, there are some designs which rely upon electrical heating, with the heating wire either wrapped round the outside of quartz tubes, through which the milk flows, or with the heating element inside the tube in direct contact with the milk. Most milk is still pasteurised in large dairies, but the past few years have seen a large increase in the number of producer–processors who pasteurise their milk on the farm. Several manufacturers now produce HTST pasteurisers specifically for these locations, most of which are plate heat exchangers.

The diversion valve is normally positioned at the end of the holding tube. There is an interesting debate over the optimum position for the temperature probe that controls diversion. If the temperature probe is positioned at the beginning of the holding tube, it is possible to detect under-processed milk quickly enough to be able to operate the flow diversion valve before the milk reaches the end of the tube. This is possible even when the response time is fairly slow. However, the probe would not indicate the extent of heat loss as the milk passed through the tube, hence it would not indicate the fall in temperature. A temperature probe positioned at the end of the holding tube does indicate the fall in temperature, but it needs a fast response time to prevent under-processed milk going forward. Most holding tubes are not thermally lagged, and there is little information presented in the literature about temperature drops in holding tubes. If temperature drops are high, then the location of the temperature probe is very important.

Hasting (1992) provides a review of some practical considerations in the design, operation and control of food pasteurisation processes. Two microbial parameters that are used are the pasteurisation unit (PU) and the p* value. With the PU, a reference temperature appropriate to the particular process is selected, for example 63 °C, together with an appropriate Z value. The process can be expressed as a PU value. PU = 30 min is equivalent to 30 min at 63 °C. The p* concept uses 72 °C as the reference temperature and Z = 8 °C. A process having p* = 1 corresponds to 72 °C for 15 s. Both these procedures allow the contribution of the heating and cooling periods to be evaluated.

Residence Times

It has long been recognised there is a distribution of residence times when liquids flow through pipes. This distribution can be measured and analysed by a variety of methods, described in more detail by Levenspiel (1972). One common method, easily applied, involves the injection of a pulse of some suitable tracer material into the tube inlet at time zero, followed by sampling and analysis of the outlet stream at regular time intervals. Figure 3 shows the traces obtained for (a) plug flow, (b) streamline flow and (c) turbulent flow.

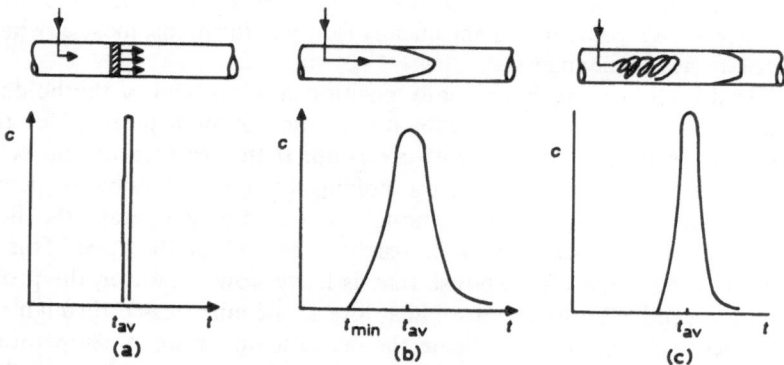

Fig. 3. Residence time distributions for (a) plug flow, (b) streamline flow, (c) turbulent flow.

Plug flow is an idealised situation where there is no velocity profile across the tube. The fluid passes through the tube in a similar fashion to a piston in a cylinder, consequently there is no spread of residence times. However, due to internal friction and/or viscosity of the fluid, there is a velocity distribution across the tube, the nature of which depends upon whether the flow is streamline or turbulent. The type of flow can be determined by evaluating a dimensionless constant known as the Reynolds number (Re), where

$$Re = \frac{V_{av}D\rho}{\mu}$$

or

$$Re = \frac{4Q\rho}{\pi\mu D}$$

where V_{av} = average velocity (m s^{-1})

D = tube diameter (m)

ρ = fluid density (kg m^{-3})

μ = dynamic viscosity (Nsm^{-2})

Q = volumetric flow rate (m^3 s^{-1}).

If the Reynolds number is less than 2000, the flow is streamline, if it is greater than 4100, the flow is turbulent. With streamline flow, there is a parabolic velocity profile across the tube with no bulk movement across the tube. The maximum velocity occurs at the centre of the tube, and is approximately twice the average velocity, where

$$V_{av} = \frac{\text{volumetric flow rate}}{\text{cross-sectional area of tube}} = \frac{4Q}{\pi D^2}$$

and

$$V_{max} = 2V_{av}$$

The average residence time (t_{av}) is based on the average velocity, and is equal to L/V_{av}. It can also be shown that

$$t_{av} = \frac{\text{volume of tube}}{\text{volumetric flow rate}}$$

When a pulse of tracer is injected, the averge residence time is the time for 50 per cent of the material in the pulse to pass through the tube. The average residence time is determined from the volumetric flow rate and the dimensions of the holding tube, both of which are easily measured. However, the minimum residence time, which is based on the maximum velocity, will equal $t_{min} = 0.5t_{av}$. Therefore, a holding tube designed to give an average residence time of 20 s would give a minimum time of 10 s under these flow conditions, and result in some of the milk being under-processed. Thus, the design of the holding tube should be based upon the minimum residence time when eradication of microorganisms is the main criterion.

When the flow is turbulent, the velocity profile is much flatter, and there is good bulk mixing across the tube. In this case

$$V_{max} = 1.2 V_{av}$$

$$t_{min} = 0.83 t_{av}$$

The distribution of residence times is also much narrower, thereby reducing the chance of over-processing. Therefore the same holding tube will give a minimum residence time of 16·6 s. The term 'holder efficiency' is often used, which is the ratio of the minimum to the average time. It is recommended that it is at least 0·8, which would correspond to turbulent flow conditions within the tube. Fouling of the tube, which reduces the diameter, will also reduce the residence times. Typical results from a $9001 h^{-1}$ HTST pasteuriser, using an injection technique with water, gives the following data

$$Re = 13,000$$

$$t_{min} = 16 \text{ s}$$

t_{av} (evaluated from holding tube size and volumetric flow rate) = 21·5 s. The distribution shown in Fig. 3 is not quite symmetrical, but there is close agreement between the average residence time (evaluated) and the time corresponding to maximum concentration of tracer in the outlet stream. However, as the distribution is asymmetrical, a long tail often results and, although the area under this tail is usually small, a small portion of fluid will have a residence time significantly longer than the average residence time. As the flow in the tube becomes more turbulent, the residence time distribution tends to be reduced. In practice, this can be achieved by making holding tubes longer and thinner, provided the pressure drops are not excessive.

Dickerson et al. (1968) describe a technique for measuring the minimum residence time in holding tubes of HTST pasteurisers. Using identical flow rates, minimum residence times for raw milk (15·8 s) and chocolate milk (15·5 s) agreed fairly closely with those for water (16·6 s). However, for more viscous products, such as condensed skim-milk (40 per cent total solids), ice cream mix (10 per cent milkfat) and cream (40 per cent fat), the discrepancies were larger. The size of the discrepancy increased as the viscosity increased, attributed to the development of streamline flow with its more parabolic velocity profile and higher maximum velocity. It may be necessary to lengthen holding tubes when processing more viscous materials to ensure that all the material is sufficiently processed; residence time distribution data become difficult to obtain at short residence times. Examples specific to UHT processing will be discussed later.

A bacteriologically effective residence time can be evaluated from the residence time distribution function, as described by Heppell (1986), which is an extension of the work of Burton *et al.* (1977). It is defined as the effective heating time that would give the same sterilisation effect found in the process. The proportion of survivors ($\log N_F/N_0$) can be calculated from a knowledge of the residence time distribution and the D_T value for the spores in question, assuming isothermal temperature conditions in the holding tube. The bacteriologically effective residence time falls between the average and the minimum residence times. Therefore, sterilisation efficiencies based on average residence times, will give an over-estimate of the lethal effect (Brown and Ayres, 1982). More detailed analyses of residence time distribution functions are given by Levenspiel (1972), Loncin and Merson (1979) and Kessler (1981).

STERILISED AND UHT MILKS

Pasteurisation conditions are not sufficient to inactivate the thermo-resistant spores in milk, but elimination of these spores would give a milk with much better keeping qualities, and one whose shelf-life would be limited by non-microbial spoilage reactions. Milk subjected to temperatures in excess of 100 °C and packaged in airtight containers goes under the general name of 'sterilised milk' or UHT (ultra-high temperature) milk. Packaging can take place either before or after heat treatment, but aseptic packaging is essential for milk packaged after heat treatment. Milk processed in this way is termed 'commercially sterile', i.e. it is not necessarily free of microorganisms, but those which survive the heat treatment are unlikely to proliferate during storage and cause spoilage of the product.

It is common practice in the heat treatment of low-acid products (pH > 4·5) to achieve at least 12 decimal reductions (10^{12} to 1) for the spores of *Clostridium botulinum*, because they are the most heat resistant of the major food poisoning organisms. Fortunately, they are hardly ever found in milk, although recently it was the cause of a very serious food poisoning outbreak in hazelnut yoghurt. The problem was eventually traced to insufficiently processed hazelnut puree used to flavour the yoghurt. However, spores which are more heat resistant than *Clostridium botulinum* may be present in milk, and are likely to cause spoilage over a long storage period (See Table I); the most heat resistant of these being *Bacillus stearothermophilus*. Many sterilisation processes for dairy products aim to achieve two decimal reductions for this spore, which requires 6 to 8 minutes heating at 121 °C. Most sterilisation procedures do not take place at constant temperature, and in many cases, the heating and cooling periods may make a substantial contribution to the total lethal effect.

A popular approach for in-container sterilisation processes has been to adopt a total integrated lethal effect value, known as the F_0 value, for the process. (Reference temperature = 121·1 °C, $z = 10$ °C). Values of L are tabulated from this equation, and summarised in standard lethality tables.

$$\log L = (\theta - 121\cdot 1)/Z \text{ or } L = 10^{(\theta - 121\cdot 1)/Z}$$

where θ = experimental temperature
  L = number of minutes at the reference temperatue equivalent to 1 min at the experimental temperature.

The temperature–time heating profile can be converted to a lethality profile where the area under the curve is equal to the F_0 value

$$\text{Total lethality } (F_0) = \int L \, dt$$

Many investigators ignore the heating and cooling sections in UHT plants, but this approach is much more valid for direct steam injection processes. The F_0 value is evaluated from the following equation

$$F_0 = 10^{(\theta - 121\cdot 1)/Z} \times \left(\frac{t_{min}}{60}\right)$$

where θ = processing temperature (°C)
  t_{min} = minimum residence time (s).

Shew (1981) suggested that a maximum spoilage rate of 1 in 10^4 containers was acceptable. Assuming that raw milk contains an average of one heat resistant spore per millilitre and UHT processing brings about eight decimal reductions, the likely survival would be one organism in every 10^5 one-litre packages. For smaller containers, the spoilage rate (per container) would be lower. If the raw milk were more contaminated, the spoilage rate would increase.

Distinction Between In-Bottle Sterilised and UHT Milk

The traditional way of producing sterilised milk involves filtering and homogenising the milk, filling into bottles, sealing them and heat treating them in batch or continuous retorts. More recently milk has been processed by using higher temperatures for shorter times, followed by aseptic packaging; when temperatures exceed 135 °C, the holding times become short enough to allow a continuous processing operation. The production of sterilised milk under these conditions has become known as ultra-high temperature (UHT) processing.

The distinction between sterilised and UHT milk can best be seen by plotting reaction kinetic data for the destruction of spores, the inactivation of enzymes and various chemical reactions on the same graph, using semi-log paper (Fig. 4). Examples are shown for the destruction of thermophilic and mesophilic spores, the inactivation of protease and a chemical reaction – superimposed on this are the conditions for sterilised and UHT milks.

Kessler (1981, 1989) presents similar reaction kinetic data for the discoloration of whole milk, evaporated milks, skim-milk and cream containing 10, 30 and 40 per cent fat respectively, loss of vitamin B_1, the formation of hydroxymethyl furfural (HMF), whey protein denaturation and losses of lysine. It can be seen that both types of process produce a commercially sterile product. However, the UHT process offers other advantages, which arise from the fact that chemical reaction rates are less sensitive to changes in temperature than the thermal inactivation of spores. Therefore, the much shorter processing times required to achieve commercial sterility result in much less chemical reaction occurring, as well as allowing a continuous processing operation.

In comparison to sterilised milk, UHT milk is whiter, tastes less caramelised and undergoes less protein denaturation and loss of heat-sensitive vitamins. Consequently, the two products are different in overall quality, and specifically in sensory, chemical and nutritional terms; note that the quality will also change during subsequent storage. However, one drawback of UHT processing may be the incomplete inactivation of the more heat-resistant, proteolytic enzymes, particularly when initially present at high concentrations. It is interesting that sterilised milk is not as

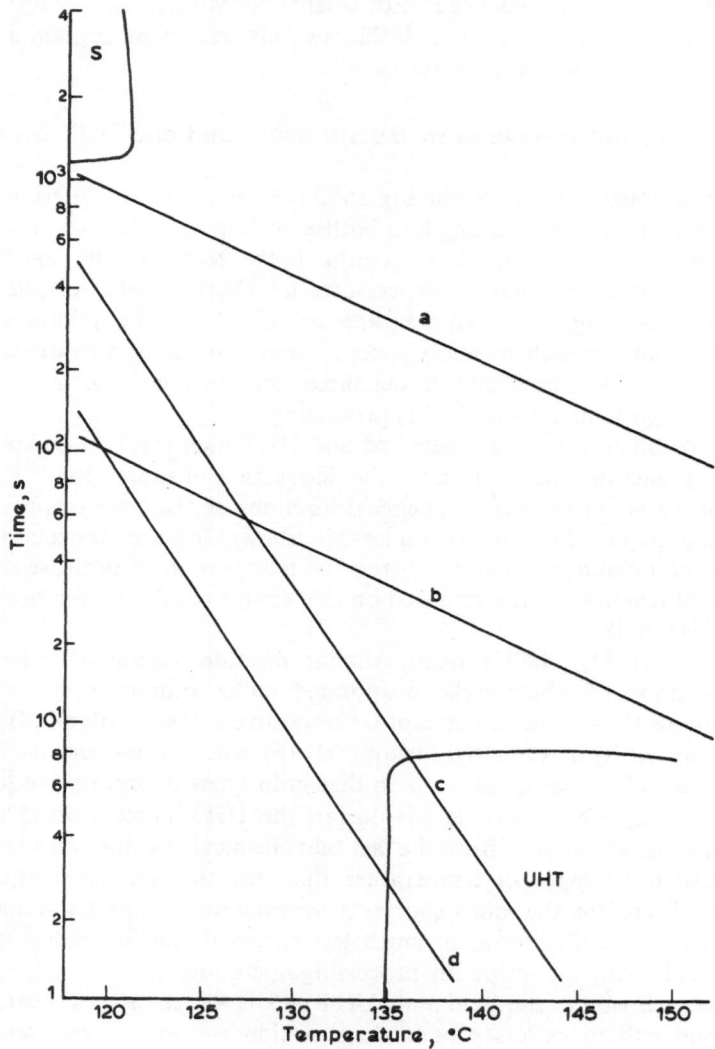

Fig. 4. Reaction kinetic data for the following reactions. (a) Inactivation of 90 per cent of protease from *Pseudomonas* (Cerf, 1981). (b) Destruction of 3 per cent thiamin (Kessler, 1981). (c) Inactivation of thermophilic spores (Kessler, 1981). Inactivation of mesophilic spores (Kessler, 1981). (S) Sterilised milk region. (UHT) UHT milk region.

susceptible to gelation as UHT milk. More recently it has been recognised that there are considerable differences between UHT milks produced not only by the methods of indirect heating and direct steam injection, but also from different equipment using the same methods. Some countries now distinguish between sterilised milk and UHT milk, whilst in other countries the term sterilised milk refers to milk produced by both methods.

Sterilised Milk (Heat Treatment Regulations, UK)

Initially, the regulations applied to milk heat treated in the container after it had been sealed. They have now been amended to incorporate milk which has been treated by continuous flow methods and then filled into sterile bottles; additional heat treatment after filling is optional. This category includes milk that is preheated in a UHT process, e.g. 137 °C for 4 s, homogenised, partially cooled, filled into bottles and further heat treated. There is some debate on the severity of the final heat treatment required. IDF (1991) considers that the effects of the two heat treatment processes may be considered additive, provided they comprise a single continuous process. The Sterilised Milk Regulations (Statutory Instruments, 1989) are much less specific, in terms of times and temperatures, than those for pasteurisation. They state that the milk should be filtered or clarified, homogenised, filled into containers of less than 1 gallon (UK) capacity and maintained at a temperature above 100 °C for sufficient time to ensure that a negative result is obtained when the turbidity test is performed. The milk should be labelled sterilised milk.

Additional regulations for sterilised milk produced by continuous flow methods state that the milk should be filled into sterile containers, using such aseptic precautions as will ensure the protection of the milk from the risk of contamination. Temperature records should be kept for the continuous heat exchanger, and a flow diversion valve should be provided. Sterilised milk, thus produced, must also satisfy the colony count for UHT milk.

The turbidity test involves the addition of 4 g of ammonium sulphate to 20 ml of milk. At this concentration, ammonium sulphate will precipitate the casein and denatured whey protein. Any undenatured protein will be left in solution. The mixture is then filtered and the resulting clear filtrate is heated. Any undenatured whey protein present in the filtrate flocculates and precipitates on heating, resulting in a positive turbidity. Sterilised milk should contain no undenatured whey protein, whereas most UHT milk should give a positive turbidity reading.

Conventional in-bottle sterilised milk is subjected to processing temperatures between 110 and 116 °C for between 20 and 30 min, depending upon the degree of caramelisation required; batch or continuous retorts are used. Sterilised milk produced by continuous processing is normally given an additional retorting period after filling, but one which is substantially lower than for in-bottle sterilisation. This additional heat treatment is generally sufficient to fully denature the whey proteins, and hence satisfy the requirement for negative turbidity. Before heat treatment, the milk is preheated and filled into glass bottles at 80 °C, or plastic bottles at 55 °C (Ashton and Romney, 1981). It may appear that the turbidity test could be used to distinguish between UHT and sterilised milks. However, it is possible to find some UHT milks which will give negative turbidity. This arises from the use of equipment with high regeneration efficiencies, and long heating and cooling periods or recirculation of milk.

In the United Kingdom, sterilised milk is a traditional product, and in 1982 it commanded 6·5 per cent of the liquid milk market (Table II), whereas in 1990 it was 4·6 per cent. In the EC it accounted for about 9·7 per cent of milk consumed in 1990. The brown colour and caramelised flavour are still popular, particularly with the elderly. It sells well in industrial areas. Its production is described in more detail by Ashton and Romney (1981). The use of continuous retorts, e.g. hydrostatic cookers or rotary retorts, is favoured for larger processing units, i.e. greater than 10,000 units per day. When using plastic bottles, consideration has to be paid to the barrier properties of the material, the processing temperatures used and pressure regulation during processing. This is to ensure that the internal pressure within the bottle is not significantly different from the pressure in the retort. Sterilised milk may have to satisfy a microbiological test. For example, UK regulations would deem the milk to be satisfactory if the plate count is not more than 10 per 0·1 ml. An interesting phenomenon is the formation of a boiling ring at the interface of the milk and headspace air within the bottle. This has always been present. More discerning consumers are now noticing this ring and suggesting that the bottles have not been properly cleaned, which is not the case.

UHT Regulations

UHT (ultra-high temperature) milk is produced by a continuous flow technique which rapidly heats the product to the UHT temperature, holds it for a short period of time and rapidly cools it. The sterile product is then packed aseptically. For this reason the term aseptic processing is often used. The following temperature–time combinations should give F_0 values of 3 and 8 respectively.

$F_0 = 3$		$F_0 = 8$	
121 °C	180 s	121 °C	480 s
131 °C	18 s	131 °C	48 s
141 °C	1·8 s	141 °C	4·8 s
151 °C	0·18 s	151 °C	0·48 s

The UK heat treatment and labelling regulations for UHT milk state the following.

The milk should be treated by the ultra-high temperature method, that is to say retained at a temperature of not less than 135 °C for not less than 1 s. [Previously this temperature was 132·2 °C.]

Flow diversion facilities should be provided if the temperature falls below 135 °C; suitable indicating and recording thermometers should be provided and records presented for three months. The milk should be packaged aseptically, securely fastened and labelled either ultra-heat treated milk or UHT milk. A sample of the milk should satisfy the UHT colony count, which is detailed in the regulations.

UHT milk can be produced using a variety of heat exchangers, in particular the plate and tubular types, where the heat transfer medium and the milk are separated. More recent equipment heats the milk by mixing it with steam. This is done either by injecting steam into the milk (injection) or by injecting milk into a steam chamber (infusion). Consequently the UHT treatment is referred to as either indirect or direct. It is not usually required to label which method has been used.

The UHT regulations for direct methods require that there should be no dilution of the milk. How this is achieved is described by Perkin and Burton (1970). The injected steam should be dry and saturated, free from foreign matter and readily removed for sampling. The boiler water should only be treated with certain listed, permitted compounds. The UHT regulations for semi-skim and skim-milk are similar. The minimum conditions for milk of 135 °C for 1 s corresponds to an F_0 value of 0·41, or a B^* value (see later) of 0·10 (assuming no contribution is made from the heating and cooling periods); this is well below the normal value of 3 recommended for low-acid products. Consequently milk thus processed will show a high level of microbial spoilage, and in practice temperatures between 137 and 145 °C and times between 2 and 4 s are used. In addition, the heating and cooling periods will contribute to the total lethality of the process.

It is also interesting to note that both the terms ultra-heat treatment and ultra-high temperature are used. My preference is for ultra-high temperature because this gives a more accurate description of the process. Ultra-heat treated implies that the product has been very severely treated. This is not so. Less chemical changes take place to produce UHT milk than to produce sterilised milk. More recently, regulations for the UHT treatment of milk-based drinks and cream have been introduced. These stipulate minimum processing times and temperatures of 2 s and 140 °C respectively. These are considerably in excess of those for milk and correspond to an F_0 value of 2·59. Similar conditions also apply to cream.

Several countries distinguish between UHT and sterilised milks. The basis for these differences has been discussed by Staal (1981) and have been summarised by Lewis (1986). Despite the low popularity of UHT milk in the UK, it is much more acceptable in many of the other EC countries; in France and Germany (1990), UHT milk accounts for 77 and 52 per cent, respectively, of the liquid milk market. Table II gives information for the amounts of pasteurised, sterilised and UHT milk sold in the EC countries for the years 1980 and 1990.

UHT Processing Plant

The short times necessary for UHT processing allow the use of continuous processes, and the equipment used for indirect UHT processing is not dissimilar to HTST pasteurisation plant (Figs. 1 and 5). Plate heat exchangers or tubular heat exchangers are used for low viscosity fluids, but for more viscous products or materials containing particulate matter, scraped-surface heat exchangers can be used. Milk is heated by regeneration and finally by steam or pressurised hot water. Homogenisation occurs either between the heating or the cooling stages or, in some cases, both. The sterile product is pumped to either an aseptic storage tank or directly to an aseptic filling system. The major differences between a pasteuriser and UHT plant are as follows.

(1) Higher operating pressures are required in order to prevent the milk boiling at the processing temperatures.

(2) An homogenisation step is required for milk containing an appreciable amount of fat. This prevents separation and hardening of the fat during storage.

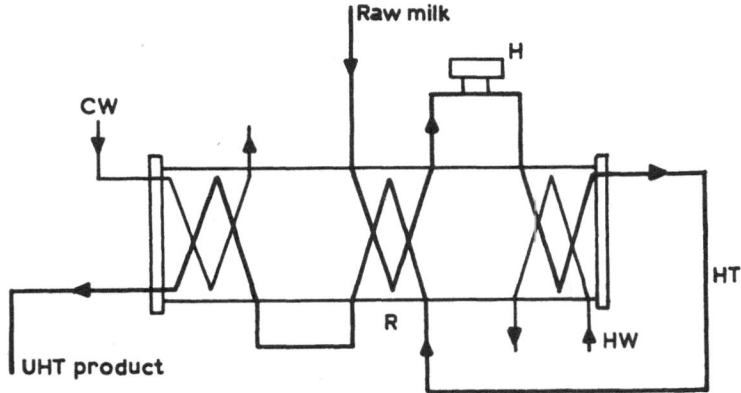

Fig. 5. UHT indirect plant. CW, Chilled Water; H, Homogeniser; HT, Holding
Tube; HW, Hot Water; R, Regeneration.

(3) Before processing, the plant needs to be sterilised downstream of
the holding tube and maintained sterile throughout processing.
The product may need to be stored in an aseptic tank and packaged into cartons aseptically. Sterilisation is achieved by circulating hot water through the plant and by ensuring that all points
downstream of the holding tube reach a minimum temperature of
130 °C for 30 min; even higher temperatures are now being used.

(4) Aseptic packaging equipment is essential and is discussed later.

(5) In general the residence times are much shorter. For example, a
direct steam injection plant was found to give an average residence
time of 3·26 s and an effective bacteriological holding time of 2·31 s.
The effective time was then used to evaluate decimal reduction
times for heat-resistant spores directly from data from a UHT
plant. There was reasonable agreement between data obtained by
this method and by capillary tube experiments.

More recent trends in design have been to use high regeneration
efficiencies, and a low temperature differential in the final stage of
heating. Most of the work on the heat resistance of spores at UHT
conditions has been done with *Bacillus stearothermophilus*. Problems
arise because decimal reduction times are very low at UHT temperatures,
and difficulties have also arisen when trying to relate data obtained from
capillary tube experiments to those from the pilot-plant. Burton *et al.*
(1977) found that corrections for temperature distributions within the

tube were necessary at temperatures above 135 °C. The determination of the total number of viable spores remaining after heat treatment is not straightforward, due to a number of factors that affect spore resistance, such as the age and storage conditions of the spores, the recovery of the spores and the presence of inhibitory substances produced by the heat treatment. Some of the discrepancies between heat resistance data obtained in capillary tubes and from an experimental steriliser have been investigated by Heppell (1986). He found that there were problems in determining the initial numbers of spores, and that the residence time distribution for milk was significantly different from that of water. Once these factors were recognised and corrected for, there was found to be a good agreement between the results. Methods for determining the heat resistance of spores at UHT conditions are reviewed by Brown and Ayres (1982), whilst Burton (1988) reviews the most recent data.

Direct Methods

One of the most interesting developments involves the direct contact of steam with milk. The mixing of saturated steam and milk results in almost instantaneous heating to the final UHT temperature, which is regulated by a combination of steam pressure and the pressure in the holding tube. The milk is normally heated (by regeneration) in a plate or tubular heat exchanger to between 70 and 80 °C before steam addition (Fig. 6). After injection, there is between 10 and 15 per cent dilution of the milk, and the resulting solids content will fall; this water is subsequently removed in the flash-cooling stage. The milk then passes through the holding tube, where it is held for the desired holding time; similar

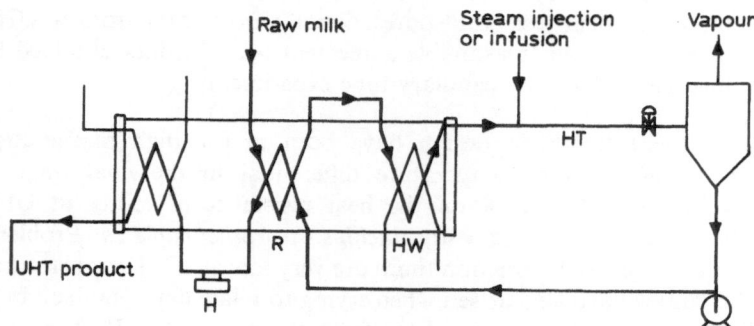

Fig. 6. UHT direct plant.

comments apply to residence time distributions as were made earlier. The dilution with steam will slightly reduce the residence time, due to the increase in the volumetric flow rate through the holding tube.

After the holding tube, the milk passes into a vessel held under vacuum conditions. The milk is well above its boiling point at these conditions, consequently there is a rapid fall in temperature. The resulting loss of energy causes some of the water to vaporise. This process is known as flash-cooling. The two most important effects are virtually instantaneous cooling, and the removal of water vapour and other volatile components. The instantaneous heating and cooling of the milk is very important because it means the milk suffers very little chemical thermal damage, since it is at a high temperature for only a very short time. The microbiological damage will depend upon the UHT temperature and holding time. One disadvantage of flash-cooling is that energy is lost from the system and cannot easily be recovered by regeneration. However, the vapour from flash-cooling can be used for regeneration purposes, but as its temperature is only about 80 °C, the amount of regeneration is limited. Therefore, the energy costs of direct systems tend to be higher than for indirect systems.

The pressure (vacuum) in the chamber will control the boiling temperature, the degree of cooling and the amount of water removed. As the pressure is reduced, the temperature falls, and the amount of water evaporated increases. The pressure should be adjusted so that the amount of water removed is the same as the amount of steam condensed during the heating process. The heat balance and a detailed analysis of the process is given by Perkin and Burton (1970). A mathematical solution relies on a knowledge of the heat loss from the plant, but unfortunately this is not easy to evaluate and varies from one plant to another. Therefore, direct UHT plants require calibrating, and the temperature conditions which give the correct water removal are determined by processing the fluid, then altering the vacuum conditions until no dilution is observed. This desired point can be checked by recirculating the product and noting whether there is any accumulation or loss as it goes through the steam addition and flash-cooling process. As a first approximation, the temperature of the milk in the vacuum chamber should be approximately 2 °C higher than the temperature of the milk before steam injection. Once these conditions have been determined, the temperatures are recorded during the processing run together with the temperature in the holding tube. This procedure is necessary to avoid dilution of the milk products. With other formulated products, e.g. milk-type drinks, reconstituted filled milks, creams and ice cream mixes, there may not necessarily be the same requirement to remove the exact quantity of water, and conditions may not be so strict.

A process could also be envisaged in which the formulated mix is made to a higher solids content than is required, and brought down to its intended solids content by steam injection; in this case, flash-cooling could be replaced by conventional cooling. There is a whole range of possible combinations, and as the amount of flash-cooling is reduced, the regeneration efficiency increases. A further advantage of flash-cooling is the loss of volatile components in the milk. Oxygen and low molecular weight volatile components are removed along with the water vapour. Of particular importance are the low molecular weight sulphur components which arise mainly from the breakdown of the sulphur-containing amino acids. Their presence in milk gives rise to a typical 'cabbage' flavour, and a pungent odour: it is this group of compounds that makes indirect UHT milk unacceptable to many people at this stage. Due to their partial removal, it is more difficult to taste the difference between direct UHT milk and pasteurised milk.

Also the low level of oxygen is beneficial in retarding oxidation reactions during storage; these will be discussed in further detail in a later section. After flash-cooling, the milk is homogenised. Homogenisation is preferred after steam injection as it results in a more stable product, but the homogeniser needs to be used under aseptic conditions; normally this is accomplished by incorporating a steam chest through which the piston rods pass (sterile block). There are also possibilities of using temperatures up to 150 °C for very short times. This will have considerable increase on the sterilisation effect. Accurate control of the residence time will be important to control the amount of chemical damage.

COMPARISON OF DIRECT AND INDIRECT PROCESSES

In the earlier literature there was a whole series of papers devoted to comparing direct and indirect heating processes. Most of the differences arise because the milk is subjected to different time–temperature profiles. Some of the differences depend on whether or not the flash-cooling process is included. The capital cost of a direct steam injection plant is usually higher than that of an indirect plant, mainly because the direct plant is slightly more complex. The energy costs of direct steam injection plants are also higher. This is because it is not possible to obtain such high regeneration efficiencies; the energy lost in the flash-cooling process is not easily recoverable. Virtually all results indicate that direct steam injection is a milder form of heat treatment as far as chemical reactions are concerned, it is more suitable for milks of poor microbiological quality, and the plant is less susceptible to fouling. It is claimed that it is possible to process milk for up to 20 h on a direct plant, compared to a maximum of between 4 and 5 h on an indirect plant. Perkin *et al.* (1973) showed that direct heat-treated milk produced twice as much sediment in the package after storing for 100 days. Both types of process reduced the rates of clotting with pepsin and rennin, but the effect of indirect processing was more severe. Lyster *et al.* (1971) used β-lactoglobulin denaturation to monitor the severity of the heat treatment. Whereas direct steam injection resulted in between 65·1 and 71·1 per cent denaturation, indirect heating resulted in between 78·8 and 83·5 per cent denaturation. These levels were calculated using temperature profiles for the two processes in conjunction with reaction kinetic data for denaturation of lactoglobulin (Lyster, 1970). The calculated values were always lower than those found experimentally and there was only fair agreement between the results. Jelen (1982) presented some further results, and some of these are given in Table III.

TABLE III
Concentration of components after direct and indirect processing

	Direct	Indirect
Whey protein nitrogen (mg/100 g)	38·8	27·6
Loss of available lysine (per cent)	3·8	5·7
Vitamin loss % B_{12}	16·8	30·1
Folic acid	19·6	35·2
Vitamin C	17·7	31·6
Hydroxymethyl furfural (μmol litre^{-1})	5·3	10·0

Difficulties do arise in comparing results when only holding times and temperatures are recorded. Kessler (1981) has adopted an interesting procedure for the assessment of UHT plants. He recognised that all UHT plants would give different time–temperature profiles as the product passed through and that, in some cases, considerably more chemical damage would result from excessive heating and cooling periods, than for the short time in the holding tube. His procedure evaluates the contributions made by the heating and cooling periods, both on microbial inactivation and chemical damage. Some typical time–temperature

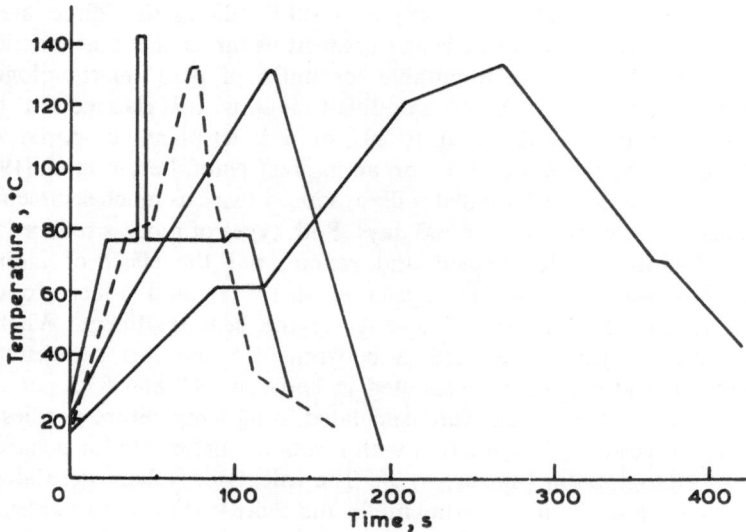

Fig. 7. Temperature–time profiles for different UHT plants. (Adapted from data in Reuter (1984).)

profiles are shown in Fig. 7. He introduced a microbiological parameter (B^*) and a chemical parameter (C^*); a temperature of 135 °C was adopted as the reference temperature. The microbiological parameter was based on the fact that commercial sterility could be achieved at 135 °C for 10·1 s. The corresponding Z value was 10·5 °C. Such a process was given a B^* value of 1·0. If the temperature–time profile is known for the product passing through the plant, B^* is evaluated in a fashion similar to F_0.

$$B^* = \int \frac{10^{(\theta - 135)/10 \cdot 5}}{10 \cdot 1} \, dt$$

where θ = temperature (°C)
  t = time (s).

Thus, a process with a B^* of 4·0 would be equivalent to 135 °C for 40·4 s. The chemical parameter (C^*) was based on the conditions required for a 3 per cent destruction of thiamin. This was chosen as it represented a typical chemical reaction in milk and corresponded to a change which would be only just detectable. The conditions required to

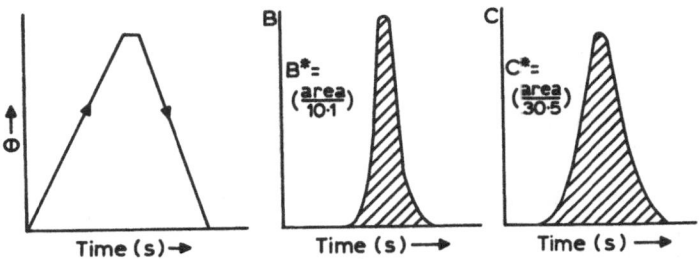

Fig. 8. Heating profile for B^* and C^* calculation.

bring this about were 135 °C for 30·5 s; the Z value was 31·4 °C. (See Fig. 8.)

$$C^* = \int \frac{10^{(\theta - 135)/31\cdot4}}{30\cdot5} \, dt$$

Conventional thermal processes use 121·1°C as a reference temperature, but on this occasion 135 °C has been selected because the extent of extrapolation required for typical UHT processing conditions is greatly reduced. It is highly desirable that B^* should be greater than one and C^* should be as low as possible. Kessler (1981) gives examples of B^* and C^* calculations for typical direct and indirect UHT processes. The following values were recorded.

Indirect UHT:	$B^* = 1\cdot25$;	$C^* = 0\cdot49$
Direct UHT:	$B^* = 2\cdot18$;	$C^* = 0\cdot30$

This simple comparison shows that microbial inactivation is greater, whereas chemical damage is less for the direct process. This approach is extremely useful as it allows a direct comparison of all UHT processes. Reuter (1984) summarised the difference in performance by describing three types of UHT plant.

First generation
Direct steam injection with rapid heating and cooling, and low overall chemical damage.

Second generation
Older, indirect processing plant with low regeneration efficiency (70–80 per cent), and shorter, non-stop operating times (6–10 h).

Third generation

Indirect processing with high regeneration efficiency (85–90 per cent), broad heating profiles and longer, non-stop operating times (14–18 h). However, compared to second generation plants, the milk is subject to a much higher degree of chemical damage.

Table IV shows the effects of heating and cooling rates on F_0, B^* and C^* values for milk held at 140 °C for 2 s (assuming linear heating and cooling profiles). It is assumed that the milk is heated from 80 °C and cooled to 80 °C. It can be seen that as the heating time period changes from instantaneous (zero) to 60 seconds, there are considerable increases in all the values. However, the relative increase in the amount of chemical damage is greater than for microbial inactivation. Thus extending the heating and cooling periods will have a more pronounced effect on the amount of chemical damage taking place. If the data for 60 s is further broken down, in order to evaluate the contributions of the heating, holding and cooling periods, the following results are obtained.

	F_0	B^*	C^*
heating 80 °C to 140 °C (60 s)	5·61	1·35	0·63
holding 140 °C for 2 s	8·20	1·94	0·73
cooling 140 °C to 80 °C (60 s)	13·81	3·29	1·37

In this case, most of the lethality and chemical damage takes place during the heating and cooling periods; in fact in this example, the material has been quite severely processed before it even enters the holding tube. There have been many attempts in the literature to distinguish between sterilised and UHT milks by chemical methods.

TABLE IV

F_0, B^* and C^* values for different heating profiles (holding at 140 °C for 2 seconds)

Heating and cooling time (s)	F_0	B^* (total)	C^*
0·1	2·61	0·60	0·10
1	2·77	0·64	0·12
10	4·45	1·04	0·31
30	8·20	1·94	0·73
60	13·8	3·29	1·37

80–140 °C (heating), hold 2 s, 140–80 °C (cooling).

TABLE V
Chemical changes in heat treated milk

	Whey protein nitrogen (mg/100 g mean)	Turbidity	Lysine loss (%)	Lactulose (mg/ml)
Raw	95·5			
Pasteurised	80·8	771	0·7–2·0	0·1
UHT direct	38·8	181	0–4·3	0·3
UHT indirect	27·6	14·2	1·7–6·5	
Sterilised (glass)	21·9	0·8	3·3–13	2·9

These have been reviewed by Burton (1983) and some results are summarised in Table V. However, the range of values is considerable and results for these different types of milk overlap. Other changes involving hydroxymethyl furfural (HMF), lysino-alanine, furosine and pyridosine were also examined. Andrews (1984) found that lactulose levels could be used to distinguish between direct and indirect UHT processes, but there was an overlap between indirect UHT milk and sterilised milk. It was observed that lactulose increased during storage in sterilised milk, but not in UHT milk. It is now obvious that some UHT milk is being heat treated as severely as some in-bottle sterilised milk. A good correlation between lactulose and C^* has been reported. In many quarters, lactulose values are taken as providing a good indicator of the amount of heat damage done. A review of the factors which affect the production of lactulose is given by Andrews (1989).

Other Reactions Taking Place in Heat-Treated Milk

A thorough review of the reactions taking place in heated milk, which do not involve oxygen, has been given by Kessler (1989). These results are presented between 100 and 150 °C, for times between 1 and 6000 s. A summary of the orders of reaction and activation energies (E_a) for some of these are given below.

	E_a (kJ mol^{-1})
thiamin (second order)	100·8
lysine (second order)	100·9
colour changes	116
HMF (zero order)	139

WHEY PROTEIN DENATURATION

The formation of activated sulphydryls has been associated with the denaturation of whey proteins, in particular β-lactoglobulin. This fraction is implicated because each dimer (36,000 daltons) contains two –SH groups and four –S–S– groups, and in addition, it accounts for over 50 per cent of the whey proteins. However, α-lactalbumin (0 and 4) and blood serum albumin (1 and 17) also contain sulphydryl and disulphide groups (SH, S–S), but the caseins appear to make little contribution. β-lactoglobulin denaturation was found to follow second order reaction kinetics (Lyster, 1970). Denaturation levels were measured by immunodiffusion, and the variation in reaction velocity constant (k) with temperature was found to be as follows

$$\log k = 37\cdot95 - 14\cdot51\left(\frac{10^3}{T}\right) : (\text{Temperature, } 68\text{–}90\,^{\circ}\text{C})$$

$$\log k = 5\cdot98 - 2\cdot86\left(\frac{10^3}{T}\right) : (\text{Temperature, } 90\text{–}135\,^{\circ}\text{C})$$

where T = temperature (K)

k = reaction velocity constant ($l\,g^{-1}s^{-1}$).

Hillier and Lyster (1979) using quantitative polyacrylamide gel electrophoresis evaluated whey protein denaturation in heated skim-milk, and were able to evaluate the reaction velocity constants for β-lactoglobulin variants A and B.

$$\text{lactoglobulin A: } \log k_A = 4\cdot25 - 1\cdot91\left(\frac{10^3}{T}\right) : (\text{Temp. range, } 100\text{–}150\,^{\circ}\text{C})$$

$$\text{lactoglobulin B: } \log k_B = 3\cdot48 - 1\cdot67\left(\frac{10^3}{T}\right) : (\text{Temp. range, } 95\text{–}150\,^{\circ}\text{C})$$

At temperatures below 95 °C, β-lactoglobulin (B) was more stable to heat, whereas above 95 °C it was less stable, but differences between the two variants were small. The irreversible step in this process was believed to involve a second order reaction between an S–S bond and an –SH group to form a new disulphide link. The second order reaction equation is

$$kt = \frac{1}{C} - \frac{1}{C_0}$$

where C = final concentration
C_0 = initial concentration.

Similar data were also presented for the denaturation of α-lactalbumin and blood serum albumin. Blanc *et al.* (1977) used several different methods for assessing β-lactoglobulin denaturation in pasteurised and UHT treated milks. These include electrophoresis, immunoelectrophoresis, immunodiffusion, rocket electrophoresis, gel filtration, differential scanning calorimetry and analysis of −SH and −S—S− groups. There was reasonable agreement between the two quantitative methods employed. High pressure liquid chromotography (HPLC) is now being used to measure whey protein denaturation.

Differential scanning calorimetry (DSC) was used to measure the denaturation temperature and heat of denaturation for β-lactoglobulin in salt solutions of different composition, and the effects of the presence of protein on denaturation reactions. From the results, it was concluded that the presence of α-lactalbumin and κ-casein would exert a protective effect on β-lactoglobulin denaturation, and that the denaturation temperature of β-lactoglobulin was dependent upon the ionic environment. Considerable differences were observed in the DSC thermograms for the different milks. Melo and Hansen (1978) found evidence for a complex between α-lactalbumin and β-lactoglobulin in model systems subject to UHT direct heating (143 °C for 8–10 s). They suggested that aggregation was due to the formation of disulphide bonds; evidence for the formation of this complex at lower temperatures (70–90 °C) has been shown by Elfagm and Wheelock (1977). Whey protein denaturation has been considered in considerable detail, both in terms of the individual proteins and total whey protein denaturation. Data is presented graphically for most of these reactions (Kessler, 1989).

NUTRITIONAL VALUE

Most of the loss in nutritional value is concerned with the vitamin and, to a lesser extent, mineral fractions. Although whey proteins are considerably denatured, there is very little loss in biological value or available lysine; Burton (1982) lists the biological values of raw milk, UHT milk and in-bottle sterilised milk as 0·90, 0·91 and 0·84 respectively. Losses of vitamins during sterilisation processes have been reviewed by Mehta (1980) and Burton (1983), and they may occur as a result of processing or during storage. On the whole, they are reasonably stable to sterilisation processes, although losses are greater during in-bottle sterilisation than during UHT production.

The fat-soluble vitamins (A, D and E) and the water-soluble vitamins of the B group (pantothenic acid, nicotinic acid and biotin) are hardly affected by UHT processes; riboflavin is stable to heat but not to light. Up to 20 per cent of thiamin and 30 per cent of vitamin B_{12} are lost in UHT processing, whereas losses during in-bottle sterilisation are up to 50 per cent for thiamin and 100 per cent for B_{12}. Losses of vitamin C occur during thermal processing, but are small in comparison to losses which can occur during storage; some values for vitamin C contents of heated milk are presented in Table VI. (Scott *et al.* 1984 a and b.) Losses of vitamin C, folic acid and vitamin B_{12} during storage are interrelated; folic acid starts to be lost once all the ascorbic acid is oxidised and vitamin B_{12} losses are also accelerated at high oxygen levels. More detailed information on losses of vitamins and minerals is given by IDF (1989).

TABLE VI
Vitamin C content of heat-treated milk

	Vitamin C ($\mu g/ml$)
Fresh milk	23·3
72 °C/16 s	19·4
80 °C/16 s	18·3
110 °C/3·5 s	17·3
140 °C/3·5 s	16·4

COLOUR OF MILK

Blanc and Odet (1981) reviewed the effects of heat on the colour of milk. Changes in casein size and the denaturation of whey protein both increase the amount of light scattering (reflectance), making milk appear whiter. However, this improvement is balanced by browning, which lowers the degree of reflectance and gives a striking increase in the green and yellow components on the Hunter or Munsell solids colour system (Francis, 1975; Clydesdale, 1975). Homogenisation is an additional, complicating factor, as most UHT milk is homogenised at some stage. In the absence of heat, homogenisation of skim-milk makes the milk very sightly lighter and gives a slight decrease in the green component. This suggests perhaps a slight modification to the size of the casein micelles. Whole milk shows a marked increase in lightness and a very striking decrease in the yellow component. This is attributed to a considerable increase in the number of fat globules and the dispersal of their coloured fat-soluble vitamins. The resulting effect is that UHT milk is usually whiter than raw milk, whereas sterilised milk is slightly browner. Non-enzymatic browning is linked to parameters such as pH, extent of heat treatment and storage temperature, as well as milk composition. In general, direct processes cause less browning than indirect heat; in UHT milk, significant browning occurs during storage, particularly at higher temperatures (i.e. greater than 25 °C).

TEXTURE OF UHT MILK

The texture of milk is related to mouth feel. Blanc and Odet (1981) summarised the factors affecting texture, and the methods of assessing some of these factors. The main defects are the separation of fat and the formation of a sediment which contains protein, fat, lactose and minerals in varying proportions. Fat may separate due to inefficient homogenisation. When fat separates, it rises and eventually hardens. Aggregation increases as the severity of the heat treatment increases and may be partially reduced by homogenisation after heat treatment. The concensus of opinion is that more sediment is produced in milk from the direct heating process, but the use of chemicals which provide additional anions, such as citrate, bicarbonate and hydrogen phosphate, can be beneficial. The total amount of sediment increases during storage, but the protein content decreases over the first five weeks and then rises again.

AGE THICKENING – GELATION

UHT milk shows a greater tendency to thicken and coagulate during storage than sterilised milk; this effect is related to the severity of the heat treatment. The gel structure is believed to be caused by casein micelles, which trap fat globules and whey proteins in a three-dimensional network. The UHT treatment of milk leads to a much larger proportion of small-sized casein micelles compared to raw or pasteurised milk. When viewed under the electron microscope, the surface of the micelles appears rough, which results in a viscosity increase of between

0.1 and 0.2 mNsm^{-2} (cP). Casein micelle size also changes during storage, more so at conditions of fluctuating temperature, and particularly before gelation. The physico-chemical and biochemical processes involved in age-gelation are summarised in Table VII.

TABLE VII
Processes involved in age-gelation

Physico-chemical

Dissociation of the casein/whey protein complexes
Cross linking due to Maillard reaction
Removal or binding of calcium ions
Conformational changes of casein molecules: breakdown of micelle structure; interaction of β-lactoglobulin and κ-casein; S-S exchange reactions; pH change; dephosphorylation of casein, and interaction of casein and carbohydrate

Biochemical

Heat resistance and reactivation of natural and bacterial proteinases
Survival of bacterial spores

Adams *et al.* (1976) concluded that the growth of Gram-negative psychrotrophs in raw milk led to detectable proteolysis, particularly of κ- and β-casein, even after two days. This breakdown subsequently had a deleterious effect on the proteins during UHT treatment, with decreased levels of the κ-, β- and $α_{s1}$-caseins and β-lactoglobulin. It was further claimed that coagulation during, or shortly after, heating increased with increasing severity of heat treatment and size of psychrotrophic population. These findings were confirmed by Law *et al.* (1977) who found that the level of psychrotropic bacteria determined whether UHT milk would gel during subsequent storage. Samples containing greater than 8×10^6 colony-forming units of *Pseudomonas fluorescens* AR11 gelled between 10 and 12 days after production when stored at $20\,°C$; below this count no gelation was observed ater 20 weeks storage. The protease from this organism caused extensive protein breakdown of κ-casein in a way similar to rennet action; β-casein was also broken down rapidly, while $α_{s1}$-casein was degraded only slowly. Chilled storage of raw milk may lead to high counts of psychrotrophic bacteria, some of which are spore-forming (van den Berg, 1981). Several species also produce significant amounts of highly heat-resistant enzymes, particularly proteases and lipases. In pasteurised milk, the effects of these enzymes are not usually noticeable, but if the count is above 10^6 ml^{-1}, significant amounts of proteases may survive UHT treatment and cause problems during storage. There is evidence to suggest that more protease inactivation occurs during in-container sterilisation (120 °C for 20 min) than during UHT processing (140 °C for 2 s) (Driessen, 1989). There is also evidence that some inactivation takes place at temperatures well below 100 °C. The clotting properties of UHT and sterilised milks also appear to be altered. Thus, while clotting times are increased and the coagulum is not so firm in hard cheese manufacture, heat treatment has been reported to increase activity of the starter, reduce clotting times and increase yield in cottage cheese manufacture using skim-milk. Chemical and physical changes during processing and storage are accentuated in concentrated milks. Muir (1984) reviews some practical methods for the production of UHT sterilised milk concentrates.

'COOKED' MILK FLAVOURS

It is widely acknowledged that consumer acceptance of and/or preference for a certain type of milk is influenced more by its flavour than by any other attribute. Flavour is a property detected by the senses, in particular taste and smell, and thereby requires taste panel work for its evaluation. Flavour changes in milk arise because of changes in its chemical constituents, and the various types of flavour defect in milk have been reviewed by Badings (1984) and Shipe *et al.* (1978). Researchers studying flavours use a combination of taste panel work and chemical analysis. The sensory aspects have been described in more detail by Piggott (1984). A detailed profiling study of heat-treated milk is described by Prasad *et al.* (1990). The flavours of particular interest in heated milk are the 'cooked' flavour and the bitter and oxidised flavours which develop during long-term storage of sterilised and UHT milk. The picture is further complicated because the 'cooked' flavour, which is developed on heating, changes rapidly during the early days of storage. The vocabulary used for description is also not straightforward. Terms, such as 'cooked', 'cabbagey', 'sulphury' and 'caramelised' are all frequently used. It is this 'cooked' flavour that is unpopular with the consumer, and is probably the principal reason why sales of UHT milk remain low in the UK (see Table II). The 'cooked' flavour is easily detectable both in liquid milk and in drinks, such as tea and coffee, where only small quantities of milk are used. One of the best accounts of the flavour changes in milk on heating and during storage is given by Ashton (1965). He recognised two phases, each with a number of distinct stages. These are summarised as follows.

Primary Phase

(a) Initial heating flavour accompanied by a strong sulphydryl or a cabbagey smell.
(b) Weaker sulphydryl or cabbage odour with residual 'cooked' flavour.
(c) Residual 'cooked' flavour with normal, acceptable agreeable flavour.

Secondary Phase

(d) Normal, acceptable to agreeable flat, acceptable flavour.
(e) Flat, acceptable to mild, oxidised flavour.
(f) Incipient oxidised flavour (or rancidity) to pronounced rancidity.

UHT milk is best consumed while in stages (c) and (d), and it can be seen that the flavour changes taking place during storage are considerable. Furthermore, such changes are very much influenced by the storage temperature, as well as dissolved oxygen levels, with a much faster progression through the above mentioned sequence of events taking place at higher temperatures. Different approaches have been taken to examine the source of the cooked flavour in heat-treated milk.

(i) Measuring the amount of sulphur components in the milk: this portion includes active sulphydryls, total $-SH$ and disulphide ($S-S$) groups: total ($-SH + S-S$) expressed as $-SH$ groups, sulphur-containing amino acids, and the low molecular weight volatile sulphur components, e.g. H_2S and CH_3SH.
(ii) Examining the denaturation levels of the proteins in milk, in particular β-lactoglobulin and α-lactalbumin as these are usually implicated as the major source of sulphur-containing components.

Another approach has been to add substances to milk before heat treatment to reduce the intensity of the 'cooked' flavour. Thus, Josephson and Doan, as early as 1939, made some observations on the source and significance of the 'cooked' flavour in milk heated above temperatures of 170 °F. They concluded that sulphydryl groups were wholly responsible for the 'cooked' flavour. These sulphydryl groups also reduced the oxidation–reduction potential of heated milk and acted as antioxidants, preventing both the oxidation of ascorbic acid and the development of tallowy and oxidised flavours. They suggested that the lactalbumin of milk, and some of the proteins associated with the fat globule membrane, were the most likely sources of the sulphydryl groups. Thus the role of sulphydryl groups was recognised at an early stage, although later work has drawn attention to active and total sulphydryl groups, disulphide groups and low molecular weight sulphur components, such as hydrogen sulphide and dimethyl sulphide. The presence of high concentrations of oxygen in the packaged milk will accelerate all the relevant reactions, i.e. the disagreeable 'cooked' flavour will disappear more quickly but the 'cardboardy', oxidised note will appear more quickly.

Milk produced by the indirect process has been found to have a very intense initial sulphydryl or 'cabbagey' smell, most likely due to a high concentration of hydrogen sulphide, which disappears rapidly within a few days. However, unless this milk is specially de-aerated, the oxygen content immediately after packaging is very high and the oxidation reactions proceed quickly. Milk produced by direct steam injection has a much reduced 'cabbagey' smell, resulting from the flash-cooling process which removes most of the volatile sulphur components and oxygen. This milk is probably already at stage (c) in Ashton's list, but if the oxygen content is still reasonably high, it may proceed to an oxidised stage quite quickly. Nevertheless, many investigators have reported that it is quite difficult to distinguish direct UHT milk from pasteurised milk.

The reaction scheme proposed by Ashton is accelerated by increasing the storage temperature. Thomas *et al.* (1975) examined the effects of high, medium and low oxygen levels on UHT milk during storage at room temperature for 150 days. High initial levels of oxygen led to rapid depletion of –SH groups and thereafter rapid depletion of ascorbic acid and folic acid; losses of these vitamins were much reduced at lower initial oxygen levels. Milks with the higher initial oxygen content were preferred for up to 13 days; thereafter, acceptability was independent of initial oxygen content. They concluded that the beneficial effect of high initial oxygen on flavour appears to be so slight, it is completely outweighed by the adverse nutritional effect. Also, the effect seems to be confined to such a short period in the early life of the milk, it probably disappears before the milk even reaches the consumer. Blanc and Odet (1981) found similar results at refrigerated storage (5 °C) and ambient storage (25 °C). At 5° C, the milk loses its sulphur flavour three weeks after production, is at its optimum period between $4\frac{1}{2}$ and 7 weeks, and starts to become stale after 8 weeks. At 25°C, the sulphur flavour is lost after 2 weeks, the milk is at its optimum between 3 and 5 weeks, and the stale flavour becomes evident after 6 weeks. Many other reactions affect the acceptability of the milk. Browning, caramelisation and the onset of gelation are also accelerated by increased storage temperatures.

Sulphydryl and Disulphide Group Determination

The occurrence of 'cooked' flavours in milk has been related to the changes in sulphydryls and disulphides in milk and, in particular, to the presence of free sulphydryl groups. Sulphydryl groups in the whey proteins, particularly β-lactoglobulin, become exposed as the molecules unfold. These may themselves contribute directly to the 'cooked' flavour, or further react to form low molecular weight compounds, such as hydrogen sulphide and dimethyl disulphide. Total sulphydryl groups, free or active sulphydryl groups and disulphide groups can be determined by the same procedure.

(1) For active sulphydryl groups, the milk is used directly.
(2) For total sulphydryl groups, the milk is reacted with a protein unfolding, denaturing agent, such as urea (Lyster, 1964).

Disulphide groups are determined by reducing the disulphide group using sodium borohydride and measuring total −SH. Patrick and Swaisgood (1976) expressed this as 'half cystine'; this is a useful approach, because it gives a measure of total sulphur. However, several chemical methods are available, and while the most popular is based on Ellman's reagent (5,5'-dithiobis-2-nitrobenzoic acid), others based on a spectro-fluorometric determination (Pofahl and Vakaleris, 1968) and a silver nitrate titration (Dill *et al.*, 1962) have been described. Lyster (1964) distinguished between active and masked sulphydryls and measured both in skim-milk heated at temperatures between 40 °C and 100 °C for 15 min. Active sulphydryl groups remained low until a temperature of 70 °C was achieved, after which they increased sharply to a maximum at 85 °C, followed by a gradual fall. Total sulphydryl groups fell throughout the temperature range, the fall being greatest from 85 to 100 °C. No information was given on disulphide activity. Milk from a UHT plant (no conditions given) had similar concentrations of free and total −SH groups. Pofahl and Vakaleris (1968) suggested that whey proteins in their native form contain a significant number of sulphydryl groups, and that this is increased on heating, probably at the expense of disulphide groups. Overall, the total amount of disulphide and sulphydryl decreased, probably due to volatilisation, and no distinction was made between active and total sulphydryl groups. Casein was found to contain hardly any sulphydryl groups, and only very small amounts were generated on heat treatment.

Koka *et al.* (1968) determined that the production of activated sulphydryl groups followed first order reaction kinetics, with reaction rate constants (min^{-1}) of 0·078 (75 °C), 0·142 (80 °C), 0·384 (85 °C) and 0·805 (90 °C) over the initial heating period, when skim-milk was heated for 15 min at different temperatures. The activated −SH groups showed a similar pattern to that observed by Lyster, with the maximum level of free −SH being observed at 100 °C, followed by a small decrease at higher temperatures. Dill *et al.* (1962) found that in direct steam injection, heat-activated sulphydryl groups increased as temperature increased from 190 to 300 °F for a time of 2 s. However, at holding times of 20 s and 150 s, the heat-activated sulphydryl groups went through a maximum, followed by a decrease which was associated with volatilisation of sulphur compounds. Patrick and Swaisgood (1976) performed work on milks heated by the UHT direct steam injection method. They measured free sulphydryl, total sulphydryl and half cystine (total sulphydryl + reduced disulphide). For direct steam injection, the concentration of reactive sulphydryl groups after heat treament approached or occasionally exceeded the concentration of total sulphydryl groups in the raw milk. Concomitant with the production of reactive sulphydryl groups was a decrease in measurable half cystine, which was about 6 per cent for milk heat treated at 150 °C for 1·5 s, compared to 17 per cent for milk heated at 100 °C for 30 min. These results suggest that both scission of disulphide bonds and volatilisation of sulphydryl groups could be occurring during heating. A similar pattern of events was observed for milk heated at 90 °C for up to 60 min. Blanc *et al.* (1977), by comparing results with Lyster's (1964), also concluded that direct steam injection caused less protein denaturation than indirect processing. The loss of sulphydryl groups during storage was accelerated by increasing the temperature – a finding in agreement with Lyster (1964) – but at 4 °C, the number of buried sulphydryl groups increased during storage, suggesting that the proteins may fold up during refrigerated storage.

VOLATILE SULPHUR COMPONENTS IN MILK

Dill *et al.* (1962) showed the existence of hydrogen sulphide in heated milk by flushing it out through lead acetate, precipitating lead sulphide and measuring it qualitatively. Thomas *et al.* (1976) described a rapid method for determining small amounts of H_2S in the headspace of milk products. Although H_2S concentration showed a good relationship to 'cooked' flavour intensity, it was not established that it was responsible. Jeon (1976) showed the existence of dimethylsulphide and isopropyl sulphide in UHT milk and observed that these disappeared less rapidly in UHT milk to which ascorbic acid had been added.

Jaddou *et al.* (1978) isolated flavour volatiles in UHT milk, by a low temperature distillation technique and identified them using gas chromatography and mass spectrometry. 'Cabbagey' defects in heated milks correlated very well with total volatile sulphur; the main components implicated were hydrogen sulphide, carbonyl sulphide (COS) methanethiol (CH_3SH), carbon disulphide (CS_2) and dimethyl sulphide ($(CH_3)_2S$). Burki and Blanc (1978) suggested that volatile sulphur components were more responsible than reactive $-SH$ groups for the early 'cabbagey' flavour of UHT milk. Jeon *et al.* (1978) analysed volatiles produced in UHT milk heated at $145\,°C$ for 3 s, aseptically packed and stored for up to 150 days. Twenty-six compounds were identified, most of which were carbonyl compounds, and it was believed that aldehydes were the most important contributors to off-flavours in UHT milk; no mention was made of sulphur-containing components. Gaafar (1987) made a detailed investigation into sulphur-containing groups, in relation to the onset of cooked flavour. He concluded that the threshold for cooked flavour corresponds to a β-lactoglobulin denaturation of 59 per cent, a hydrogen sulphide concentration $3\cdot4\,\mu g\,l^{-1}$ and a reactive sulphydryl concentration of $0\cdot037\,mmol\,l^{-1}$.

FLAVOUR IMPROVEMENT

There have been several attempts to improve the flavour of UHT milks. Potassium iodate (10–20 ppm) has been used to reduce the amount of 'cooked' flavour in milk, by causing the oxidation of any exposed sulphydryl groups (Skudder *et al.* 1981), and it was also found that the presence of potassium iodate reduced denaturation of α-lactalbumin, probably by not allowing sulphydryl disulphide exchange reactions to occur. The presence of free $-SH$ groups encourages the formation of protein aggregates on heated surfaces and the addition of L-cysteine before UHT processing resulted in a massive increase in the amount of deposit, as well as a most unacceptable 'cabbagey' or 'sulphury' flavour. Unfortunately, the addition of iodate (at these levels) results in the formation of bitter components about 14 days after processing, possibly due to increased instability of proteases within the milk, or suppression of natural protease inhibitors. Badings (1977) reported that the 'cooked' flavour could be reduced by adding L-cystine to milk prior to heating, both in direct and indirect processes. For milk treated by the indirect process, the amount of hydrogen sulphide produced immediately after processing was $82 \cdot 5\,\mu g\,kg^{-1}$, which was approximately eight times greater than that produced by direct steam injection. The addition of 30 and 70 mg cystine per kilogram of milk reduced the hydrogen sulphide level to $9 \cdot 5$ and $1 \cdot 7\,\mu g\,ml^{-1}$ respectively for indirect milk. Hydrogen sulphide disappeared very quickly during the first 24 h of storage, even at $3\,°C$, with changes being quicker in the indirect milk, presumably due to its higher oxygen content. They also concluded that addition of L-cystine produced no inclination toward oxidation or other flavour defects; it was later proposed that the hydrogen sulphide was removed by L-cystine with L-cysteine as the reaction product.

Swaisgood (1977) patented a process for removing the 'cooked' flavour from milk ($30–35\,°C$) by immobilised sulphydryl oxidase attached to glass beads, and he suggested placing the reactor downstream of the UHT holding tube. The flow rate has to be limited to $40\,ml\,h^{-1}$ and there may also be problems of sterilising the reactor contents before processing. Another patent application claimed to improve flavour by preheating milk to between 70 and $90\,°C$, centrifuging it at high speed to remove many of the microorganisms and spores, then holding it at between 35 and $40\,°C$ for a period of 10 to 20 min. Thereafter, the milk is UHT treated at between 100 and $140\,°C$ for short periods of time up to 10 s. It claims to improve the keeping quality of the milk by using centrifugation to remove the vast proportion of bacteria before heat treatment, thus preventing them from causing it to deteriorate during storage. A patent (US Patent 1989) describing the use of caraway seeds as a means of reducing the cooked flavour has recently been filed. Most attention has been paid here to the 'cooked' flavour, but Mehta (1980) has reviewed the factors affecting the onset of the stale or oxidised flavour, which appears after the 'cooked' flavour has disappeared, together with the volatile components responsible for it. Methyl ketones were the largest class of compounds isolated, athough aldehydes were thought to make the most significant contribution to this off-flavour (Blanc and Odet, 1981).

HEAT EXCHANGER FOULING

One of the most significant problems in UHT processing is fouling of the heat exchanger surface. This results in a narrowing of the flow passage, greater pressure drops and a reduction in the overall heat transfer coefficient. Either the pressures will become too high for safe operation or the UHT temperature will fall to a value that gives a large increase in the spoilage rates. At this point, the plant must be closed down, cleaned and re-sterilised. Therefore, to ensure long processing runs, it is important to understand the factors which affect fouling.

Milk starts to form deposits at 80 °C, and the extent of the deposition can be monitored by weighing, or indirectly by monitoring the pressure drop. In practice, there is a time period over which there is no significant loss of pressure and this time can vary significantly. The pressure drop then increases in a parabolic fashion up to a value where the limiting pressure drop is achieved: the plant is then closed down and cleaned. Deposit formation is also dependent upon the temperature of the milk, and the temperature for maximum deposit formation is 110 °C. The deposit that forms between 80 °C and 105 °C is a white voluminous precipitate. It is predominantly proteinaceous and tends to block the flow passages. Above 110 °C, the deposit is finer, more granular and predominantly mineral in origin. As the build-up is in the final section, it tends to reduce the efficiency of heat transfer; consequently temperatures begin to fall. Pressurised hot water sets tend to induce less fouling than steam heating and high velocities help to reduce deposits, but their use is limited by holding time and pressure drop considerations.

In addition, forewarming of the milk at between 80 and 85 °C for between 5 and 10 min will reduce both the amount and nature of the deposit, which is then predominantly a mineral one. Such forewarming may be ideal before evaporation to produce high heat powders, or before UHT processing followed by an in-container sterilisation process. It is also important to stop the air coming out of the milk, as the presence of bubbles increases deposit formation. The application of a reasonable back pressure, approximately one atmosphere (14·7 psi, 0·1 MPa) above that required to prevent boiling, should keep the air in solution. The quality of the milk plays an important role in the formation of deposits.

Ageing the milk at 4 °C for between 10 and 24 h improves the heat stability, perhaps due to lipolysis, because the presence of carreic acid, and later stearic acid, inhibits the fouling process. Milks with high contents of β-casein are more prone to deposit formation, but the addition of pyrophosphates to milk brings about a significant improvement.

There is no simple heating test to assess whether a milk is likely to foul or not. Alcohol can be used indirectly, to assess quickly the susceptibility of raw milk to fouling before it is processed. Equal volumes of milk and alcohol solutions are mixed together. If the milk flocculates in less than 74 per cent alcohol, it is likely to cause fouling problems during processing. Factors affecting alcohol stability have been reviewed by Horne (1984). Zadow (1971) reported that milk started to become unstable to UHT treatment using direct steam injection, below a pH of 6·62; at a pH below 6·50, virtually complete precipitation was observed. Reduction in the calcium ion concentration through the addition of EDTA, phosphate or citrate salts had a slight effect on stability, but very little sediment was observed above pH 6.7. Skudder (1984) observed that it is extremely useful to measure pH, as a slight increase in pH improves processing times, and similar results were obtained with reconstituted milk powder with a plate-type UHT plant (Zadow, 1978). The natural range of pH values is probably not greater than 0·15 units, but a change in initial pH from 6·67 to 6·54 vastly reduces the processing time attainable; note that during UHT processing, the pH of milk falls to well below 6·0. At pH 5·3–5·4, the milk becomes grossly unstable, but the addition of NaOH to increase to pH by 0·1 unit, prior to heat treatment, improves stability. The addition of trisodium citrate (4×10^{-3} M) raises the pH by 0·7, increases processing times, and appears to have a slightly beneficial effect on quality, due to an increase in viscosity of the product. Reducing the pH increases the amount of fat in the deposits formed from whole milk. The surface finish and temperature differentials across the plate are also important. The presence of cold spots results in calcium phosphate deposits remaining on the plate after in-place cleaning with sodium hydroxide followed by orthophosphoric acid. Tissier *et al.* (1984) found two major deposit peaks, one at 90 °C (predominantly protein-50 per cent) and one at 130 °C (predominantly mineral-75 per cent). The major protein contributing to the lower temperature peak was β-lactoglobulin (62 per cent), while in the second peak, β-casein (50 per cent) and α_{s1}-casein (27 per cent) were dominant. The effects of plant design have been evaluated by Lalande *et al.* (1984). Grandison (1988) surveyed fouling throughout a complete year, finding considerable seasonal variations. However, he was not able to correlate the severity of fouling with any chemical or physical parameters. Burton (1988) has reviewed the factors responsible for fouling from dairy products in more detail, and fouling models in relation to milk products have been discussed by Fryer (1989) and De Jong et al. (1992).

UHT goat's milk has recently appeared on the UK market. The author's experience is that this is a very difficult product to produce, as it is extremely susceptible to fouling, as well as producing a heavy deposit after only a few hours storage. Its alcohol stability is between 50 and 60 per cent. Zadow et al. (1983) have studied the stability of goat's milk to UHT processing and concluded that either pH adjustment to well above 7·0 or addition of 0·2 per cent disodium phosphate before processing, were necessary to reduce sedimentation. The higher content of ionic calcium was thought to be responsible for these problems in goat's milk. Buffalo's milk is also UHT processed in India. This does not provide a severe fouling or sedimentation problem, despite its much higher total solids and low alcohol stability (compared to cow's milk).

ASEPTIC PACKAGING

UHT technology relies on the ability of the processor to transfer the sterile product into a container under aseptic conditions. Obviously the container itself will need sterilising before filling; cans (Dole) are sterilised by superheated steam, whilst most other packages are sterilised with hydrogen peroxide at a concentration of between 20 and 35 per cent and a temperature between 80 and 85 °C. Residence times of several seconds are required. Care should be taken to ensure that all hydrogen peroxide is removed, as it is a strong oxidising agent. Plastic pots, with heat-sealed foil lids, are also available for individual portions (Bosch) and retail packs (Fresh-fill). The oxygen permeability of the plastic is important and may well influence the shelf-life of the product. Laminated packages, comprising several layers and including paper, plastic and aluminium foil, thereby offering a complete barrier, are popular with milk products. The most popular shape is the block or brick. The package is either formed from a single roll of material (Tetra Pak) or preformed blanks (Combibloc). Some of these have a headspace, whereas others do not.

More recently, a can with a plastic body and metal ends has appeared on the market (Milk Can) for milk and flavoured milk drinks. Again, this is sterilised with hydrogen peroxide and the product is designed to sell alongside and compete with popular soft drinks. It is also possible to package UHT products in bulk, using bag-in-the-box systems, ranging from a few litres through to one metric tonne. The bags and connections are sterilised by gamma irradiation, and a more recent development allows the removal of samples without contaminating the remaining product. Some of these aseptic systems are discussed in more detail in Burton (1988) and Reuter (1989).

CONCLUDING REMARKS

In terms of microbial quality and reducing spoilage rates, the emphasis is now toward that of prevention. For sterilisation processes, where very low spoilage rates are required (less than 1 in 10,000), it is now recognised that quality control which relies solely on end-product sampling is not effective in controlling the process. Low sampling rates only highlight gross defects in processing and finding zero defects gives no grounds for complacency and could permit poor quality batches to be released for sale. On the other hand, large sampling rates are very time-consuming, wasteful and costly. There is also a chance that satisfactory batches will be rejected. More details on sampling plans are given by ICMSF (1986) and Cerf (1989). One approach, now widely used, is that of Hazard Analysis Critical Control Points (HACCP) (ICMSF, 1988). Here the philosophy is to identify where hazards may occur, from raw materials, during the different processing stages, packaging and subsequent handling and storage. Critical control points are then established. These are points in the production process where the hazards can be effectively controlled. Loss of control permits the realisation of the potential hazard as an unacceptable food safety or spoilage risk. Thus in UHT and sterilisation processes, attention is paid to raw material quality, UHT temperatures and times, plant sterilisation, including the aseptic tank and filling machines, and quality is ensured by setting out to keep the process under control. Such an approach is combined with the use of codes of practice and staff education programmes (IFST, 1991).

In theory, the organoleptic quality can be improved by the use of even higher temperatures for shorter times. For such processes, the holding period becomes the controlling factor, both in terms of microbial inactivation and chemical damage. Thus very minor variations in holding time give large changes in these factors and could lead to a non-uniform product. The principles of hazard analysis are also applied to pasteurisation processes, in order to improve the safety and keeping quality of the products. Another important goal is the production of a low heat-treated product with a long shelf-life under refrigerated storage.

Process Development

One interesting recent development is concerned with the direct conversion of electrical energy to heat using the fluid as an electrical resistance and conducting medium (ohmic heating). The fluid is placed in a non-conducting tube that has electrodes at each end. As an alternating current is passed through the fluid, heat is generated. The conversion efficiency is greater than 90 per cent. The major advantages are that even heating results, there are no temperature gradients, and none of the usual limitations due to conduction and convection arise. In addition, it is suitable for continuous processing, and there is no requirement for a hot, heat-transfer surface, which will reduce fouling. Its big advantage is that liquids containing particulate matter can be processed, and will not be subjected to the high shear rates found in scraped-surface heat exchangers. In principle, the technique is similar to microwave heating, but it is claimed to have a lower capital cost and a higher conversion efficiency. However, it is unlikely to offer any obvious quality or economic advantages for milk, although it may be useful for milk-based desserts. Ohmic heating is described in more details by Biss *et al.* (1989). Other developments will aim towards obtaining longer processing runs. They will also attempt to improve the quality of heat-treated products by a better understanding of the physical, chemical and biochemical reactions which occur during heating and subsequent storage.

REFERENCES

Aboshama, K. and Hansen, A. P. (1977). *Journal of Dairy Science*, **60**, 1374.
Adams, D. M., Barach, J. T. and Speck, M. L. (1976). *Journal of Dairy Science*, **59**, 823.
Andrews, A. T., Anderson, M. and Goodenough, P. W. (1987). *Journal of Dairy Research*, **54**, 237.
Andrews, G. R. (1984). *Journal of the Society of Dairy Technology*, **37**, 92.
Andrews, G. R. (1989). *IDF Document*, **238**.
Ashton, T. R. (1965). *Journal of the Society of Dairy Technology*, **18**, 65.
Ashton, T. R. and Romney, A. J. D. (1981). *IDF Bulletin*, **130**.

Badings, H. T. (1977). *Nordeuropaeisk Mejeri-tidsskrift*, **43**, 379.

Badings, H. T. (1984). *Dairy Chemistry and Physics* (Ed. P. Walstra and R. Jenness), John Wiley, New York.

Badings, H. T. and Nester, R. (1978). In: *Brief Communications, XX Int. Dairy Congress*, Paris.

Biss, C. H., Coombes, S. A. and Skudder, P. J. (1989). In: *Process Engineering in the Food Industry* (Ed. R. A. Field and J. A. Howell), Elsevier Applied Science Publishers, London.

Blanc, B. and Odet, G. (1981). *IDF Bulletin*, **133**.

Blanc, B., Baer, R. and Ruegg, M. (1977). *Schweizetische Milkwirtschafiliche Forschung*, **6**, 21.

Bramley, A. J. and McKinnon, C. H. (1990). In: *Dairy Microbiology*, Volume 1 (Ed. R. K. Robinson), Elsevier Applied Science Publishers, London.

Brown, K. L. and Ayres, C. A. (1982). In: *Developments In Food Microbiology – 1* (Ed. R. Davies), Applied Science Publishers, London.

Burki, C. and Blanc, B. (1978). In: *Brief Communications, XX Int. Dairy Congress*, Paris, 1978.

Burton, H. (1982). In: *CRC Handbook of Processing and Utilization in Agriculture* (Ed. I. A. Wolfe), Chemical Rubber Company, New York.

Burton, H. (1983). *IDF Bulletin*, **157**.

Burton, H. (1988). *UHT Processing of Milk and Milk Products*, Elsevier Applied Science Publishers, London.

Burton, H., Perkins, A. G., Davies, F. L. and Underwood, H. M. (1977). *Journal of Food Technology*, **12**, 149.

Busse, M. (1981). *IDF Bulletin*, **130**.

Cerf, O. (1981). *IDF Bulletin*, **130**.

Clydesdale, F. M. (1975). In: *Theory, Determination and Control of Physical Properties of Food Materials* (Ed. C. Rha), D. Reidel, Dordrecht.

De Jong, P., Bouman, S. and Van der Linden, H. J. L. V. (1992). *Journal of the Society of Dairy Technology*, **45**, 3.

Dickerson, R. W., Jr., Scalzo, A. M., Read, R. B., Jr. and Parker, R. W. (1986). *Journal of Dairy Science*, **51**, 1731.

Dill, C. W., Roberts, W. M. and Aurand, L. W. (1962). *Journal of Dairy Science*, **45**, 1332.

Driessen, F. M. (1989). *IDF Bulletin*, **238**.

EC Dairy Facts and Figures (1991). Economics Division, Milk Marketing Board.

Elfagm, A. A. and Wheelock, J. V. (1977). *Journal of Dairy Research*, **44**, 367.

Fox, P. F. (1982). In: *Developments in Dairy Chemistry – 1* (Ed. P. F. Fox), Applied Science Publishers, London.

Francis, R. J. (1975). In: *Theory, Determination and Control of Physical Properties of Food Materials* (Ed. C. Rha), D. Riedel, Dordrecht.

Fryer, P. J. (1989). *Journal of the Society of Dairy Technology*, **42**, 23.

Gaafar, A. M. M. (1987). PhD Thesis, University of Reading, UK.

Grandison, A. (1988). *Journal of the Society of Dairy Technology*, **41**, 43.

Griffiths, M. W. (1986). *Journal of Food Protection*, **49**, 696.

Gould, G. W. (Ed.) (1989). In: *Mechanisms of Action of Food Preservation Procedures*, Elsevier Applied Science Publishers, London.

Hasting, A. P. M. (1992). *Journal of Food Control*, **3**, 27.

58 *M. J. Lewis*

Heppell, N. (1986). *Journal of Food Technology*, **21**, 385.
Hillier, R. M. and Lyster, R. L. J. (1979). *Journal of Dairy Research*, **46**, 95.
Horne, D. S. (1984). Report of the Hannah Research Institute, **89**, Hannah Research Institute, Ayr, Scotland.
ICMSF (1986). *Micro-organisms in Foods 2, Sampling for microbiological Analysis: principles and specific applications*, Blackwell Scientific Publications, Oxford.
ICMSF (1988). *Micro-organisms in Foods 4, Application of the hazard analysis critical control point (HACCP) system to ensure microbiological safety and quality*, Blackwell Scientific Publications, Oxford.
IDF (1986). *IDF Bulletin*, **200**.
IDF (1989). *IDF Bulletin*, **238**.
IDF (1991). IDF Document No. 222.
IFST (1991). *Food and Drink – Good Manufacturing Practice*, 3rd edn, Institute of Food Science and Technology, London.
Jaddou, H. A., Pavey, J. A. and Manning, D. J. (1978). *Journal of Dairy Research*, **45**, 391.
Jelen, P. (1982). *Journal of Food Protection*, **45**, 878.
Jenness, R. (1982). In: *Developments in Dairy Chemistry – 1* (Ed. P. F. Fox), Applied Science Publishers, London.
Jeon, I. J. (1976). PhD Thesis, University of Minnesota, St Paul, Minnesota.
Jeon, I. J., Thomas, E. L. and Reineccius, G. A. (1978). *Journal of Agricultural and Food Chemistry*, **26**, 1183.
Josephson, D. V. and Doan, F. J. (1939). *The Milk Dealer*, **29**, 35.
Jukes, D. J. (1987). *Food Legislation of the UK – A Concise Guide*, Butterworths, London.
Kessler, H. G. (1981). *Food Engineering and Dairy Technology*, Verlag A. Kessler, Freising.
Kessler, H. G. and Horak, F. P. (1984). *Milchwissenschaft*, **39**, 451.
Kessler, H. G. (1989). In: *Developments in food preservation*, Volume 5 (Ed. S. Thorne), Elsevier Applied Science Publishers, London.
Koka, M., Mikolajcik, E. M. and Gould, I. (1968). *Journal of Dairy Science*, **51**, 217.
Kosaric, N., Kitchen, B., Pandial, C. J., Sheppard, J. D., Kennedy, K. and Sargant, A. (1981). *CRC Critical Reviews in Food Science and Nutrition*, **14**, 153.
Lalande, M., Tissier, J. P. and Comev, G. (1984). *Journal of Dairy Research*, **51**, 557.
Law, B. A., Andrews, A. T., and Sharpe, M. E. (1977). *Journal of Dairy Research*, **44**, 144.
Levenspiel, O. (1972). *Chemical Reaction Engineering*, John Wiley, New York.
Lewis, M. J. (1986). In: *Modern Dairy Technology*, Volume 1, (Ed. R. K. Robinson), Elsevier Applied Science, London.
Loncin, M. and Merson, R. L. (1979). *Food Engineering, Principles and Selected Applications*, Academic Press, New York.
Lund, D. (1975). In: *Principles of Food Science, Part 2. Physical Principles of Food Preservation* (Ed. O. Fennema), Marcel-Dekker, New York.
Lyster, R. L. J. (1964). *Journal of Dairy Research*, **31**, 41.

Lyster, R. L. J. (1970). *Journal of Dairy Research*, **37**, 233.

Lyster, R. L. J., Wyeth, T. C., Perkin, A. G. and Burton, H. (1971). *Journal of Dairy Research*, **38**, 403.

Mehta, R. S. (1980). *Journal of Food Protection*, **43**, 212.

Melo, T. S. and Hansen, A. P. (1978). *Journal of Dairy Science*, **61**, 710.

Muir, D. D. (1984). *Journal of the Society of Dairy Technology*, **37**, 135.

Muir, D. D. (1990). In: *The Microbiology of Milk* (Ed. R. K. Robinson), 2nd edn, Elsevier Applied Science, London.

Patrick, P. S. and Swaisgood, H. E. (1976). *Journal of Dairy Science*, **59**, 594.

Perkin, A. G. and Burton, H. (1970). *Journal of the Society of Dairy Technology*, **23**, 147.

Perkin, A. G., Henschel, M. J. and Burton, H. (1973). *Journal of Dairy Research*, **40**, 215.

Piggot, J. R. (Ed.) (1984). *Sensory Analysis of Foods*, Elsevier Applied Science, London.

Pofahl, T. R. and Vakaleris, D. G. (1968). *Journal of Dairy Science*, **51**, 345.

Prasad, S. K., Thomson, D. M. H. and Lewis, M. J. (1990). In: *Trends in Food Product Development*, (Eds Yan, T. C. and Tan, C.), Proceedings of the World Congress of Food Science and Technology, Singapore.

Reuter, H. (1984). In: *Engineering and Food*, Volume 1 (Ed. B. M. McKenna), Applied Science Publishers, London.

Reuter, H. (Ed.) (1989). *Aseptic Packaging of Foods*, Technomic Publishing Company Inc., Lancaster, USA.

Schroeder, M. J. A. (1984). *Journal of Dairy Research*, **51**, 59.

Schroeder, M. J. A., Cousins, C. M. and McKinnan, C. H. (1982). *Journal of Dairy Research*, **49**, 619.

Scott, K. J., Bishop, D. R., Zechulko, A., Edwards-Webb, J. D., Jackson, P. A. and Scuffam, D. (1984a). *Journal of Dairy Research*, **51**, 37.

Scott, K. J., Bishop, D. R., Zechulko, A., Edwards-Webb, J. D., Jackson, P. A. and Scuffam, D. (1984b). *Journal of Dairy Research*, **51**, 51.

Scott, R. (1970). *Process Biochemistry*, **5**(5), 39.

Shew, D. I. (1981). *IDF Bulletin*, **133**.

Shipe, W. F., Bassette, R., Deane, D. D., Dunkley, W. L., Hammond, E. G., Harper, W. J., Kleyn, D. H., Morgan, M. E., Nelson, J. H., and Scanlan, R. A. (1978). *Journal of Dairy Science*, **61**, 855.

Skudder, P. J., Thomas, E. L., Pavey, J. A. and Perkin, A. G. (1981). *Journal of Dairy Research*, **48**, 99.

Society of Dairy Technology (1983). *Pasteurising Plant Manual*, Huntingdon.

Solberg, P. (1981). *IDF Bulletin*, **130**.

Staal, P. F. J. (1981). *IDF Bulletin*, **133**.

Stall, P. F. J. (1986). *IDF Bulletin*, **200**.

Swaisgood, H. E. (1977). US Patent 4 053 644.

Thomas, E. L., Burton, H., Ford, J. E. and Perkin, A. G. (1975). *Journal of Dairy Research*, **42**, 285.

Thomas, E. L., Reineccius, G. A., De Waard, G. J. and Slinkard, M. S. (1976). *Journal of Dairy Science*, **59**, 1865.

Tissier, J. P., Lalande, M. and Corrlev, G. (1984). In: *Engineering and Food*, Volume 1 (Ed. B. M. McKenna), Applied Science Publishers, London.

UK Dairy Facts and Figures (1991). The Federation of UK Milk Marketing Boards.

United States Patent (1989). US Patent 4 851 251.

van Boekel, M. A. J. S. and Walstra, P. (1989). *IDF Bulletin*, **238**.

Van den Berg, M. G. (1981). *IDF Bulletin*, **130**.

Walstra, P. and Jenness, R. (1984). *Dairy Chemistry and Physics*, John Wiley, New York.

Zadow, J. G. (1971). *Journal of Dairy Research*, **38**, 393.

Zadow, J. G. (1978). In *Brief Communications, XX Int. Dairy Congress*, Paris, p. 711.

Zadow, J. G., Hardham, J. F., Kocak, H. R. and Mayes, J. J. (1983). *Australian Journal of Dairy Technology*, **38**, 20.

Chapter 2

Developments in Cream Separation and Processing

C. Towler

New Zealand Dairy Research Institute, Palmerston North, New Zealand

Cream consists of a concentration of the fat in milk, with the fat existing mainly as globules protected by a membrane. As such, cream can have a variety of compositions and is normally defined according to fat content or function. Fat content may range from 10 per cent (half-cream) to 80 per cent plus (plastic cream). Cream for butter manufacture would normally contain approximately 40 per cent fat. The United Nations Food and Agricultural Organization and World Health Organization (1977) have suggested the following standards for market cream.

Pasteurised, sterilised and ultra-high temperature (UHT) treated cream	≥18% milkfat
Half-cream	10–18% milkfat
Whipping cream	≥28% milkfat
Heavy whipping cream	≥35% milkfat
Double cream	≥45% milkfat

The physico-chemical properties of cream are very much influenced by the state of dispersion of the milkfat globules and the globule membrane that surrounds them. Mulder and Walstra (1974) have prepared a comprehensive treatise on the chemistry of the milkfat globule, including the properties of various systems that incorporate milkfat. The fat globules in milk or cream are not of uniform size and vary from 0·5 to 10 μm in diameter. Figure 1 shows a typical Gaussian distribution curve for milkfat globules in cow's milk. As the concentration of milkfat globules is altered in cream, changes that have a marked effect on the rheology and physical state take place. Temperature changes also have a marked effect as the different lipid components undergo changes of state. Not only is the magnitude of change important, but also the rate of change, as this affects crystallisation patterns. The non-fat milk solids also play an important part in the properties of cream, and additives such as salts, proteins, emulsifiers and hydrocolloid stabilisers all affect cream properties. The properties of cream are also affected by physical handling such as pumping, aeration and agitation as they affect the disintegration and agglomeration of the globules. It will be the function of this chapter to explain the principles of cream manufacture from milk, along with further treatments that may be applied to give cream products for direct consumption, or cream that may be used for manufacture of other dairy products.

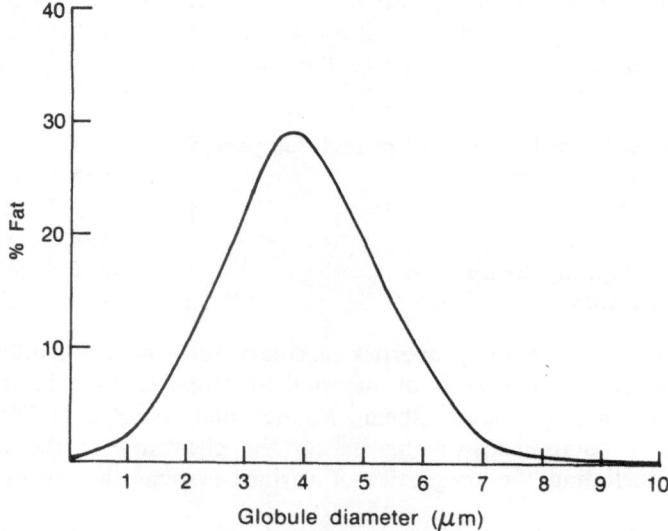

Fig. 1. Typical distribution curve of fat globules in milk expressed as a percentage of total fat taken up by globules with a particular diameter.

SEPARATION AND DEVELOPMENTS IN SEPARATORS

The separation of milk is a process whereby an essentially fat-free portion (skim-milk) is separated from a fat-rich portion (cream). The process is physical, relying on the density difference between the milkfat in the globules and the aqueous phase in which they are dispersed. If milk is allowed to stand, fat rises and the familiar process of 'creaming' is observed with a fat-rich fraction collecting at the surface. The upward gravitational force (f_u) on a fat globule is given by

$$f_u = 4\pi r^3 g(\rho_s - \rho_f)/3$$

where r = radius of globule
         g = acceleration due to gravity
         ρ_s = density of serum (skim-milk)
         ρ_f = density of fat globule.

The rise of the globule is inhibited by frictional force (f_f) which is given by Stokes' Law

$$f_f = 6\pi\eta r v$$

where η = fluid viscosity of serum
         v = velocity of globule.

When the fat globule is rising at a constant terminal velocity, then

$$f_f = f_u$$
$$6\pi\eta r v = 4\pi r^3 g(\rho_s - \rho_f)/3$$
$$v = 2r^2 g(\rho_s - \rho_f)/9\eta$$

Thus, the velocity with which a fat globule rises is directly proportional to the square of its radius and the density difference between the globule and the serum, and is inversely proportional to the viscosity of the serum. The densities of the milkfat and serum and the viscosity of the serum can be manipulated by altering temperature, but the radius of a globule is fixed. Table I gives the theoretical velocities of fat globules of 1, 2 and 5 μm diameter at different temperatures, with the relevant time it would take to rise 200 mm; this is the approximate depth of milk in a 600 ml bottle. These figures are only of academic interest as small globules tend to remain dispersed through thermal currents and Brownian motion, whereas larger globules tend to coalesce leading to more rapid separation. The figures do illustrate, however, that separation by gravity is most inefficient.

TABLE I

Velocity of different diameter fat globules at different temperatures, with the time taken to rise 200 mm

Temperature (°C)	Density of serum (kg m^{-3})	Density of fat (kg m^{-3})	Viscosity of serum (N m^{-2} s)	Diameter of globule (µm)	Velocity of globule (µm s^{-1})	Time to travel 200 mm (days)
5	1037	961	2.96×10^{-3}	1	0.014	165
				2	0.056	41
				5	0.350	6.6
20	1034	930	1.79×10^{-3}	1	0.032	73
				2	0.127	18
				5	0.791	2.9
35	1029	908	1.17×10^{-3}	1	0.056	41
				2	0.225	10
				5	1.410	1.6
50	1022	898	8.5×10^{-4}	1	0.080	29
				2	0.318	7.3
				5	1.99	1.2
65	1015	888	6.5×10^{-4}	1	0.106	22
				2	0.426	5.4
				5	2.66	0.9

Centrifugal Separators

The other factor that may be increased to increase sedimentation is the force acting on globules. This can be achieved through the use of centripetal force in a rotating vessel. The resultant globule velocity can then be given as

$$v = 2r^2(\rho_s - \rho_f)R\omega^2/9\eta$$

where R = radial distance of globule from axis of rotation
ω = angular velocity (radians s^{-1})

or
$$v = 2r^2(\rho_s - \rho_f)4\pi^2 RN^2/9\eta$$

where N = rotational frequency (revolutions s^{-1}).

Table II gives the velocity of fat globules 0·1 m from the axis of rotation in a centrifuge rotating at different speeds with milk at different temperatures. Figures 2 and 3 illustrate the effect of the various parameters on the velocity. These indicate that the velocity of globules can be increased substantially by increases in centrifuge speed or milk temperature, but the size of the fat globules is a critical factor; small globules have only a limited velocity at high rotational speeds and high milk temperatures.

Gustaf de Laval devised the first continuous centrifugal separator, the principle of which is illustrated in Fig. 4. Whole milk was fed in through the top of the bowl to a distributor in the base which brought the milk to rotational speed while being channelled into the bowl itself. The fat globules moved in toward the axis of rotation to form a cream layer, whilst the denser skim-milk flowed to the outside of the bowl and was channelled out. The cream was taken out of the inner bowl as overflow, and separation was controlled by the flow of incoming milk. In later models, the fat content of the cream was controlled through restriction in the skim-milk flow. Allowing more skim-milk to flow out reduced cream flow, and the fat content was increased. Such separators had limited capacity, as fat globules had a reasonably large distance to travel before reaching the cream layer. If high milk flows were used, then separation was inefficient with fat globules escaping with the skim-milk flow. This difficulty was resolved by provision of a number of separation zones through a 'disc stack'. The discs were conical in shape (cone angle approximately 60°) with holes in them to channel milk through (rising channels). Identical discs were stacked one on top

C. Towler

TABLE II

Velocity of fat globules of various diameters in a centrifuge rotating at different speeds at different temperatures

Temperature (°C)	Globule diameter (μm)	Rotational frequency of centrifuge (rpm × 10³)	Velocity of globule (mm s⁻¹)
5	1	3	0·014
		5	0·039
		7	0·077
	2	3	0·056
		5	0·156
		7	0·307
	5	3	0·352
		5	0·978
		7	1·916
35	1	3	0·057
		5	0·158
		7	0·309
	2	3	0·227
		5	0·630
		7	1·235
	5	3	1·418
		5	3·938
		7	7·718
65	1	3	0·107
		5	0·298
		7	0·583
	2	3	0·429
		5	1·190
		7	2·333
	5	3	2·678
		5	7·440
		7	14·582

of the other with spacers (caulks) fitted on the upper surfaces to provide a gap between adjacent discs. Such discs are still features of modern separators. Figure 5a shows how fat globules have only a limited distance to move before being directed inwards on the upper surface of a disc, while skim-milk flows outwards on the lower surface of the adjacent disc. Figure 5b shows the arrangement of a complete disc stack with relevant flows. Figure 5c is a photograph of an actual disc showing the holes for

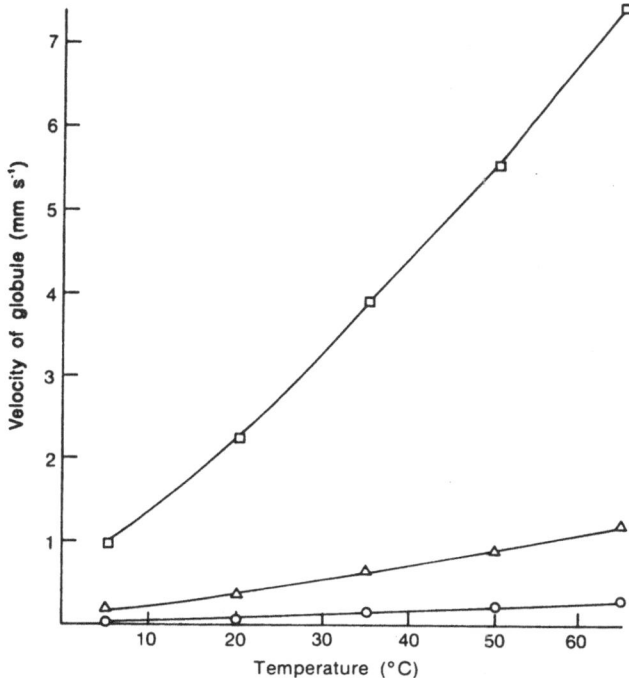

Fig. 2. The dependence of the velocity of fat globules of different diameters on temperature in a centrifuge rotating at 5000 rpm; the globules are 0·1 m from the axis of rotation. ○: globule 1 μm diameter; △: globule 2 μm diameter; □: globule 5 μm diameter.

channelling milk and the spacers (caulks) that maintain the required separation between discs.

Paring Discs (Centripetal Pumps)

A further development came with the provision of 'paring discs' at the outlet of separators. The paring disc converts the rotational energy of exiting milk and cream into linear kinetic energy and acts as a stationary centripetal pump. Figure 6 shows a cutaway diagram to illustrate the action of a paring disc. The paring disc has these advantages.

(1)   The generated pressure can be used to push the exiting cream or skim-milk through heat exchangers.

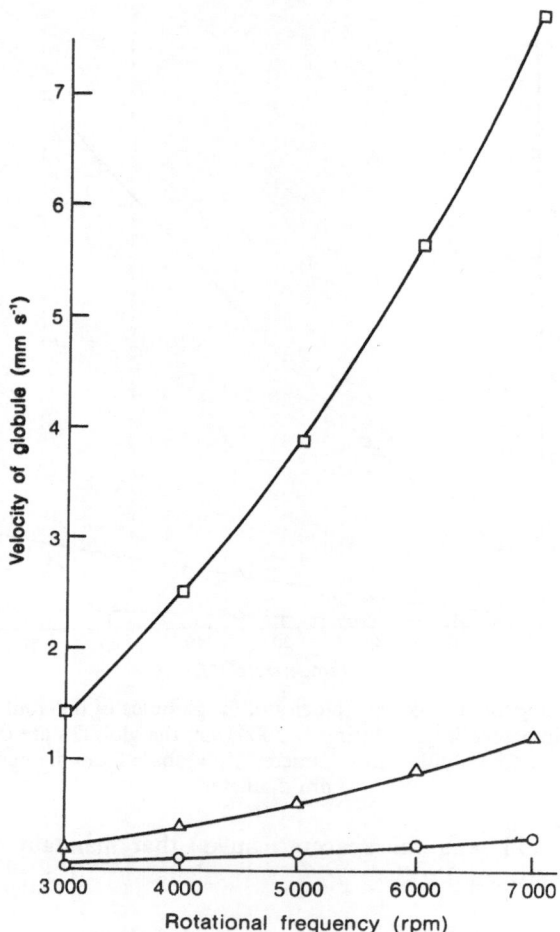

Fig. 3. The dependence of the velocity of fat globules of different diameters on the speed of rotation of a centrifuge at 35 °C; the globules are 0·1 m from the axis of rotation. ○: globule 1 μm diameter; △: globule 2 μm diameter; □: globule 5 μm diameter.

(2)   Variable flow restrictive devices on the outlets can be used to generate back pressures which will control flow. In this way a fairly accurate control of fat content in the cream can be ensured.

Separators with milk feed at atmospheric pressure and paring discs at the

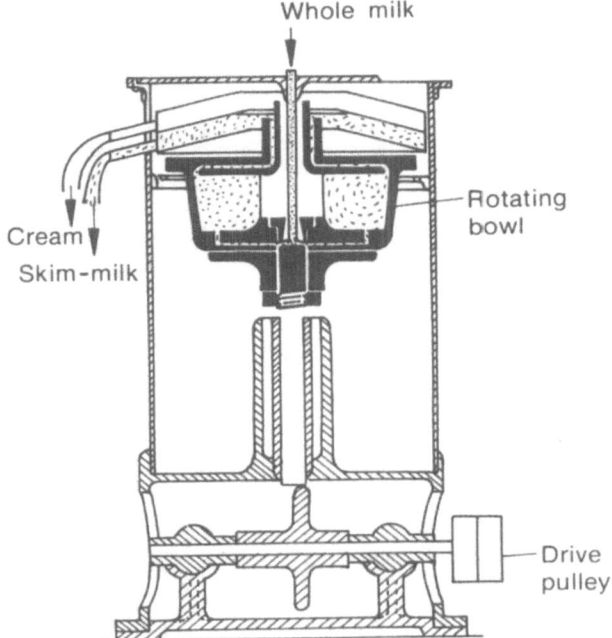

Whole milk

Rotating bowl

Cream

Skim-milk

Drive pulley

Fig. 4. An early model separator (courtesy of Alfa-Laval).

outlet are known as semi-open types; those mentioned previously with straight discharges are open types.

Hermetic Separators

Entrainment of air in milk inhibits separation. Although it might be expected that low density air would migrate rapidly with the fat globules, the presence of naturally occurring surface-active agents in the milk gives a somewhat stable structure to the air bubbles. This led to another development in separators, the airtight or hermetic separator. The development was based on the provision of hermetic seals which effectively isolated the separator from the atmosphere. In contrast to open or semi-open separators, the milk is introduced into the separator from below via a hollow spindle in the central shaft. The milk gradually reaches the rotational speed of the separator, in contrast to the older open types in which rapid acceleration took place on reaching the distributor. Separation takes place as in a normal separator through the

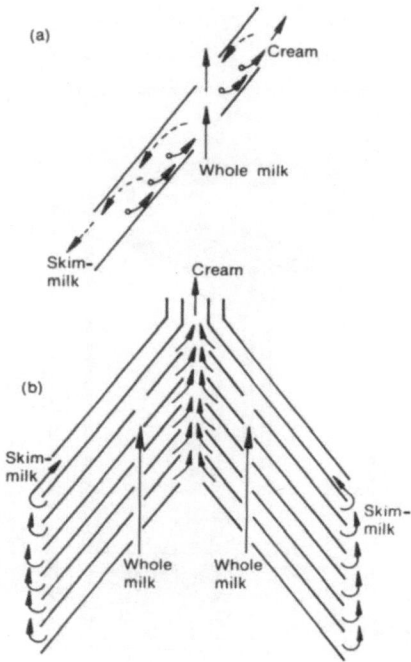

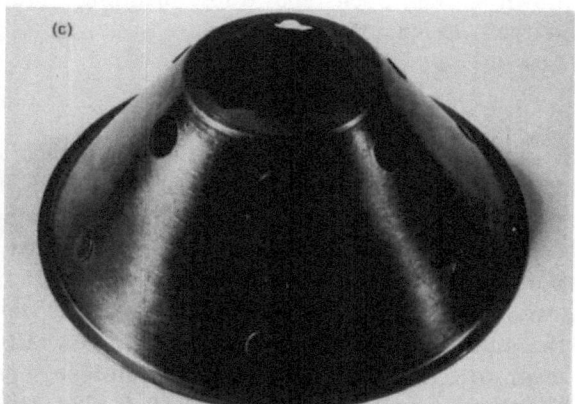

Fig. 5. (a) Flow of cream and skim-milk in the space between discs in a centrifugal separator. (b) A disc stack. (c) Photograph of a separator disc showing holes for channelling of milk and spacers (caulks).

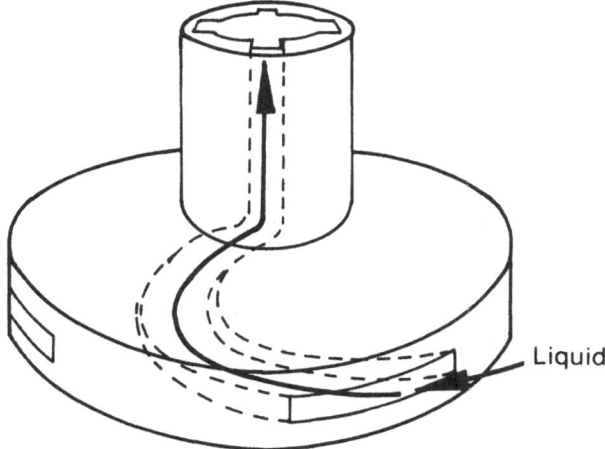

Fig. 6. A paring disc functioning as a centripetal pump.

disc stack. Efflux of cream and skim-milk takes place through hermetic seals under a moderate pressure. If a higher pressure discharge is required to feed heat exchangers, then pumps or paring discs may be built into the outlets. Often a paring disc is used on the skim-milk outlet without a hermetic seal. Such separators are known as 'semi-hermetic'. The feeding of hermetic separators is important as correct flow and discharge pressure must be achieved. The effectiveness of a hermetic separator does depend on the whole milk being air-free as it is fed to the separator, and it is important that air entrainment be avoided. The importance of hermetic sealing is that air will not be mixed into the product as a result of the separation process.

Self-desludging Separators

Milk does contain solid particles, including contaminating dirt and cellular material from blood and bacteria. This dense solid material collects on the outside of the spinning bowl and, if left, inhibits the efflux of skim-milk and stops the separation process. This limited the running time of separators and led to the development of a mechanism for automatically removing the solid material without having to interrupt the operation of the machine. Such separators are variously called 'self-desludging' or 'self-cleaning' separators. The diagram of a self-desludging separator (Fig. 7) shows the principle of operation. Slots are cut in the

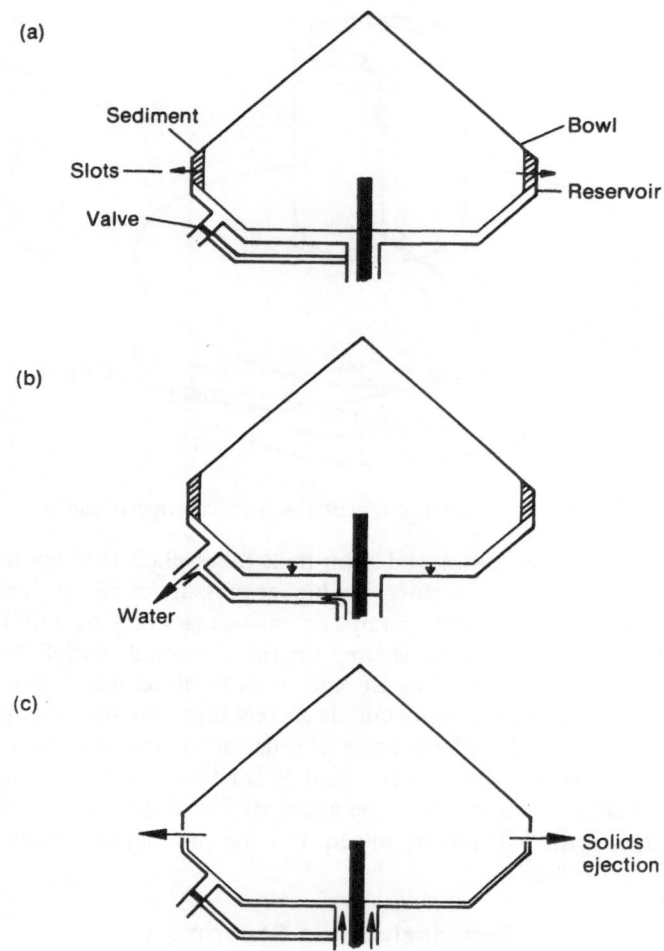

Fig. 7. (a) Self-desludging separator in normal operation with bowl bottom elevated by hydraulic pressure of water in reservoir. (b) Activation of hydraulic-ally operated valve causes reservoir to empty and bowl bottom moves down. (c) Bowl bottom depressed opens up slots and solid sediment is ejected. Valve closes to allow reservoir to refill moving bowl bottom up again.

outside of the bowl and are kept closed by a sliding bowl bottom (piston), elevated by hydraulic pressure of running water in a reservoir under-neath. When desludging is required, a hydraulically operated valve opens up and the reservoir drains to allow the bowl bottom to fall, thus opening the slots. The outward pressure on the sediment forces it out and it is ejected into an outer bowl (hood). The waste material has rotational motion which allows it to be taken tangentially out of the outer bowl into a cyclone for suitable disposal. The valve then closes to allow the reservoir to refill, and the slots are closed off. The action is very rapid, the slots are open for less than one-fifth of a second. Discharges are normally programmed to occur at regular, predetermined intervals that depend on the volume of the sediment space in the separator bowl and the condition of the milk. The above mechanism is not exclusive; several other methods of operating the sliding bowl have been devised. A system that uses less water has the sliding piston only covering about half of the bowl bottom. This gives considerably shorter opening and closing times and allows much better control of partial desludgings. Such systems require a water metering device to control the lower volumes required (Lehmann and Zettier, 1987). Non-self-desludging separators are often distinguished as 'solid-bowl' centrifuges as they do not have the separate 'floating' bottom.

Factors Affecting Separation

The previous section has covered the major advances in separator design and the principles involved in separation. At this stage it is pertinent to look at the different processing conditions that have an influence on separation and how these should be viewed in a practical situation.

Normally the object of separation is to attempt to recover all the fat in the whole milk within the cream fraction, with the minimum amount of fat being retained in the skim-milk. Skimming efficiency is assessed as the fat content of the skim-milk. The word 'normally' is used as, in fact, it is dependent very much on the fate of the various fractions. If the skim-milk is merely going to be remixed with cream to produce milk with a standardised fat content, then the skimming efficiency is not so vital. If the skim-milk is to be converted into skim-milk powder or casein, then it is important that the fat content be low, to meet the various specification and functional requirements of these products. It is particularly important if a casein whey is to be subjected to ultrafiltration or other membrane processing, as the fat can have an inhibiting effect on the flux rate through a membrane. Figures 2 and 3 illustrate the theoretical effects of globule size, milk temperature and rotational speed of a centrifuge on the velocity of fat globules, but, for practical purposes, other factors have to be taken into account. The most comprehensive treatises on separation are generally of Russian or German origin, Lipatov (1976) having written

a book on the subject. Other pertinent literature is to be found in Lang
and Thiel (1955), Alfa-Laval (1980), Kessler (1981), Lehmann and Zettier
(1987) and Rothwell (1989).

Temperature

An increase in milk temperature leads to an increase in density difference
between skim-milk and milkfat and reduces the viscosity of the skim-
milk, so that, in theory, an increase in temperature should lead to an
increase in separation efficiency. In practice, a higher temperature can
lead to disruption of fat globules and, as noted earlier, this has a very
major effect on the rate of separation. King *et al.* (1972) showed that fat
losses in skim-milk were slightly higher with a separation at 72 °C than
with a separation at 54·5 °C, but separation at 32 °C gave a higher fat loss
than separation at 54·5 °C or 72 °C. The temperature history of the milk
has an important bearing on separation efficiency; the modern practice of
farmers cooling to approximately 5 °C and holding for collection leads to
some disruption of globule agglomerates, and there is an increase in
water binding at the fat globule membrane interface (Lehmann, 1982). In
practice it would appear that most plants in New Zealand have settled on
temperatures around 50 °C, and 50–55 °C is generally recognised as being
optimal for skimming efficiency (Lehmann and Zettier, 1987), and is in
agreement with manufacturers' recommendations. Separation temperatures can have an effect on free fat levels. Te Whaiti and Fryer (1975)
found greater increases of free fat levels in cream with separation at 70 °C
than with separation at 35 °C. However, separation at low temperature
(less than 10 °C) will give little increase in free fat levels as the fat is mostly
in the solid form. Free fat formation is thus less when the fat is all in
liquid form or all in solid form, and the range 10–40 °C is the most
critical. Free fat formation is, however, largely a factor of mechanical
treatment, and can be minimised by preventing air incorporation and
taking care in pumping cream.

   If separation is practised at high temperature, then the separation step
can be included as part of the pasteurisation holding time. This approach
can reduce heat exchange requirements as the separate cream and
skim-milk streams need not be pasteurised, although it will mean a
somewhat more complex regenerative cooling system to get the cream
and skim-milk to suitable final temperatures. The choice will depend very
much on the fate of the cream and the skim-milk. The major effects of a
high separating temperature are protein denaturation and phospholipid
migration to the skim-milk; this latter point is of great importance in

cream properties, and will be referred to later. Phospholipids in whey also inhibit flux during membrane processing.

Cold milk separators that will operate at temperatures less than 10°C are available. These allow separation of milk as it is received at the factory, and, although fat losses to skim-milk are somewhat higher, they do sometimes allow substantial savings in energy and capital costs. In some cheesemaking operations, heat treatment of the milk is undesirable and cold milk separators offer some advantages. In addition, cold milk separators produce cream with a greater phospholipid content; this gives better whipping properties. The major modification in a cold milk separator is a wider disc spacing than in a conventional model (approximately double) to allow adequate flow of the more viscous cold cream.

Bowl speed

The velocity of a fat globule is proportional to the square of the rotational speed, so an increase in bowl speed will have a very major effect on separation efficiency. An increase in bowl speed, however, requires an increase in energy input, and a more robust design to withstand the large forces at the bowl periphery. The separator also generates more noise. For this reason bowl speeds have not increased significantly, as skimming efficiency is quite adequate at moderate speeds of 4000–6000 rev min^{-1} (rpm). Early design separators operated with bowl speeds of 3000 rpm. However, it is important that the bowl speed be maintained during operation, and many separators are fitted with tachometers to ensure that rotational velocity is consistent. Momentary deceleration does take place during a desludging operation, so it is important that it be carried out within a minimum time period, as separation efficiency is adversely affected.

Disc configuration

Lang and Thiel's (1955) review on centrifugal separators in the dairy industry summarised several studies on the effect of disc spacing on separation efficiency. In theory, the smaller the space between the discs, the higher should be the efficiency of separation as the fat globules have less distance to travel before being 'captured' on the disc surface. However, flow patterns must be taken into consideration; it is important that laminar flow conditions exist for maximum efficiency of separation. Any turbulence will result in fat globules being remixed with the milk stream, increasing the possibility of their being swept out with the skim-milk. For relatively narrow spacings (less than 0·2 mm) it is found that separation efficiency is independent of disc spacing, because the disc spacings take equal volumes of milk in a given time. Thus the tangential force moving milk through the disc space is created by the friction of the disc on the milk. For a narrow disc spacing, there is more surface contact with the milk giving a proportionately greater flow outwards than for a wide spacing. Thus, within certain limits, separation efficiency is independent of the disc spacing. These limits are dependent on flow factors (i.e. rate of milk flow and viscosity) and bowl rotational speed. For example, a cold milk separator requires a larger disc spacing than a hot milk separator because of the higher viscosity of the cream.

The disc angle has an effect on the distance fat globules have to travel within the disc spaces and the velocity of the components when they reach the disc surface. The number and size of discs determines the volume of milk that can be separated in a given time. Equations that relate disc configuration to separation and flow parameters can be found in the account of the subject by Kessler (1981).

It is important that discs are not distorted through physical damage from stress, or dismantling and cleaning operations. The disc spacers (caulks) are also of importance in ensuring that excessive turbulence does not occur as fluid passes by. Thus, the assembly of a disc stack containing the necessary multitude of discs makes it necessary to use a large amount of physical pressure. This is to force the discs into the necessary configuration, where the discs are in intimate contact with the spacers to get the required interdisc spaces. An O-ring locking device is also used to ensure that this spacing is maintained.

The placement of channels in the discs is important, and should reflect the likely proportion of flow between light and dense phases. For milk separation, the channels are displaced toward the axis of rotation as the cream flow is less than the skim-milk flow. In practice, the larger fat globules separate within a relatively short distance after entering the disc space and a distinct 'line' is seen on the discs when the stack is disassembled. The line denotes the zone of separation and this should ideally coincide with the placement of the channels. A separator for anhydrous milkfat (AMF) production from molten butter has channels that are very close to the disc periphery as there is proportionately more fat than aqueous phase.

Flow
The flow of incoming milk and the relative flows of exiting skim-milk and cream are factors that must be carefully controlled in order to

achieve good separation efficiency with adequate throughput. Much more efficient separation can be achieved with lower input feeds, as the milk has more time within the separator to allow fat globules to segregate. However, it is important that the flow should not be so low as to allow significant air entrainment to fill available bowl space, as separation efficiency will be adversely affected.

In practice, as high a flow rate as is possible will be aimed for, i.e. minimum processing time, but separating efficiency cannot be compromised as this will lead to high fat losses. In extreme cases, if incoming milk flows are too high, 'flooding' of the separator occurs and virtually no separation takes place. The flow of cream and skim-milk must, naturally, equal the flow of incoming whole milk. In a fully open separator with no control on discharge flow, the fat content of the cream must be controlled by the flow of the incoming milk. In commercial separators, however, the flow of skim-milk and cream can be controlled by back pressure, and this will control the fat content of the cream. The fat content of the cream is dependent on its final use, as will be referred to later in standardisation. Modern separators are normally supplied with flow controllers on the cream and skim-milk lines to adjust for maximum separating efficiency, and to maintain the desired fat content in the cream. It is important that, once separating conditions have been stabilised, constant back pressure be maintained in the lines and normally the flow controllers incorporate this function.

Lehmann and Zettier (1982) have described a 'Soft-Stream' system for a non-hermetic separator in which the diameter of the inlet chamber is approximately 1·4 times the diameter of the feed pipe, and the cross-sectional area is greater than the total cross-sectional area of the feed ducts into the disc stack. The introduction of milk into the expanded area in the chamber results in more gentle treatment. The reduced area of the distributor outlets acts as a natural throttle and keeps the inlet chamber full of milk. Some gas is present in milk because of natural dissolution; Lehmann and Zettier (1987) have shown the levels that are present at various points in the milk treatment and separation processes. Any gas liberated from the milk escapes through a drill-hole in the feed ducts and is circulated back to an annular chamber at the top of the inlet chamber to be released with the cream or reintroduced into the liquid stream. This internal gas circuit prevents large bubbles of gas entering the disc spaces to disrupt the separation process. Figure 8 illustrates the important features of a 'Soft-Stream' system.

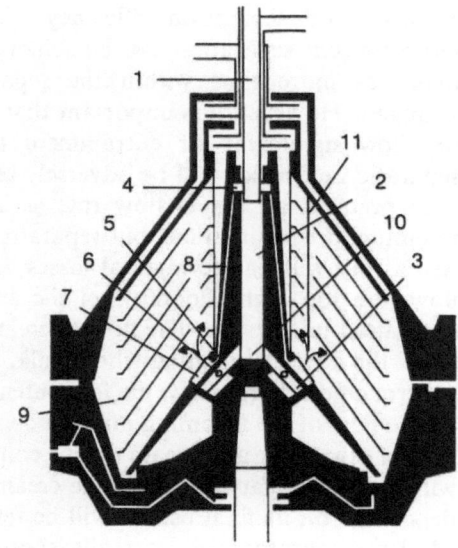

1   Feed tube
2   Inlet chamber
3   Outlets
4   Annular chamber
5   Feed ducts
6   Drillhole
7   Chamber
8   Drillhole
9   Disc stack
10 'Soft-Stream' inlet
11 Rising channels

Fig. 8. Cross-section of a separator with a 'Soft-Stream' inlet system (courtesy of Westfalia Separator).

Raw milk

The state of the raw milk has a large bearing on separation efficiency. Figures 2 and 3 illustrate that, whatever conditions prevail in separation, the velocity of a fat globule is very dependent on its size, and a high proportion of small fat globules will lead to a reduced separation efficiency. In practice, fat globules less than approximately 1 μm represent the lowest limit of centrifugal separation. In normal milk, that would represent a level of approximately 0·04 per cent fat in the skim-milk. Fat contents in skim-milk in commercial operation are normally in the range 0·04–0·06 per cent with cream being taken off at approximately 40 per cent fat, so this represents a good separation efficiency. There are several factors that may affect the fat globule size distribution.

(1) Breed of cow.
(2) Stage of lactation.
(3) Temperature history of the milk.
(4) Handling of milk through agitation, pumping, aeration, etc.

Conclusions on Separation

Figures 9 and 10 show diagrams of commercial separators of various types. Further advances in separators have been confined to increases in capacity to fit the needs of larger processing units and refinements to cleaning-in-place (CIP) procedures, whereby separators can be washed and rinsed without stripping down at regular intervals. Mechanical advances have concentrated on energy efficiency and noise reduction. The latter has been achieved through production of double-walled skins with water in the gap. This water, together with water used in the self-desludging mechanism, also exerts a cooling action. Sand-filled hoods can also be used for noise reduction.

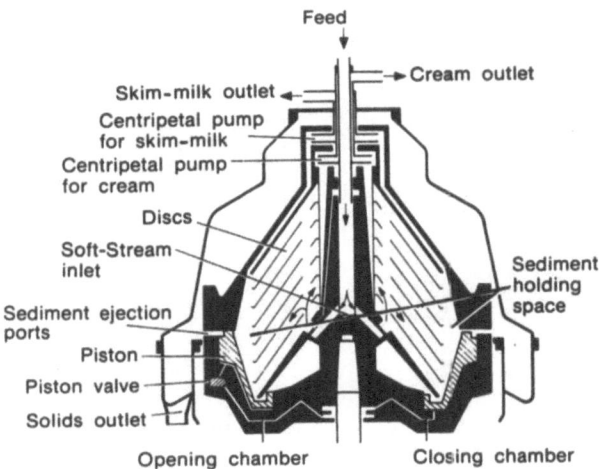

Fig. 9. Cross-section of a semi-open separator (courtesy of Westfalia Separator).

More recent major advances have been the automation and control of the separation process. Separation is normally included as an integral part of milk reception and pasteurisation processes, all of them subject to microprocessor control. The microprocessor controls the operation of valves and pumps. Information is fed back from devices that measure temperature, pressure, flow and level in tanks. Under such control, milk can be received, preheated, separated, pasteurised and cooled, with cleaning to follow. This can be carried out using a variety of conditions and a number of modes of operation through different programs that can be fed into the microprocessor from a keyboard, disc or tape. In the

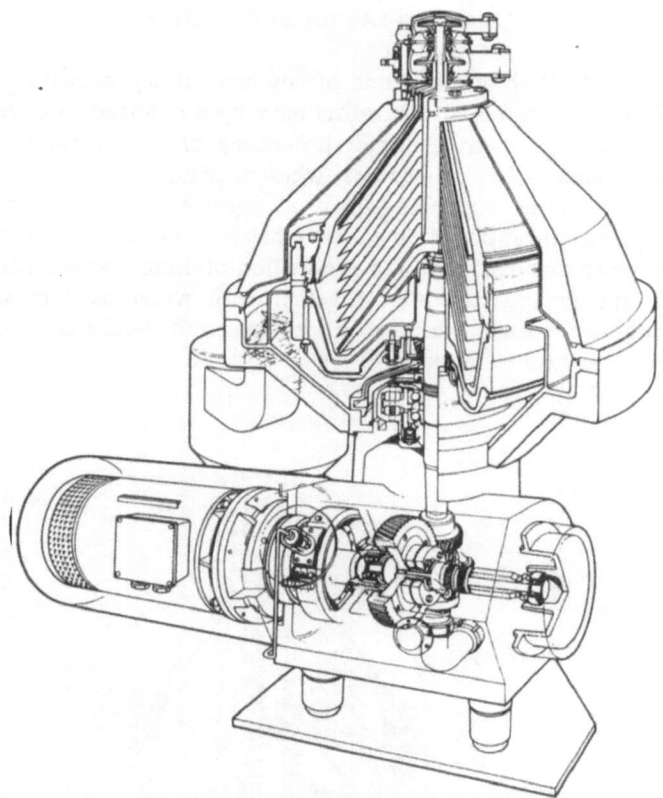

Fig. 10. Cutaway diagram of a hermetic separator including drive system (courtesy of Alfa-Laval).

separator, microprocessor control applies in particular to the incoming milk flow, back pressures on skim-milk and cream, and the desludging frequency. More sophisticated control systems will be referred to in the next section on standardisation and Chapter 9 gives a much more detailed account of automated control systems in general for dairy factories. Special disc-bowl separators have been designed to perform functions other than the separation of skim-milk and cream. Some of these functions are listed below.

(1) Cream concentrators to produce plastic cream (60–80 per cent fat) from normal cream of approximately 40 per cent fat content.

(2) Clarifixators for inverting phase and separating aqueous serum from liquid milkfat in AMF (or butteroil) manufacture.
(3) Clarifiers merely to remove solid material from milk or whey.
(4) Bactofuges which decrease the bacterial count in milk by centrifugal action – such centrifuges have a high rotational speed.
(5) Nozzle-bowl centrifuges for continuously ejecting dense phase concentrates.
(6) Desludging type separators for separating fine particulate matter from slurries, e.g. lactalbumin manufacture.
(7) Cream cheese separators for separating cream cheese (which forms the light phase) from whey.

These modified separators have different disc designs and various systems of feeding and ejecting components dependent on the consistency and relative flows of material. The large desludging separators may also be belt-driven to cope with the extra mechanical strain associated with frequent desludging operations. West (1985) has written a general account of disc-bowl separators.

STANDARDISATION

The important factor in standardisation of cream is the fat content. In market cream, if the fat content is higher than the specified requirement, then financial returns will suffer. If the fat content is low, then the cream may not meet regulatory requirements and the cream may also lack desired functional properties, such as viscosity or whipping properties. In cream for butter manufacture, although accurate standardisation is not so important, the fat content should be such as to give ease of churning and low fat losses to the buttermilk. Too low a fat content will result in a high buttermilk volume; this will not return as much as the skim-milk that would otherwise have resulted from better separation control. Standardisation is also an important facet of market milk production, and much of the technology is applicable to cream and milk.

To achieve accurate standardisation, the fat content should ideally be controlled as it leaves the separator. For small-scale operations, however, standardising in bulk is the norm. The separator is adjusted to give a slightly higher fat content than required and a suitable diluent such as skim-milk is added to obtain the required fat content. Such dilutions can be calculated from first principles, or by using Pearson's square. A very important consideration in this operation is the measurement of fat content. Normally a reasonably quick measurement that precludes the accurate Werner-Schmidt or Rose-Gottlieb methods is required. The Babcock or Gerber methods, which rely on volumetric measures of extracted fat, are quicker, if a little less accurate. Most dairy factories, however, possess Milkotesters®, Milkoscans® (A.N. Foss Electric, Denmark) or instruments that will give rapid measurements of good accuracy provided they are set in the correct mode with a properly prescribed diluent in the cream. Once the fat content of the cream has been determined, it is necessary to add the correct proportion of diluent (e.g. skim-milk or whole milk) to get the required fat content. Tanks may be fitted with load cells, but are normally fitted with volumetric measures of contents, and as such the density of the cream should be taken into account as higher fat contents give lower densities. For bulk standardisation, it is more appropriate to consider some examples.

Example 1. A 5300 l quantity of cream has a fat content of 42·3 per cent. A fat content of 40 per cent is required. How much skim-milk should be added? If we assume that the skim-milk contains 0 per cent fat, then

$$\text{Quantity of fat in cream} = 5300 \times 0·423 \times \rho \text{ kg}$$
$$= 2241·9\rho \text{ kg}$$

where ρ is the density of the cream.

The total amount of cream that would have this quantity of fat as 40 per cent of its content would be

$$2241·9\rho \times 100/40 = 5605\rho \text{ kg}$$
$$= 5605\rho/\rho' \text{ l}$$

where ρ' is the new density of the cream.

To a first approximation, $\rho = \rho'$; hence the total quantity of cream required is 5605 l and skim-milk (305 l) should be added to the cream to get the required fat content. Cream close to 40 per cent fat content has a density of approximately 1 kg l^{-1}. So for practical purposes litres volume can be roughly taken as kilograms weight.

Example 2. If in the above example whole milk with 4·2 per cent fat content were to be used as a diluent, then the Pearson square method becomes much easier to use. The essentials of the method are shown in Fig. 11. The fat content of the cream used is set in the top left-hand corner

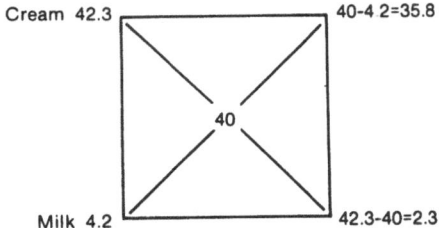

Fig. 11. Pearson's square as a means of calculating proportions of components for standardising.

of a square (42·3 per cent) with the fat content of the diluent at the bottom left-hand corner (4·2 per cent). The required fat content is then placed at the intersection of the square's diagonals (40 per cent) and the proportions of the cream and the whole milk are simply obtained by subtraction along the diagonals.

$$\text{Cream required} = 40 - 4{\cdot}2 = 35{\cdot}8 \text{ parts}$$

$$\text{Whole milk required} = 42{\cdot}3 - 40 = 2{\cdot}3 \text{ parts}$$

Thus the amount of whole milk added to 5300 l of cream should be

$$5300 \times 2{\cdot}3/35{\cdot}8 \, l = 341 \, l$$

i.e. the cream should be made up to 5641 l with whole milk.

These are only approximations as strictly speaking the calculation should be based on weight rather than volume. However, if the change in fat content between the original and the standardised is not too great, then the approximation of equal density can be made. The relative inaccuracy of the tank volume measurements and any air entrained in the cream will also give a false volume measurement. Because of these inaccuracies, final analytical checking of fat content is advisable.

In high volume factories, it is more convenient to produce cream of a required fat content directly from the separator. This can be done purely by adjusting the flow of skim-milk and cream via back pressures for a fixed flow of incoming milk. Provided the fat content of the milk is known, the back pressures on the cream and skim-milk lines can be adjusted with special valves to provide the required proportional flow of skim-milk and cream. A further refinement is a branch from the cream line with a flow adjuster to feed a proportion back into the skim-milk and produce a milk with a standard fat content. The important factors in such an operation are that standard conditions prevail during separation and that flow regulators are accurately calibrated. Flow measuring devices that assist these operations are available.

More sophisticated operations require an accurate measurement of milkfat content as the cream leaves the separator. A convenient method of ascertaining the fat content of cream is to measure its density, which is a function of the fat content. For example, in the temperature range 40–80 °C density (kg m^{-3}) can be computed using an equation from Phipps (1969).

$$D = 1038 \cdot 2 - 0 \cdot 17T - 0 \cdot 003T^2 - \phi(133 \cdot 7 - 475 \cdot 5/T)$$

where   T = temperature in °C
        ϕ = fractional fat content.

As cream density is largely influenced by fat content, the use of an in-line densitometer can provide a constant monitor of fat content. For standardisation, a constant density is required, and in this regard the unsophisticated principle of maintaining a constant volume at a constant weight can be used. A U-shaped tube of liquid is counterbalanced by weights, and any movement of the tube through losing balance can provide a signal to a controller to restore the balance. A diagram of such an instrument can be found in the Alfa-Laval Dairy Handbook (1980), and Pato (1978) has also described incorporation of density measurement as a means of direct standardisation.

The densitometer is sensitive to vibration, and air incorporation in the cream will greatly affect its accuracy; other mass flow meters that relate density and total flow of liquid are available. Less sophisticated systems that incorporate a Milkotester® are available and, although not as accurate as a densitometer, are less sensitive to environmental changes. There have been very significant advances in control systems, which incorporate microprocessors to receive signals from measuring devices and transmit them to operate controls. Figure 12 illustrates a typical arrangement for producing standardised cream. Figure 13 shows an extended system for producing standardised milk and cream. The major perturbations for such systems are changes of milk feed silos, which may change milk composition as well as feed pressure to the separator, and desludgings of the separator. The efficiency of a system depends very much on the rapidity of response to such changes and the re-establishment of the required conditions through design and tuning of the feedback control loops.

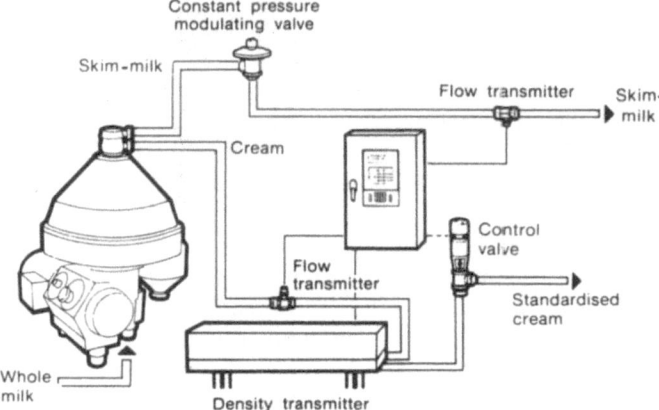

Fig. 12. System for automatic production of standardised cream (ALFAST courtesy of Alfa-Laval). Key: ————, electrical signals; – – – –, pneumatic signals.

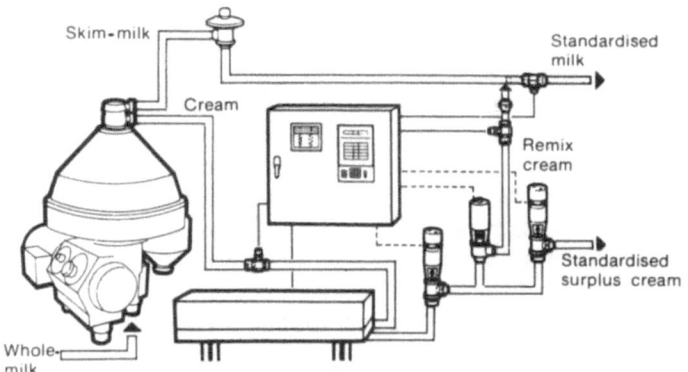

Fig. 13. System for automatic production of standardised cream and milk (ALFAST courtesy of Alfa-Laval). Key as for Fig. 12.

PASTEURISATION AND VACREATION

Heat treatment of cream is necessary to destroy organisms and enzymes that may be pathogenic or cause spoilage. Lactic acid, produced by bacteria, causes souring and coagulation of cream. Proteolytic enzymes may produce bitter peptides and also cause coagulation. Lipolytic enzymes will break down the lipids to produce free fatty acids which give a rancid flavour. Cream can be pasteurised by conventional means and the principles of heat treatment of milk given in Chapter 1 can be applied to cream. Cream is more viscous and somewhat more susceptible to mechanical breakdown, and this should be borne in mind when choosing equipment for cream pasteurisation, e.g. positive pumps would be preferred to centrifugal pumps. Whereas 72 °C for 15 s might be a typical regime for pasteurising milk, a higher temperature of about 80 °C would generally be used for cream. A higher temperature than 80 °C may impair cream quality, possibly through activation of bacterial spores (Rothwell, 1989). Eibel and Kessler (1986) have given an account of how heat treatment of cream affects shelf-life. They recommended that temperatures be kept below 105 °C as higher temperatures severely disrupt the milkfat globule membrane.

Pasture feeding of animals can produce flavour taints through herbage-derived substances dissolved in the fat. As most of the tainting substances are relatively volatile, a process was devised in New Zealand both to pasteurise the cream and to remove the volatiles through what is essentially a steam distillation process. The piece of equipment is known as a Vacreator®, which was the trade name adopted for the Murray Vacuum Pasteuriser (present manufacturers and agents NDA Engineering Group, Auckland, New Zealand). The process is known as vacreation. Vacreation has been used in a number of countries, and not only improves the flavour of creamery butter, but also extends the shelf-life significantly when compared with butter derived from plate-pasteurised cream. In the Vacreator, steam is intimately mixed with cream and the condensed vapour plus volatiles are removed by flash evaporation under vacuum. Figure 14 shows a diagram of a Vacreator consisting of five vessels. The typical pressure and temperature conditions pertaining to each vessel are shown on the diagram. Raw cream is preheated in a tubular heat exchanger by vapours exiting from vacuum vessels 3 and 4. The cream is mixed with steam and vapours. It exits from vessel 1 and passes into vessel 3, where the pressure is reduced slightly and the cream and vapour are separated. The cream is then mixed with steam and vapour, exits from vessel 2 and the mix is passed into vessel 4 for separation. The vapours from vessels 3 and 4 are combined and passed through the preheater, before passing to a water jet condenser which provides vacuum and condenses the remaining condensable vapours. A spring-loaded baffle valve applies a back pressure to vapour from vessel 3, so that the pressure difference required to transfer cream between vessels is maintained.

The cream from vacuum vessel 4 passes into an internal cream pump and is pumped to vessel 1, where it meets fresh incoming steam. The

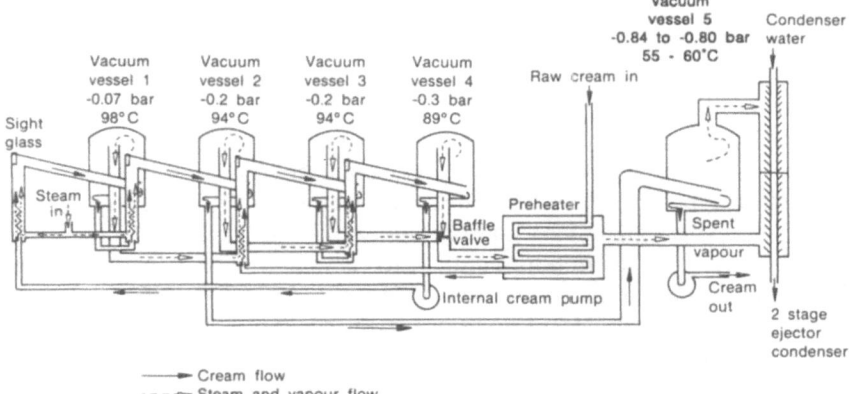

Fig. 14. Schematic diagram of a Vacreator®.

cream separated in vessel 1 is mixed with fresh steam again before passing into vessel 2. The cream exiting from vessel 2 passes into vessel 5 which acts purely as a flash cooler, with the vacuum removing water vapour and the associated latent heat. The cream exits at approximately 55–60 °C, and the vapours are removed in the condenser. The somewhat complicated system is called 'weaving flow', and is essentially counter-current with the cleanest cream meeting the cleanest steam.

The somewhat complex nature of flows has been found to be necessary because of the tendency of cream to foam under vacuum treatment, so that the separation of the liquid and vapours becomes difficult. The liquid and vapours are separated via a cyclonic centrifugal action with the cream being fed tangentially into each vacuum vessel at a slight downwards angle. The vapours are removed through a centrally mounted pipe. If foaming is severe, liquid gets carried over with the vapour stream resulting in product loss. The flow of steam assists in providing the cream with sufficient kinetic energy to flow through the system, but loss of energy occurs in the separation process which necessitates the use of the internal cream pump to push the cream to the final two stages. The flow from vessel to vessel is also controlled through the pressure differential between the vessels, but the high operating temperatures (and thus low vacuums) mean that transfer of cream by pressure differences is limited to two vessels in series. The vacuum levels, temperatures and flows of cream and steam thus require very careful control to ensure that excessive product loss through foaming or flooding does not occur. The modern

Vacreator is now equipped with microprocessor control to assist in achieving optimum temperature conditions during operation.

The amount of taint removed is proportional to the quantity of steam used. In the spring or during prolonged periods of wet weather, feed growth and the proportion of green feed in the diet results in increased levels of taints, and consequently high steam flows during vacreation are required, typically 0.25–0.3 kg steam (kg cream)$^{-1}$. During drier parts of the season, less green feeds are consumed so that less steam is required, typically 0.18 kg steam (kg cream)$^{-1}$, to remove the lower levels of taint. Taints resulting from poor quality cream may also be removed by vacreation, but high steam flows are required.

The major disadvantage of vacreation is its energy usage through the relatively large quantities of steam required, as the design of the Vacreator is such that vessel flooding will prevent operation at steam flows less than 0.15 kg steam (kg cream)$^{-1}$. Some heat is recovered through the preheater. Also available are thermorecompressors, which will generate low pressure steam from waste heat recovered from the vapours. The low pressure steam can be fed to the raw steam entering the Vacreator, but the cost of thermorecompressors warrants their use only with large, high-throughput units. The high energy usage of the Vacreator has led to some companies investigating flash-pasteurisation which incorporates a limited vacuum treatment. Such a process is acceptable for treating cream with a low taint level, but is generally unsuitable for cream with a high taint level if the cream is to be used for producing butter. Further experimental work in New Zealand has indicated that steam:cream ratios can be reduced if the proportion of steam entering vessels 1 and 2 is more carefully controlled. Control is by a valve that limits the quantity of steam passing to vessel 2, which is at lower pressure. With such control, the differential pressure between the two vessels can be maintained, and flooding of vessel 1 is eliminated even at steam flows as low as 0.09 kg steam (kg cream)$^{-1}$ (Cant, personal communication).

Steam quality is of utmost importance in Vacreator treatment. It must be of culinary standard and filtered because of the intimate contact with the cream. This limits the use of certain chemicals for the treatment of boiler feedwater.

Steam is injected into cream during vacreation at a velocity of approximately 140 m s^{-1} (500 km h^{-1}), and this violent treatment causes disruption of fat globules with an increase in the proportion of fat present as small globules (<2 μm). Vacreation will also increase the number of large fat globules (>10 μm) due to agglomeration resulting from foaming or flash-boiling (Dolby, 1953, 1957). The increase in the number of small fat globules can lead to higher losses of fat to buttermilk in butter manufacture. The introduction of low velocity steam diffusers alleviates this, and is possible when the steam split into vessels 1 and 2 is carefully controlled (Cant, personal communication).

New versions of the Vacreator will have the water ejector condenser replaced by a plate heat exchanger surface condenser and liquid-ring mechanical vacuum pump. Cream throughput will also be increased from the present maximum of 10,000 kg h^{-1} to 18,000 kg h^{-1}. A photograph of a Vacreator is shown in Fig. 15. The four primary vacuum vessels lie in a

Fig. 15. Photograph of a Vacreator®.

line with the preheater lying horizontally just underneath and to the rear of these. The flash vacuum vessel and condenser are situated behind the preheater.

CONSUMER CREAM PRODUCTS

Towler (1982a) has presented a review of consumer cream products, and more recently Mann (1989) has provided a literature review of market cream. The variable composition of cream and the properties relating to its emulsified state give a wide variety of forms that have different functions for use as food. The following presents a resumé of important consumer cream products.

(1) *Half-cream and single cream (10–18 per cent fat):* These creams would normally be used as pouring creams for use in desserts and beverages.

(2) *Coffee cream (up to 25 per cent fat):* The function of coffee cream is to provide an attractive appearance to the coffee with an appropriate modification in flavour. The hot, acid conditions in coffee provide an alien environment for cream, and protein precipitation may occur (feathering) with the release of free fat and a reduction in 'whitening' effect. The incidence of feathering and methods of alleviating it will be referred to later.

(3) *Cultured (sour) cream:* Inoculation of cream with lactic acid bacteria will induce souring through the conversion of lactose to lactic acid with subsequent coagulation of the protein. The solids-not-fat (SNF) content of the cream is thus as important as the fat dispersion in determining the texture of the final product. Preheating given to the cream will modify the protein in such a way as to alter the protein coagulum, and a higher protein content will lead to a firmer coagulum. Homogenisation of the cream will also increase its viscosity, and will inhibit creaming during the period of fermentation. The flavour of the finished product very much depends on the types of culture used and the acidity of the final product. Sour cream is used in a number of meat dishes, vegetable dishes and confectioneries. Smetana is a popular sour cream product in eastern Europe. Normally cultured cream would have less than 25 per cent fat, although crème fraîche, a lightly cultured product, may have a fat content of approximately 40 per cent.

(4) *Whipping cream (30–40 per cent fat):* When cream is subjected to mechanical beating, air is incorporated and fat globules collect at the air–water interface and the milkfat globule membrane becomes disrupted. If there is sufficient solid fat present, a firm structure will develop through bridging of adjacent fat globules. Anderson and Brooker (1988) have provided an account of the structure development in whipped cream including some aspects of the microstructure. There are several factors of importance for whipping cream.

(a) The amount of beating required to form a stable aerated structure (whipping time). This is dependent on the apparatus used to whip the cream, and the state of the cream itself. If whipping is continued beyond the optimum time, phase inversion and separation of fat will take place as in churning.

(b) The overrun. This expresses the percentage volume increase of the cream due to air incorporation. Thus, if a certain quantity of cream is whipped

$$\text{Overrun} = [(V_w - V_u)/V_u] \times 100\%$$

where V_w is volume of whipped cream
V_u is volume of unwhipped cream.

That is, if a quantity of cream doubles in volume on whipping, then the overrun is 100 per cent.

(c) The stiffness of the whipped cream. A stiff whip is particularly important for cream fillings in cake or for 'piped' cream decorations.

(d) Serum leakage from the cream. This will occur with partial churning (overwhipping) and leads to an unattractive 'pool' around the whipped cream, or 'sogginess' if applied to cakes.

There are several variables that affect the whipping properties of cream.

(e) Fat content of the cream. A high fat content will lead to a whipped cream with a low overrun but a firm structure and little tendency to synerese (provided it is not overwhipped). Conversely a cream of low fat content will form a high overrun whip with a soft structure. A certain minimum fat content is required to form a whip, and this would normally be about 30 per cent fat for a cream without specific additives or treatment.

(f) Temperature of the cream. The formation of a stable whipped structure depends on a sufficient quantity of solid fat, and the cream must be below 10 °C to whip satisfactorily; though there will be differences according to the lipid make-up of the milkfat in the cream. If cream is cooled to below 10 °C, warmed and then cooled again, an increase in viscosity results and the whipping time of the cream is reduced. The process is known as rebodying and was patented by Bergman and Svedberg (1934). The exact mechanism of the change has not been elucidated, but it is postulated to be due to changes in the milkfat globule membrane, although fat crystallisation changes could also be involved as marked changes in butter hardness can occur with the same temperature cycling. Time is also important when considering the aspect of temperature. It is necessary to hold fresh cream for several hours at low temperature before it will whip. This time allows fat crystallisation to take place and there may be other essential changes to the membrane.

(g) The fat globule size distribution and membrane structure. Homogenisation of cream has a deleterious effect on whipping properties, resulting in a marked increase in whipping time and a considerable reduction in whip firmness. Homogenisation results in a net increase in surface area of the fat globules, and the new membrane incorporates proportionately more protein from the serum. The new membrane is more stable to mechanical beating, and it is thus much harder to form a whipped structure.

Addition of natural membrane material (phospholipid) or emulsifiers can lead to an improvement in whipping properties. Separation of milk at low temperatures gives a cream with a high phospholipid content which will lead to better whipping properties (Thomé and Eriksson, 1973). Acidity in cream will reduce the solubility of protein in the membrane and this leads to reduced whipping times. Kammerlehner (1973a,b; 1974) has presented much useful information on factors that affect the properties of whipping cream, and has described how a low fat whipping cream can be produced using a ripening process. Mann (1987) has produced a literature review on whipping creams and whipped creams. The handling of milk and cream through transportation, separation, pasteurisation and other movements is quite critical in obtaining a cream that is of suitable quality as a whipping cream. Any disruption of globules can lead to a deterioration in whipping properties (Brooker, 1990). Temperature changes have a critical bearing on the distribution of phospholipids, which are also important in the whipping properties of the resultant cream. Aerosol cream represents a variation on whipped cream whereby cream (usually UHT) is packed in a can with nitrous oxide gas under pressure. Release of pressure through a valve causes the cream to be propelled out and dissolved gas volatilises to form a high overrun whipped cream structure, depending on the volume ratio of gas to cream. Such a whipped cream does not have the permanency of a mechanically whipped cream and shrinkage occurs. Aerosol creams are thus more suitable for desserts and confectioneries that will be consumed immediately. Machines are also manufactured for bulk or continuous dispensing of whipped cream. Batch machines have bowls with gas injection systems. Continuous machines normally consist of a refrigerated tank for holding the cream. A pump with an air-bleed then takes the cream and passes it through an insert designed to provide a tortuous path for the cream, resulting in much turbulence and shear (i.e. a simple static mixer) which whips the cream before it is dispensed out of a nozzle. The extent of whipping is controlled through the proportion of air that is bled into the cream.

(5) *Double cream (>48 per cent fat):* Double cream is marketed predominantly in Europe and represents an extra rich product for addition to desserts. Whipping will produce a very dense whipped cream which can be used in gateaux.

(6) *Clotted cream (>55 per cent fat):* This is traditionally produced in southwest England. The traditional process involves batch heating of gravitationally separated cream. The heating process induces rapid fat rise. The fat agglomerates on the surface and protein denaturation also takes place. Superficial skimming gives a thick spreadable product with a distinct cooked flavour. More modern methods for high volume manufacture use heat exchangers to get the cream to the required temperature followed by mechanical separation to the required fat content.

(7) *High fat creams:* Several high fat spreadable creams are found as indigenous products around the world. Examples are Gammer cream (Iraq) and Kajmac (Yugoslavia). High fat creams can be simply produced by passing normal cream through a second separation. Such a separator should have a relatively wide disc

spacing, and the distribution channels should be approximately half way down the disc faces. As the fat content in the cream rises and the globules pack together more tightly, the stability of the emulsion is reduced and phase inversion takes place very easily. A very high fat cream (70–80 per cent fat) is known as plastic cream.

(8) *Confectionery, butter or mock creams:* Such products are low moisture products containing high concentrations of sugar or other sweeteners, and are used as cake and bun fillings. They are, in fact, phase-inverted creams with the aqueous phase emulsified in the fat. The fat base is aerated by mechanical beating. Sweetener is usually added later.

PRESERVATION AND PACKAGING OF CREAM

As cream is a high moisture product, it is perishable and, without special care to preserve it, enjoys only a limited life without spoilage. Pasteurisation of cream extends the shelf-life to some extent, and this can be accomplished with conventional equipment. Pasteurised cream is packaged in cartons and bottles for local consumption. Unit volumes are normally in the range approximately 100–1000 ml. Bulk packaging may be used for catering or institutional use. Normally this is in the form of plastic (e.g. polythene) bags contained in plastic crates or cardboard cartons. Unit portions range from 5–25 l. Creaming is the major defect of pasteurised cream that has not been homogenised. Cream plugs are formed by the free fat which 'welds' the globules together. In severe cases the cream may totally solidify into a gel (Te Whaiti and Fryer, 1975). The free fat content may result from disruption of the membranes at high temperatures, or it may result from mechanical treatment through pumping or air incorporation. The rate of cooling of the cream and the composition of the fat are other factors that affect plug formation (Streuper and Hooydonk, 1986). For extended shelf-life there are several options, each presenting different technical difficulties in terms of maintaining the required properties of the cream during processing and storage.

Sterilisation

Applying sufficient heat to destroy all microbes and enzymes in cream, with steps taken to prevent further contamination, will result in a shelf-life dependent only on physico-chemical changes that may occur as a result of temperature changes or time. The simplest method of sterilisation is to package the material, then heat the complete package and material to confer sterility. The can and glass bottle have been the traditional containers for such operations, but other retortable plastic materials are now available for packaging. It is normal to give the cream a preheat treatment before packaging to destroy bacterial spores. Sterilisation takes place in a retort or hydrostatic steriliser using temperature–time regimes of 110–120 °C for 10–20 min. This severe heating induces gross changes in the cream with protein denaturation, Maillard browning and fat agglomeration all taking place to modify texture and flavour. A calcium sequestering agent, such as sodium citrate or a sodium phosphate, may be added to make more casein available for stabilising the emulsion. The unit packaging volumes have to be relatively small (< 400 ml) because of the restriction on heat transfer with larger volumes. Normally cream of approximately 23 per cent fat content (often called 'reduced cream') is the base cream for in-can sterilised cream manufacture. It enjoys a substantial market as a dessert adjunct or ingredient in a number of food items, such as dressings and sauces. It is well known to be a good base for party dips. Such a low-fat cream will not whip. In-bottle sterilised coffee cream has been marketed in Europe for many years. Higher fat creams are more difficult to deal with as their higher viscosity reduces heat transfer, and it is difficult to stabilise them against coagulation. Kieseker and Zadow (1973a) have described the effects of various parameters on the properties of in-can sterilised whipping cream.

The adoption of UHT sterilisation followed by aseptic packaging has led to the availability of a wider range of functional creams that can be stored for some months. The high temperature (135–150 °C) short time (3–5 s) treatment does not induce as much chemical change as in-container sterilisation, but other changes, such as creaming and fat agglomeration, will take place on storage. Steps must be taken in the processing of UHT creams to alleviate these problems.

Coffee cream tends to feather more after a period of storage, and Anderson *et al.* (1977) have shown that the index of feathering is related to a progressive increase in the proportion of calcium and casein associated with the fat phase of the cream. An increase in casein content (to provide more buffering capacity to the acid in coffee) and a reduction in calcium content markedly increase resistance to feathering during storage (Cheeseman *et al.*, 1978). The gravitational separation of fat in the stored cream is inhibited through homogenisation and the extent of homogenisation has a marked effect on the whitening of the coffee cream (Towler, 1982b). Geyer and Kessler (1989) have shown that resistance to feathering can be induced by coating the fat globules with denatured whey protein. Abrahamsson et al. (1988) have shown that homogenisation conditions markedly affect resistance to feathering. Optimum stability is achieved with two-stage homogenisation (20 MPa/5 MPa), both upstream and downstream of the UHT process.

The production of UHT whipping cream presents problems that require compromises to create a long shelf-life and adequate functional attributes. Homogenisation, which inhibits creaming, has a deleterious effect on whipping properties, although Graf and Müller (1965) have shown that adjustment of conditions to give fat globule clusters of 15–20 μm will result in a good whipping cream. It is essential that homogenisation be downstream of the UHT process to reform membranes damaged by the heating process (Fink and Kessler, 1987). This necessitates the use of an aseptic homogeniser; some tubular systems use a high pressure feed pump with remote homogenising valves after the heating tubes. Additives can markedly improve the stability and properties of UHT whipping cream. Stabilisers, such as hydrocolloids, gums and gelatin, inhibit fat rise and agglomeration of fat. Emulsifiers aid the whipping properties of the creams; some substantially increase overrun through their surface activity, whereas others will enhance fat globule interactions to decrease whipping times and form stiffer whips. Their incorporation is limited by the off-flavours they impart. Kieseker and Zadow (1973b) have reported the effect of several factors on the properties of UHT whipping cream. They showed that whipping improved with separation at low temperature and addition of calcium, but the cream had poor storage stability, whereas separation at high temperature and addition of calcium sequestrants led to creams with good storage stability but poor whipping properties. Towler (1988) has described the effect that additives have on the properties of UHT whipping cream.

The storage temperature of UHT whipping cream is particularly important in determining its shelf-life. Storage at 30 °C will induce considerable agglomeration of fat within a relatively short time, dependent on the amount of homogenisation and added stabilisers. Unstabilised UHT whipping cream can have a shelf-life of several weeks if stored at 5 °C. However, it is important that homogenisation conditions are adjusted carefully to obtain the necessary resistance to creaming while at the same time maintaining whippability. Stabilised creams have a shelf-life of several months at 5 °C. The rebodying process is particularly effective in improving the whipping properties of UHT homogenised whipping cream, but excessive temperature cycling is deleterious to shelf-life.

UHT cream is available in other forms such as single, reduced and double creams. A number of different packaging options are available, and aseptic canning was probably the first to be utilised with cream. Now plastic (e.g. polythene), paper and foil laminate cartons and plastic (e.g. polystyrene or polypropylene) form-fill-seal packages are most widely used. Unit sizes range from 7·5 ml (coffee cream) to 1000 ml. Aerosol cream is packaged in lacquered aluminium or tin-plate cans. Figure 16 shows several forms of packaged cream, both pasteurised and UHT. For aseptic packaging in preformed pots or with laminates, the packaging material is first treated with hydrogen peroxide solution, later removed by squeezing or natural drainage. Residual solution is removed by heat; the evaporating and decomposing peroxide sterilises the material. The sterilised product must then be filled in a sterile environment. With the Tetra Pak® system, the laminate is in the form of a continuous tube and the evaporating peroxide above the filler forms a natural aseptic barrier. Individual packages are formed by heat sealers and cutters at the base of the filler. Figure 17 shows the principle of the

Fig. 16. Photograph of creams in different packages.

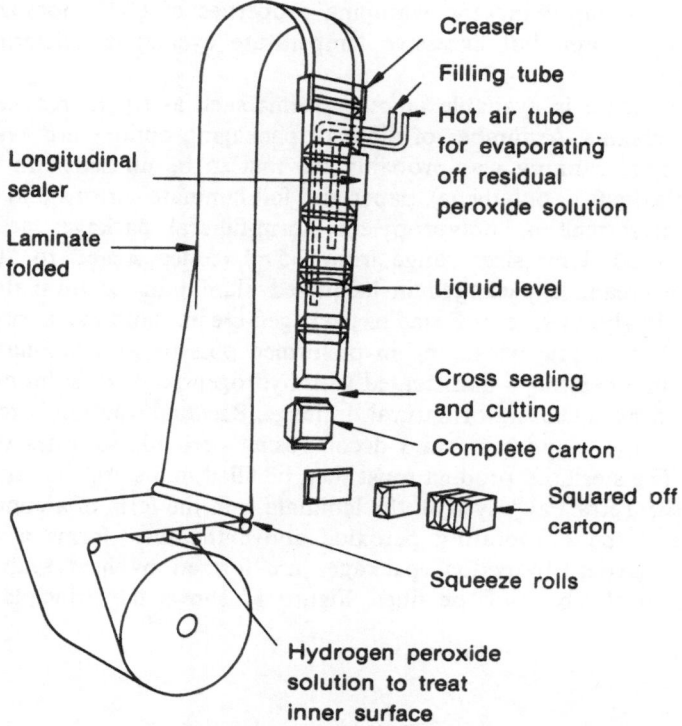

Fig. 17. The principle of the Tetra Pak® filling system.

Tetra Pak system. Most other systems incorporate laminar air flow cabinets for filling operations.

Bulk packaging is also available for UHT cream. Such packaging is normally in the form of a plastic (e.g. polythene) bag contained within a cardboard carton (bag-in-box). Unit volumes are in the range 5–1000 l. The larger volumes require something more substantial than cardboard for external support. An innovative bulk packaging system is the Intasept® which precludes the use of chemical sterilants for the packaging material (Anderson, 1985).

Freezing

Freezing of cream inhibits bacterial spoilage but also leads to destabilisation, and gross separation of fat and serum results on thawing. Such cream is suitable for reprocessing and bulk frozen cream has been exported from New Zealand for a number of years. The cream may be used as an additive for 'cream soups', where flavour is the prime consideration, or in recombined milk and ice cream where homogenisation is an essential part of the process. Cooper (1978) has described a suitable formulation for the preparation of a functional whipping cream from thawed frozen cream.

Freezing of cream has been practised in Europe for storage and subsequent utilisation in butter manufacture, which alleviates fluctuations in seasonal production in terms of both quantity and quality of final product. Frozen cream may also be used to boost the fat content of market milk when the natural fat content drops below the legal minimum. In New Zealand, cream is bulk packaged in containers of 20–25 l capacity. Plastic containers or bag-in-box systems have been used. The containers are placed in racks which pass through a blast-freezing chamber.

In Europe, plate or rotary-drum freezers are commonly used, and Rabich (1969, 1971) has described the problems of freezing cream by these methods. In plate freezing, the plates contain circulating refrigerant and are arranged vertically, in parallel, with bottom and end seals to form a series of moulds with hydraulic pressure maintaining the plates in place. The cream is poured into gaps between plates, and surface freezing is instantaneous at the precooled plate surface; this is essential to prevent adhesion of the cream to the plate. The cream freezes progressively toward the centre with refrigerant in the plates absorbing the heat, so that finally slabs of frozen cream are formed. The slabs are removed for packaging and subsequent storing by separating the plates.

In drum freezing, a rotating drum containing recirculating refrigerant is immersed in a vat of cream to form a frozen film. The frozen cream is then removed from the drum with a knife, and a flaked product is obtained. Such a process gives somewhat more rapid freezing than plate freezing, is less damaging to the cream and is continuous. The major disadvantage is that the flakes have a lower density than the slabs when packaged in bulk, and more freezer space is required for storage. Figure 18 illustrates the principle of drum freezing. It is essential that cream is adequately pasteurised to destroy enzymes, as many are still active at the low temperatures used in frozen storage. A temperature less than $-18\,°C$ is also recommended for long-term storage of frozen cream, as the rate of deterioration is inversely related to temperature of storage.

Without the use of additives, the stability of cream through a freeze–thaw cycle can only be ensured by the use of rapid freezing. It is only in

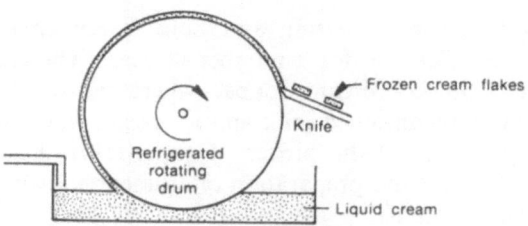

Fig. 18. Diagram of a drum freezer.

rapid freezing that the ice crystals are small and rupture of fat globule membranes is minimised. Löndahl and Johansson (1974) first reported the development of the Pellofreeze® machine (Frigoscandia, Sweden), which provides rapid freezing of cream contained as a thin sheet between two continuous stainless steel belts sprayed with low temperature glycol. The resultant frozen sheet is then broken up into flakes. This frozen cream thaws to form a fluid product that can be used as a normal consumer cream. Whipping cream and double cream are marketed in such forms; the cream flakes are packaged in heat-sealed plastic bags as used for frozen vegetables. A frozen novelty machine can be used to freeze cream at a rate that gives a reasonable consumer product, and Chase (1981) has described the marketing of such a cream.

Rapid freezing can also be achieved through a cryogenic process using the latent heat of low temperature boiling liquids to remove heat. Liquid nitrogen, which has a boiling point of $-196\,°C$, will freeze cream very rapidly. Work with a freezing tunnel has shown that a satisfactory frozen cream for direct consumer use can be produced by such a system; Fig. 19 illustrates the principle of the freezing technique. Cream is poured into suitable containers on a continuously rotating belt which, in turn, feeds the containers into an insulated tunnel. Liquid nitrogen is introduced at a point towards the other end of the tunnel and a series of fans distributes the cold nitrogen gas through the tunnel. Freezing is thus accomplished

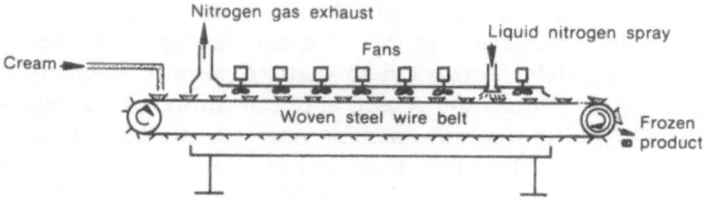

Fig. 19. Cryogenic freezing tunnel.

by an essentially counter-current flow of cold nitrogen gas, and the frozen cream is removed on exiting the tunnel. Taylor (1984) and Platt (1986) have described another cryogenic process. Cream is pumped into a circulating stream of liquid nitrogen; the cream breaks up into droplets and its crust frozen in less than 10 s. The frozen 'pea-like' droplets are separated from the liquid nitrogen with a perforated surface and freezing is completed in a turbulent gas stream.

Experiments on cryogenic freezing of cream have shown that the state of the cream is equally as important as the rapidity of the freezing process in the prevention of destabilisation. The preservation of the natural milkfat globule membrane is very important in getting freeze–thaw stability. Homogenisation adversely affects freeze–thaw stability as does partial churning or separation at high temperature (Towler, unpublished data). Additives, such as emulsifiers and stabilisers, assist in the freeze–thaw stability of cream if rapid freezing is not possible. Low molecular weight carbohydrates, such as glucose and sucrose, give protection against freezing, hence sweet creams can be frozen successfully and stored. The storage of frozen cream for consumer purposes demands not only a low temperature, but also a constant temperature, as temperature cycling may lead to the formation of larger ice crystals with resultant damage to the fat globules.

Drying

Removal of water from cream gives a product with an extended shelf-life. Commercial production has been practised on a limited scale for some years. The most common method of producing dry products from liquids is through spray drying, and Hedrick (1967) has described the manufacture and utilisation of spray dried cream. The particular problem in the production of spray dried cream is the high fat:SNF ratio in the finished product. The fat is in a liquid state at the temperature at which it would exit a spray drying chamber, so it is essential that the fat globules are encapsulated and protected by the non-fat solids. For cream, this normally requires the addition of extra protein, usually sodium caseinate, and the addition of a suitable carbohydrate (e.g. lactose, dextrose, maltodextrin or sucrose) to act as a carrier. Cooling of the powder is necessary to solidify the fat and prevent caking, which will occur if the thin protective membranes are ruptured. Cyclone collection of the powder results in mass caking of powder on the cyclone walls, and bag filters soon become impervious if fat is deposited. A Filtermat® drier (Damrow Co., Wisconsin, USA) has a continuous woven belt for collecting powder from the primary chamber and cool air can then be directed on to the powder in subsequent (fluid bed) stages of the drier. This drier is used extensively to dry powders with high fat contents; these powders are usually based on vegetable fats. Lundqvist (1978) has described a two-stage Filtermat drier; modern versions have three stages.

The high fat content of cream powder renders it susceptible to deteriorations which impair the flavour. It is essential that the cream is given sufficient heat treatment to destroy lipases. Oxidation of the fat is also a potential problem, and addition of antioxidant extends shelf-life. Storage of cream powders should be at low ambient temperatures, as an elevated temperature gives high proportions of liquid fat, resulting in caking problems, as well as more rapid flavour deterioration. The addition of free-flow agent is recommended to help prevent caking.

Cream powder provides a milkfat concentrate in a free-flowing form and as such has a number of potential uses. It can be used as an ingredient for dried soup, dessert, ice cream or packet cake mixes. Cream powders with a variety of fat contents are available, but difficulties in production limit upper levels to around 80 per cent. New Zealand currently markets products with 55 per cent and 70 per cent fat contents.

Dried cream has limited functionality when reconstituted; unlike the natural product, the globules do not reform. To obtain a functional product, incorporation of emulsifiers is necessary and homogenisation conditions are of particular importance (Griffin et al., 1970; Cooper & Peacock, 1979; Kieseker et al., 1979a, b). Prasad and Gupta (1988) have written a review of cream and butter powders.

Recombined Cream

A cream can be reformed from milkfat concentrates and recombined skim-milk or buttermilk by application of heat followed by homogenisation. For a good recombined whipping cream, it is better to incorporate phospholipids through the use of buttermilk or alpha-serum powder (the by-product from the manufacture of AMF from cream). Alternatively emulsifiers may be used. Towler and Stevenson (1988) have described the use of emulsifiers in recombined cream. The homogenisation conditions are critical in obtaining a cream with the desired functional characteristics. Whipped toppings and coffee whiteners are recombined emulsions designed to be cream substitutes. The use of vegetable fat sources results in much cheaper products, and, with a selection of different softening points, functional attributes can be much more easily tailored to meet the requirements of the product.

Cream Liqueurs

Cream liqueurs have become very popular alcoholic beverages and represent a substantial volume market for cream. The alcohol acts as a preservative against bacterial spoilage and the resultant products have long shelf-lives if suitable steps are taken to prevent creaming and destabilisation of the emulsion. Homogenisation is thus important, with sodium caseinate and sodium citrate additions to provide emulsion stability. Banks and Muir (1988) have described the factors of importance in the stability of cream liqueurs.

CONCLUDING REMARKS

The word cream has long been associated with a premium product. Milkfat possesses a unique flavour and its properties make it a favoured ingredient in many foods. In cream, the milkfat is protected and preserves a special flavour. New methods of packaging and preservation are going to lead to newer forms of cream for consumer use. Although improved forms of fabricated and simulated products may ensue, natural cream is bound to retain a special place in the eye of the consumer.

ACKNOWLEDGEMENTS

The author thanks his colleagues at the New Zealand Dairy Research Institute for assistance in preparing this chapter, particularly Dr Euan Cant for his contribution to vacreation. The valuable assistance of Alfa-Laval, Lund, Sweden, and Westfalia, Oelde, Germany, (through their New Zealand agents APV Baker) is also gratefully acknowledged.

REFERENCES

Abrahamsson, K., Frennborn, P., Dejmek, P. and Buchheim, W. (1988). *Milchwissenschaft*, **43**, 762.
Alfa-Laval (1980). *Dairy Handbook*, Alfa-Laval AB, Dairy and Food Engineering Division, Lund, Sweden.

Anderson, I. M. (1985). *Food Technology in Australia*, **37**, 399.

Anderson, M. and Brooker, B. E. (1988). In: *Advances in Food Emulsions and Foams* (Ed. E. Dickinson and G. Stainsby), Elsevier Applied Science, Barking, Essex, pp. 221–55.

Anderson, M., Brooker, B. E., Cawston, T. E. and Cheeseman, G. C. (1977). *Journal of Dairy Research*, **44**, 111.

Banks, W. and Muir, D. D. (1988). In: *Advances in Food Emulsions and Foams* (Ed. E. Dickinson and G. Stainsby), Elsevier Applied Science, Barking, Essex, pp. 257–83.

Bergman, T. V. and Svedberg, H. A. (1934). US Patent 1 944 541.

Brooker, B. E. (1990). *Food Structure*, **9**, 223.

Chase, D. (1981). *Milk Industry*, **83** (6), 17.

Cheeseman, G. C., Anderson, M. and Wiles, R. (1978). British Patent 1 526 862.

Cooper, H. R. (1978). *New Zealand Journal of Dairy Science and Technology*, **13**, 202.

Cooper, H. R. and Peacock, I.C. (1979). *New Zealand Journal of Dairy Science and Technology*, **14**, 291.

Dolby, R. M. (1953). *Journal of Dairy Research*, **20**, 201.

Dolby, R. M. (1957). *Journal of Dairy Research*, **24**, 372.

Eibel, H. and Kessler, H. G. (1986). *Deutsche Molkerei-Zeitung*, **107** (16), 478, 480, 482.

Fink, A. and Kessler, H. G. (1987). *Deutsche Molkerei-Zeitung*, **108**, 1376.

Geyer, S. and Kessler, H. G. (1989). *Milchwissenschaft*, **44**, 423.

Graf, E. and Müller, H. R. (1965). *Milchwissenschaft*, **20**, 302.

Griffin, A. T., Amundson, C. H. and Richardson, T. (1970). *Food Product Development*, **4** (7), 49, 52, 56.

Hedrick, T. I. (1967). *American Dairy Review*, **29** (9), 36, 120, 122, 124.

Kammerlehner, J. (1973a). *Deutsche Molkerei-Zeitung*, **94**, 1516, 1518, 1520, 1521, 1637.

Kammerlehner, J. (1973b). *Deutsche Molkerei-Zeitung*, **94**, 1742.

Kammerlehner, J. (1974). *Deutsche Molkerei-Zeitung*, **95**, 1758, 1789, 1820.

Kessler, H. G. (1981). In: *Food Engineering and Dairy Technology*. Verlag A. Kessler, Freising, Germany, pp. 67–69.

Kieseker, F. G. and Zadow, J. G. (1973a). *Australian Journal of Dairy Technology*, **28**, 108.

Kieseker, F. G. and Zadow, J. G. (1973b). *Australian Journal of Dairy Technology*, **28**, 165.

Kieseker, F. G., Zadow, J. G. and Aitken, B. (1979a). *Australian Journal of Dairy Technology*, **34**, 21.

Kieseker, F. G., Zadow, J. G. and Aitken, B. (1979b). *Australian Journal of Dairy Technology*, **34**, 112.

King, D. W., Russell, R. W., McDowell, A. K. R. and Dolby, R. M. (1972). *New Zealand Journal of Dairy Science and Technology*, **7**, 4.

Lang, F. and Thiel, C. C. (1955). *Dairy Science Abstracts*, **17**, 85.

Lehmann, H. R. (1982). *Deutsche Milchwirtschaft*, **33**, 172.

Lehmann, H. R. and Zettier, K.-H. (1982). *Deutsche Molkerei-Zeitung*, **38**, 1270.

Lehmann, H. R. and Zettier, K.-H. (1987). *Separators for the dairy industry*, Technical scientific documentation No. 7, 3rd rev. edn, Westfalia Separator AG, Oelde, Germany.

Lipatov, N. N. (1976). *Separieren in der Milchindustrie [Separation in the Dairy Industry]*, VEB Fachbuchverlag, Leipzig, Germany.

Löndahl, G. and Johansson, S. (1974). In: *Brief Communications, XIX International Dairy Congress*, Volume IE, pp. 649–50.

Lundqvist, H. J. (1978). *Nordeuropaeisk Mejerei-Tidsskrift*, **44**, 116.

Mann, E. J. (1987). *Dairy Industries International*, **52** (9), 15.

Mann, E. J. (1989). *Dairy Industries International*, **54** (11; 12), 24, 27; 15.

Mulder, H. and Walstra, P. (1974). *The Milkfat Globule. Emulsion Science as Applied to Milk Products and Comparable Foods*, CAB International, Wallingford, England.

Pato, T. (1978). In: *XX International Dairy Congress*, Volume E, pp. 624–5.

Phipps, L. W. (1969). *Journal of Dairy Research*, **36**, 417.

Platt, G. (1986). *Milk Industry*, **88** (2), 26.

Prasad, S. and Gupta, S. K. (1988). *Agricultural Reviews*, **9** (2), 81.

Rabich, A. (1969). *Deutsche Milchwirtschaft*, **20**, 645.

Rabich, A. (1971). *Dairy Industries*, **36**, 207.

Rothwell, J. (1989). (Ed.) *Cream Processing Manual*, 2nd edn., The Society of Dairy Technology, Huntingdon, England.

Streuper, A. and Hooydonk, A. C. M. van (1986). *Milchwissenschaft*, **41**, 547.

Taylor, R. I. (1984). *Institution of Chemical Engineers Symposium Series*, No. 84, pp. 231–40.

Te Whaiti, I. E. and Fryer, T. F. (1975). *New Zealand Journal of Dairy Science and Technology*, **10**, 2.

Thomé, K. E. and Eriksson, G. (1973). *Milchwissenschaft*, **28**, 502.

Towler, C. (1982a). *New Zealand Journal of Dairy Science and Technology*, **17**, 191.

Towler, C. (1982b). In: *Brief Communications, XXI International Dairy Congress*, Volume 1, Book 2, p. 114.

Towler, C. (1988). *New Zealand Journal of Dairy Science and Technology*, **23**, 109.

Towler, C. and Stevenson, M. A. (1988). *New Zealand Journal of Dairy Science and Technology*, **23**, 345.

United Nations Food and Agricultural Organization/World Health Organization (1977). *Milchwissenschaft*, **32**, 278.

West, J. (1985). *Chemical Engineering*, **92** (1), 69.

Chapter 3

Production of Butter and Dairy Based Spreads

R. A. Wilbey

Department of Food Science and Technology, University of Reading, UK

The evolution of batch buttermaking practices with the gradual replacement of wooden churns by stainless steel was overtaken in the 1950s by the development of continuous processes. McDowall (1953) provided a standard reference for batch buttermaking.

The various methods of continuous buttermaking developed at that time were described by Wiechers *et al.* (1950). The Fritz process has since become the most common in commercial use, while the reseparation of cream to fat levels above 80 per cent followed by phase reversal was not generally adopted except in the manufacture of anhydrous milk fat (AMF). AMF may be used as the fat source for manufacture of recombined butter and spreads.

The structure of butter has been reviewed by King (1964), Mulder and Walstra (1974) and more recently by Mortensen (1983). Accounts of the composition and variability of the constituents of the milk fat may be found in either the latter publication or that by Walstra and Jenness (1984).

For the purpose of this discussion, the term 'dairy based spreads' is used for those products with more than half their ingredients derived from milk, and with a continuous lipid phase. These products are water in oil (w/o) emulsions unlike milk and cream which are oil in water (o/w) emulsions. This definition provides a wider scope than the draft general standard for yellow fat spreads discussed within the IDF (Forman, 1990) where dairy spreads would contain only milkfat.

PRINCIPLES OF BUTTERMAKING

The buttermaking process, whether by batch or continuous methods, consists of the following steps:

(i) preparation of the cream, with or without ripening;
(ii) destabilisation and breakdown of the o/w emulsion;
(iii) aggregation and concentration of the fat particles;
(iv) formation of stable w/o emulsion;
(v) packaging;
(vi) storage and distribution of the product.

Preparation of Sweet Cream

The separation of sweet (unripened) cream for buttermaking follows the general principles described in the earlier chapter. The fat content of the cream should be standardised to the optimum level for the buttermaking equipment to be used. A typical level for continuous buttermakers would be 40 per cent fat.

In the UK the legal minimum for heat treatment of cream for retail sale is 72 °C for 15 s (or equivalent), compared to the IDF recommendation of 80 °C for 15 s for creams of $\geqslant 35$ per cent fat. Cream for manufacture of butter to be sold into intervention must be subjected to a high temperature short time (HTST) heat treatment at 71·1–76·7 °C for at least 15 s, or held momentarily at $\geqslant 79.4$ °C (UK Intervention Board, 1990). In the USA a minimum heat treatment of 74 °C for 30 min or 85 °C for 15 s is required (Bodyfelt *et al.*, 1988). In practice a heat treatment at 74–76 °C for 15 s or equivalent would be satisfactory. More severe heat treatments should be avoided, as the higher the temperature the greater the migration of copper from the milk serum into the milk fat globules. Increasing the level of copper associated with the milk fat makes it more susceptible to the development of oxidative rancidity and can reduce the shelf-life of the butter. Excessive heat treatment leads to the formation of cooked flavours, more noticeable in sweet cream butters.

After pasteurisation the cream should be cooled to 4–5 °C, using regeneration and chilled water cooling. (For butter to go into intervention the cream should be cooled to below 7·2 °C.) The plate heat exchanger design should avoid high pressure differentials as the high shear rate associated with this could result in premature damage to the milkfat globules. This damaged cream would be unstable and liable to cause problems on subsequent handling prior to buttermaking. Cream may also be damaged by incorrect size of pumps, particularly centrifugal pumps. It is recommended that the cream for buttermaking be handled by positive displacement pumps; though these too can cause destabilisation of the cream emulsion (Kessler, 1980). Cream pasteurisation technology is reviewed by Bøgh-Sørensen (1992).

Where any rubber components in a plant have contact with cream or butter they must be fat resistant, e.g. nitrile rubber. Rubber compounds that are not fat resistant will break down after several days contact and may contaminate the product.

Direct cooling of the cream by evaporation may be used as an alternative to indirect cooling. In direct cooling, the cream is heated to a higher temperature, e.g. 90 °C, then passed into a vessel at low pressure, about 20 kPa (0·2 bar). Evaporation of water from the cream cools the cream and has a flavour stripping action – it removes some of the volatiles from it. When processing cream of indifferent quality, flavour stripping can be advantageous and can compensate for the additional processing cost. Evaporative cooling does create additional shear forces on the fat globules, resulting in smaller fat globules. Some of the fat globules appear to survive in the continuous lipid phase of the butter, where the smaller globule size may contribute to an improved texture.

Care is needed in this form of cream treatment to avoid excessive shear forces which would create too many small fat globules. This would result in higher losses of fat in the buttermilk. Vacreation is more severe. It employs direct steam injection heating in addition to evaporative cooling, and should only be used where there are taint problems with the cream.

During pasteurisation all of the fat becomes liquid; crystallisation commences on cooling. Most of the latent heat of crystallisation of the milkfat is released before the cream leaves the cooler, but there is a significant release of heat in the 2 h following the heat treatment. The amount of heat released will depend on the time–temperature characteristics of the process and on the fat content of the cream. Allowance must be made for the resulting rise in temperature, e.g. 2 °C, either by cooling the cream to a lower temperature or by continuing to cool in the storage vessel. The latter option is usually preferable.

The pasteurised and cooled cream should be held for a minimum of 4 h, preferably overnight, to permit sufficient crystallisation of the milkfat. The fat usually crystallises in mixed crystals of the α and β' forms, varying with the cooling conditions used (Mulder and Walstra, 1974). Some equilibration may be expected on storage. This storage is often referred to as ageing the cream.

Cream ageing tanks of 10–20,000 l capacity may be used for smaller-scale operations. The tanks are usually vertical with cooling panels or coils built into the walls, as illustrated in Fig. 1. Agitation is by slow-moving gate-type agitators which may also contain refrigerant. These ageing tanks are suitable for both cultured and sweet creams.

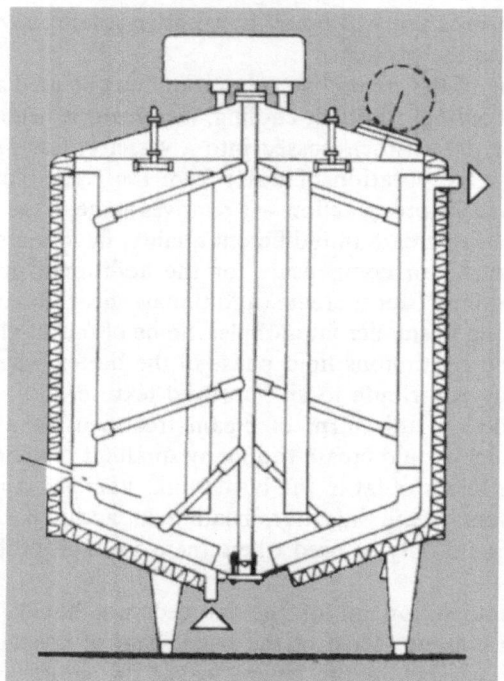

Fig. 1. Sectional drawing of cream processing tank (courtesy APV Pasilac AS, Denmark).

At the larger scale of operations, this size of vessel would require frequent changes during a product run. The feed to the larger continuous butter-makers is commonly 10,000 l h^{-1} and may be in excess of 20,000 l h^{-1}. Any change in the supply would require compensating adjustments to be made to the running of the buttermaker. Thus for large-scale butter-making, variation in cream supply may be minimised by ageing the cream for a days production in a single silo. The silo should be equipped for intermittent mixing to prevent separation and stratification without excessive damage to the cream. Cooling of cream in the silo may be achived either by refrigeration panels in the silo wall, or by recirculation through an external heat exchanger. Where smaller tanks have been used for cooling and ageing the cream, then the creams should be blended in the silo at least 2 h before buttermaking commences. The salvaging of homogenised cream by blending into cream for butter production should be avoided as this leads to higher fat losses in the buttermilk.

Ripened Cream

It can be argued that sour cream was the original form of cream for buttermaking; a consequence of microbial activity in raw milk during lengthy gravity separation, aided by lack of refrigeration. The introduction of rapid separation, pasteurisation and cooling techniques required the re-introduction of suitable bacteria to produce the ripened cream.

Pasteurisation of cream for ripening is commonly carried out at higher temperatures than appropriate for sweet cream butter, e.g. 90–95 °C for 15 s or 105–110 °C with no hold (Boutonnier and Dunant, 1985). Severe heat treatment denatures whey proteins, particularly lactoglobulins, exposing sulphydryl groups which act as antioxidants and can enhance starter growth. Any cooked flavours may be masked by the aroma produced by the starter culture. The cream may also be subjected to vacuum treatment during cooling. The fat content of cream for ripening is normally 36–40 per cent. This fat content is lower than for sweet cream butter.

In the simplest situation the cream may be cooled to the ripening temperature and the culture of starter organisms added at 1–2 per cent. A typical fermentation would be for 12–18 h at 20 °C. For cream ripening a mixed starter system should normally be used, comprising *Lactococcus lactis*, including the *cremoris* sub-species and the *diactylactis* biovariant, plus *Leuconostoc mesenteroides* sub-species *cremoris*. The mixed culture is needed to produce the desired combination of butter flavour and acidity.

Diacetyl is a major flavour component in the ripened cream, a large number of other compounds also contributing at levels both above and below the individual sensory thresholds where synergistic interaction may occur (Mick *et al.*, 1982). Biosynthesis of diacetyl is not significant above pH 5·2. Thus, stopping the fermentation by cooling the cream at pH 5·1–5·3 results in a mild flavour; whereas continuing the fermentation to pH 4·5–4·7 results in higher levels of both diacetyl and lactic acid, giving a more pronounced flavour.

The fermentation may be stopped by cooling the cream. This may be achieved by chilling the walls of the culture vessel, or by passing the cream through an external heat exchanger. Allowance must be made for the continuation of the fermentation during cooling and the commencement of cooling should be offset accordingly.

Starter cultures vary considerably in their sensitivity to low temperatures. Cooling to less than 10 °C is usually adequate for short-term retardation of starter metabolism, but for longer-term storage of ripened cream then a lower temperature of 3–4 °C is desirable. Cooling and storage is also needed to permit further crystallisation of the milkfat.

Modifications to cream treatment

The slow cooling of the cream after it is ripened leads to the formation of larger fat crystals than when cream is cooled to 5 °C immediately after pasteurisation. This can result in ripened creams producing firmer butters than the equivalent sweet cream. Control of the final texture of the butter by varying the fermentation and cooling conditions has been developed to counteract this effect and to produce a more consistent product throughout the year with better spreading properties. The first method was published by Samuelsson and Pettersson (1937), and is often referred to as the Alnarp process.

To produce a softer butter from winter cream with higher melting milkfats, the cream should first be cooled to 8 °C and held for 2 h to promote rapid crystallisation with the formation of fine crystals. The starter culture should then be added and the cream gently warmed to 19 °C and held for a further 2 h, after which the cream should be cooled to 16 °C to complete the fermentation. Fermentation under these conditions usually takes 14–20 h, at the end of which the cream should be cooled to 12 °C before churning.

A firmer butter may be produced from summer cream (with relatively lower melting fat) by initially cooling the cream to 19 °C. Starter culture should be added and after 2 h the cream cooled to 16 °C and held for a further 3 h before cooling to 8 °C and holding overnight. This method requires a higher level of starter addition to compensate for the shorter fermentation time.

These process conditions have been modified extensively to suit individual requirements. A number of series of processes have been proposed based on steps in the iodine value of the milkfat.

A more fundamental approach by Frede *et al.* (1983) used the melting and solidification curves of the milkfat, obtained by differential scanning calorimetry (DSC), in establishing the process conditions. Examples of melting and solidification curves are shown in Fig. 2.

For instance, in summer the cream should first be cooled to a fermentation temperature above the upper solidification point of the milkfat, starter culture added and the cream held at that temperature till the desired pH is attained. The ripened cream should then be cooled to 6 °C and held for 3 h before bringing the temperature up to the butter-making temperature, which should be below that of the low melting peak.

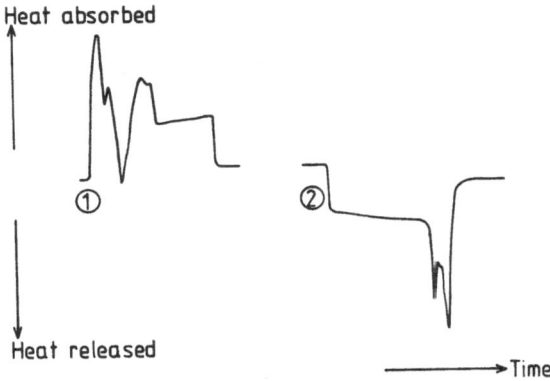

Fig. 2. Example of (1) heating and (2) cooling curves for milkfat – obtained by differential scanning calorimetry.

Conversely a softer butter would be prepared, e.g. from winter cream, by first cooling the freshly pasteurised cream to 6 °C and holding for 3 h. The cream should then be warmed to 2–3 °C above the melting point of the lower main fraction (typically 16–21 °C). The starter should be added and the cream held for at least 2 h until the desired pH has been attained. The cream should then be cooled to a temperature between the solidification points of the two main fractions before buttermaking, this being a compromise between spreadability of the butter and losses of fat in the buttermilk. As with the original Alnarp process, the starter culture addition varies 1–7 per cent depending on starter activity, temperature and the fermentation time available. DSC techniques, however, are not easily applied in a factory laboratory environment, where the simpler nuclear magnetic resonance (NMR) techniques may be more appropriate.

These thermal processes can be applied to sweet cream as well as cultured cream but add to the cost of processing and require greater control of process hygiene. For instance, using the more saturated winter milkfat, the pasteurised cream could be subjected to 2 h holds at 8 °, 19 ° then 15 °C followed by 12 h at 8·5 °C before warming up to 12 °C and feeding into the buttermaker (Berntsen, personal communication). A modified 'Alnarp' treatment using a plate heat exchanger to simplify and shorten the processing was reported by Dixon (1970).

Transfer of Cream to the Buttermaker

Optimum temperatures for ageing the cream are normally lower than that needed for efficient churning. The aged cream is also susceptible to damage if mishandled, which may result in blocked pipework and excessive fat losses in the buttermilk.

Cream should be transferred using a variable speed positive pump, conservatively rated so that its speed does not normally exceed half the maximum. Feed pipes should also be sized to prevent starvation of the pump, allowing flow rates of $0·2–0·4 \, ms^{-1}$ for sweet cream but lower rates for the more viscous ripened creams.

The most economic churning temperature for the cream may be achieved by passing the aged cream through a heat exchanger. Plate heat exchangers are normally used for this duty. There should be a low pressure drop, typically less than 50 kPa (0·5 bar), across the heat exchanger, which should also be operated with a small temperature differential (1–2 °C) between the cream and the water. The cream outlet temperature must be controlled to within $\pm 0·25$ °C of the target. Minimising the temperature differential avoids localised overheating and thus disturbance of fat crystallisation during reheating. It also reduces the need for recycling cream from the heat exchanger at start-up and during stoppages.

A consistent feed rate is important in maintaining steady buttermaking conditions. One method is to interpose a balance tank between the ageing silo and the pump. This is necessary when a cream recycling line is installed and when smaller silos are used, giving frequent changes of cream during the day. An alternative method when the reheating of the cream uses small temperature differentials, is to use an in-line flowmeter with a control loop back to the variable speed pump. Pipework, valves and fittings should be designed and installed to transfer the cream with the minimum of stress.

Thus, the supply of cream to the buttermaker should be consistent in these terms

(i) chemical composition	(fat, pH)
(ii) physical characteristics	(viscosity, fat crystallisation)
(iii) temperature	(max $\pm 0\cdot25$ °C)
(iv) feed rate	(max $\pm 0\cdot5$%)

Though it is a general rule to avoid damage to the fat globule before it gets to the buttermaker, some controlled destabilisation can be carried out. Simon-Freres produced a turbo-cream feed unit which injects oil-less filtered compressed air into the cream line at a rate of approximately $5 \mathrm{N\,m^{-3}}$ air per 1000 l of cream. The air–cream mixture was passed through a static mixer to produce a foam, thus destabilising the cream emulsion immediately before it entered the buttermaker. The principal claims for this pretreatment are reduced power consumption and fat losses at buttermaking, together with greater flexibility in the types of cream and tolerance of sub-optimal process conditions. However, great care must be taken to ensure that the unit is maintained in a satisfactory hygienic condition.

Conversion of the Cream to Butter

Once the cream is in the butterchurn then the fat globules must be disrupted under controlled conditions to destabilise the oil in water (o/w) emulsion and bring about agglomeration of the milkfat. The aged cream contains approximately equal proportions of solid and liquid fat in the globules. The solid fat is in the form of crystals, possibly lining the globule membrane, with liquid fat occupying the core of the globule. Larger fat crystals, limited by the diameter of the globule, may also be formed. The size of the milkfat crystals is related to the heat treatment profile and the rate of cooling, slow cooling produces a smaller number of large crystals. The crystals make the globules less elastic so that the membrane will be disrupted when the globule is subjected to mechanical stress. In batch churns, illustrated in Fig. 3, the mechanical stress is created by rotating the partly filled churn so that the cream is lifted up the ascending wall of the churn then cascades to the base.

Air bubbles also become entrained in the cream to form an unstable foam. Liquid fat escaping from the disrupted fat globules spreads over the water–air interface in the cream, together with whole and disrupted fat globules. Coalescence of the air bubbles leads to a reduction of the

Fig. 3. Batch churning of butter (courtesy APV Pasilac AS, Denmark).

surface: volume ratio, with concentration of the liquid and globular fat at the surface. Further concentration takes place at desorption of the air. This results in an association of the globules into granules bound together by the hydrophobic liquid fat fraction. Walstra and Jenness, (1984) give a more detailed account. Though air is normally present in buttermaking operations, destabilisation of the cream emulsion by shear forces may take place in the absence of air – a common accident in cream handling as well as being used in the preparation of AMF. The continuing agitation of the destabilised emulsion builds up the aggregates to form visible butter grains. Approximately half of the membrane material is lost from the grains into the aqueous portion to form the buttermilk.

Whereas the batch process carries out destabilisation relatively slowly with a large mass of cream, the continuous buttermakers (Fig. 4) operating on the Fritz principle destabilise small quantities at a fast rate. Modern designs limit the noise level to 75–80 dB(A). Cream is fed into the continuous buttermaker at the top of the first churning cylinder, where breakdown of the cream emulsion is carried out by a multibladed dasher. This dasher is driven through a variable speed gearbox and is normally operated at about 1000 rpm. The action of the dasher both aerates the cream and damages the globules in a period of 1–2 s. The speed of the dasher blades (Fig. 5 top) can be adjusted to obtain the desired buttergrain – faster speeds will give larger buttergrains. Excessive speed results in large grains with too much buttermilk retention. There is a greater formation

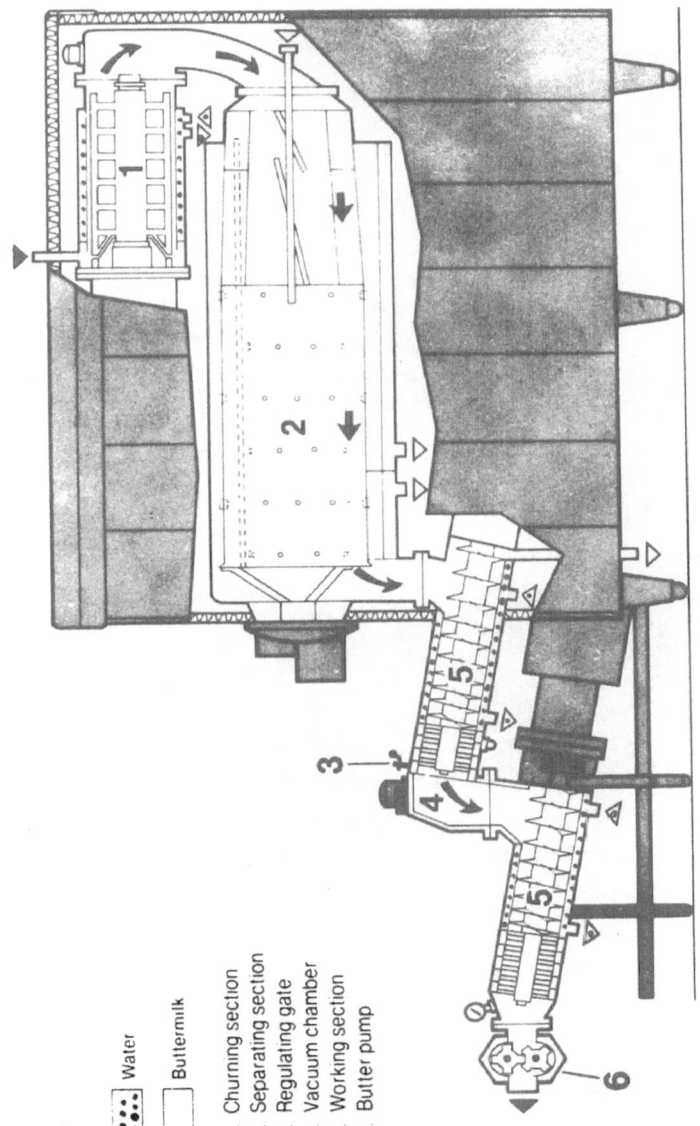

Fig. 4. Section through a modern continuous buttermaker (courtesy of APV Pasilac AS, Denmark).

Water

Buttermilk

1. Churning section
2. Separating section
3. Regulating gate
4. Vacuum chamber
5. Working section
6. Butter pump

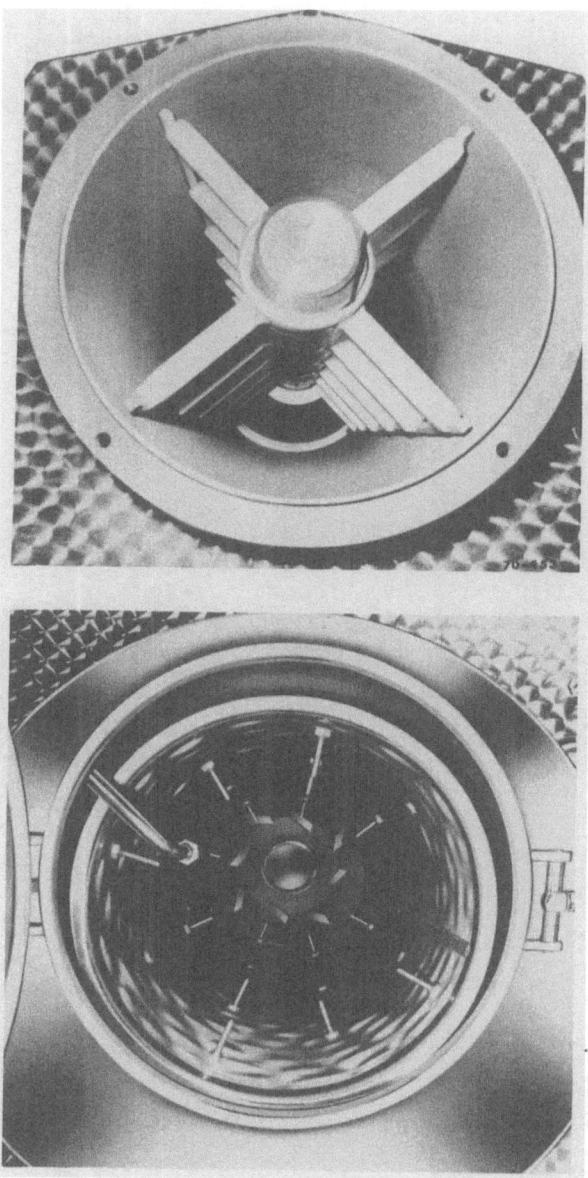

Fig. 5. Views of churning (top) and separating (bottom) sections (courtesy APV Pasilac AS, Denmark).

of colloidal liquid fat particles and this gives higher fat losses in the buttermilk.

The mixture of buttergrains and buttermilk drops from the first cylinder into the back of the second cylinder where the buttergrains are consolidated. This second cylinder is built as a rotating drum, larger than the first cylinder and rotating slowly (e.g. up to 35 rpm) so that the buttergrains pass along the rotating surface. The relatively gentle tumbling action brings about further consolidation and aggregation of the buttergrains while most of the buttermilk drains away through the perforations in the second part of the drum (Fig. 5 bottom). The cylinders in the first and second stages may also be mounted concentrically (Fig. 6).

The buttermilk is pumped away, but some may be cooled and then recycled to the second cylinder to help cool the buttergrains. Washing of the grains in the later stage of the cylinder may also be carried out, with separate collection of the washwater. This additional process is not desirable under normal conditions as the disposal of the washwater adds to costs and the replacement of non-fat milk solids by water results in an increase of the more expensive milkfat to keep within the maximum moisture content (typically $\leqslant 16$ per cent).

The moist grains of butter fall from the second cylinder into the working section of the buttermaker. This section uses contrarotating augers to consolidate the grains of butter into a heterogeneous mass, expelling more buttermilk from the grains as they are squeezed together and carried up to a series of perforated plates and mixing vanes. The expelled buttermilk, together with buttermilk coming off the end of the second cylinder, forms a pool at the base of the working section. The level of this pool should be held constant as this will affect the moisture content of the butter. Fat granules in the buttermilk float to the surface of the buttermilk and are reincorporated into the butter. A spinning disc clarifier may also be incorporated into the buttermaker to aid fat recovery. Butter grains may also be recovered by running the buttermilk over a fine sieve before pumping away to further treatment. Better recovery can be achieved by using a vibrating sieve. Vibrating sieves do not get blocked so easily and they agglomerate the recovered butter fines. Some fat may also be recovered from the buttermilk by using a milk separator at approximately half its normal capacity. Recovery of the fat is not complete as some phospholipids (from the globule membrane) and the colloidal fat droplets remain in the buttermilk to give a fat content of approximately 0·5 per cent. The presence of the membrane material in the

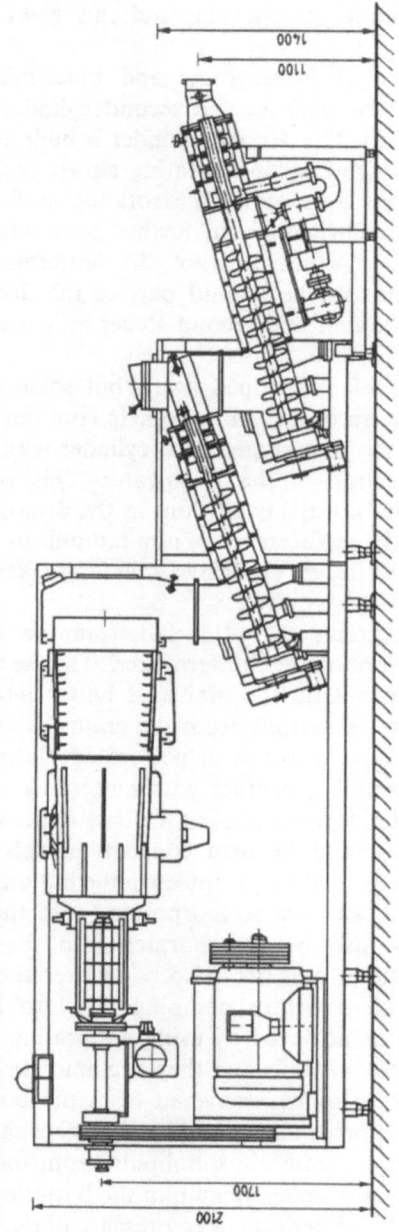

Fig. 6. Coaxial arrangement of beater and drum (courtesy GEA Ahlborn, Germany).

buttermilk confers useful properties but can also cause fouling problems when spray drying sweet buttermilk concentrates.

Compacted buttergrains are fed from the auger through the series of alternating perforated plates and impeller blades, examples of which are shown in Fig. 7. The amount of work done on the butter may be increased by reducing the size of the holes in the plates, by increasing the number of plates and impellers and by varying the angle of some of the impeller blades. Blades with no pitch angle generate the highest shear conditions on the butter, but rely on the other blades and the auger to move the butter through the orifice plates. The shear forces not only further the consolidation of the buttergrains, but break up the droplets of buttermilk remaining in the matrix of fat and fat globules. This forms a dispersed aqueous phase of what is now a water in oil (w/o) emulsion. The droplets of aqueous phase should ideally have a diameter of less than 10 μm.

Partially worked butter may contain more than 5 per cent entrained air. This can be reduced by passing the butter through a vacuum section

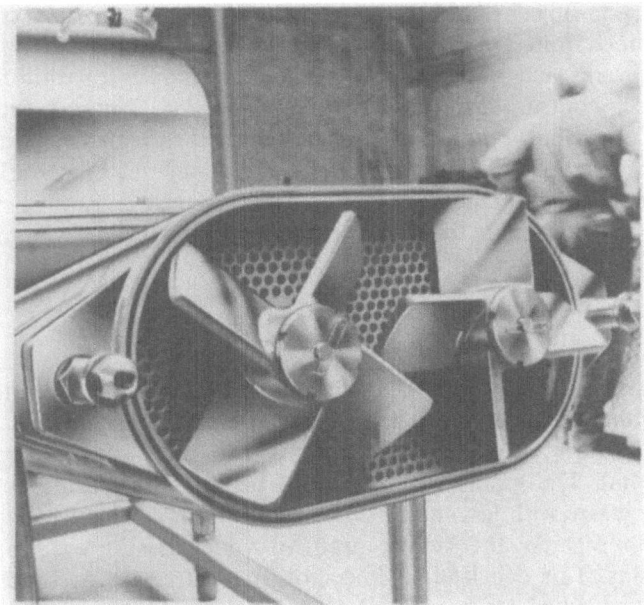

Fig. 7. View of perforated plate and mixing vanes in working section (courtesy APV Pasilac AS, Denmark).

to obtain a denser, finer textured product with less than 1 per cent air incorporation. A second set of augers removes the butter from the vacuum section and forces it through a further set of orifice plates and blades to complete the emulsification before discharge from the butter-maker. This second working section may be driven from the same drive as the first section or may have an independent drive. Independent drives can enable a larger vacuum stage to be incorporated and give greater flexibility in setting up optimum working conditions for the butter, e.g. run the second stage at 2–3 times the speed of the first stage. However, the greater flexibility should be balanced against the need to monitor and control a further set of variables.

The Fritz principle may be applied in slightly different ways. In the Simon-Freres Contimab machines, where de-emulsification and granule formation are carried out in a single larger cylinder, the cylinder may have a standard sand-blasted finish or inside there may be an expanded metal screen. Such screens are for use with creams needing a more rigorous churning effect, e.g. sweet cream with less than 42 per cent fat or cultured cream with less than 35 per cent fat. The beater blades operate at 600–800 rpm, slower than normal, and the blades vary in size to increase the clearance from 3–4 mm up to 6–7 mm at the discharge. The mixture of buttergrains and buttermilk falls into the second, separation section which uses contrarotating augers to convey and compact the butter-grains. Cooled buttermilk may also be recycled to cool the buttergrains before compaction. The action of the augers squeezes buttermilk out of the grains as they are compacted. Unworked butter is then extruded through an orifice plate to drop into the working sections, where surplus buttermilk may also be drained off.

Salted Butter

Batch produced butter may be salted by adding salt to the buttermaker after washing and draining. The salt must be of high quality e.g. BS 998: 1969, with low levels of lead (max. 1 ppm), iron (max. 10 ppm) and copper max. 2 ppm). The grains of salt should be fine, all passing through a 1.4 mm test size with less than 0·2 per cent retention on an 850 μm sieve.

The salt sets up an osmotic gradient which draws water from the buttergrains. This can lead to free moisture defects if the butter is not adequately worked. Coarse grains of salt that do not completely dissolve will give a gritty product. Fine milling of the salt enables the salt to be more evenly dispersed and to dissolve faster in the butter.

Continuous buttermaking increases the need for a fine salt. The grain size should not exceed 50 μm. Salt must be dosed as a continuous stream into the buttermaker, preferably into the first working section. The solubility of salt is not very temperature sensitive and approximates to 26 per cent w/w at ambient temperatures. This restricts the use of a saturated brine to low salt addition rates, less than 1 per cent salt in the final product. Allowance must be made in running the buttermaker to produce buttergrains with a lower moisture content, to compensate for the water in the brine. For a butter with not more than 16 per cent moisture and a 1 per cent salt content, an addition of saturated brine at 3·8 per cent requires a moisture content of less than 13·2 per cent in the unworked butter before salt addition.

In the UK a salt level of about 2 per cent is common. This requires an excessive addition rate of brine. The brine should then be made up as a saturated suspension, normally containing 50 per cent salt but possibly up to 70 per cent salt for manufacture of extra salted butter containing 3–4 per cent salt.

The brine suspension should be prepared by dissolving finely milled salt (particle size less than 50 μm) in potable water. During and after the initial dispersion the mix must be vigorously agitated to ensure the suspension is kept homogeneous. Dispersion of the salt should be well before the brine is used (preferably at least 2 h before use) to ensure that the brine is saturated and the remaining salt crystals are very small. A rise in the temperature of the brine during agitation has little effect on the solubility of the salt, but will aid incorporation of the brine into the butter.

The brine slurry should be metered to the buttermaker by positive displacement pumps. The metering accuracy of the pumps may be improved by minimising the level variation in the brine tank. The presence of coarse grains of salt in the brine will obstruct the valves and result in inaccurate dosing as well as increased wear. The brine must not be allowed to stand in the pump or pipework as the salt will come out of suspension and cause blockages. At any stoppage the brine slurry must be recycled till either the buttermaker is restarted or the dosing equipment washed out.

A 2 per cent addition raises the average salt level in the aqueous phase of the butter to over 11 per cent, apparently sufficient to inhibit most spoilage microorganisms. However as not all the aqueous phase is salted, the keeping quality of the butter is largely determined by the efficiency of the working, the overall handling and process hygiene.

The dosing system may also be used for adding water to unsalted or slightly salted butters to maximise yields. Good bacteriological quality of the water is essential to avoid contamination, especially with psychrotrophs which are frequently present in the mains water supply. The water should be treated by filtration, ultraviolet light or pasteurisation.

Alternative Processes for Ripened Butter

The ripening process adds to production costs and causes copper to migrate into the fat, increasing the susceptibility of the butter to oxidative rancidity. Addition of salt to the ripened butter causes further destabilisation. Lactic buttermilk also presents a disposal problem as it can seldom be converted economically into other dairy products and so must often be sold at a nominal sum for animal feed.

Several methods have been proposed for overcoming problems with lactic buttermilk. The most important was developed at the Netherlands Institute Voor Zuivelondazoek and is referred to as the NIZO process (1976).

The NIZO process churns sweet cream to yield sweet buttermilk, suitable for further processing, and unripened buttergrains with a low moisture content (13–13·5 per cent). Starter culture and a concentrated lactic acid preparation are added to the unworked butter which may then be worked and packed in the normal way. The characteristic flavour and aroma of the butter, which is similar to traditional products, develops on storage.

The starter cultures should include both *Lactococcus lactis* biovariant *diacetylactis* (e.g. strain 4/25) and *Leuconostoc mesenteroides* sub-species *cremoris* (e.g. strain Fr 19). The leuconostocs are able to metabolise acetaldehyde produced by the lactococci and hence avoid the development of a 'green' flavour. In the original method, the starter organisms were mixed together before injection into the butter, but this was altered to permit separate aeration of the lactococcal culture. Separate aeration increases the diacetyl level in that culture to about 40 ppm, and this avoids problems of reduced diacetyl production on storage below 6 °C. The addition of starter at up to 2 per cent of the butter is not sufficient to reduce the pH of the butter to less than pH 5·3 as required by Dutch regulations. This was overcome by adding a 'culture concentrate' together with the leuconostoc culture.

Culture concentrate is a lactic acid preparation, produced by fermentation of a lactose depleted whey by *Lactobacillus helveticus*. The fermented

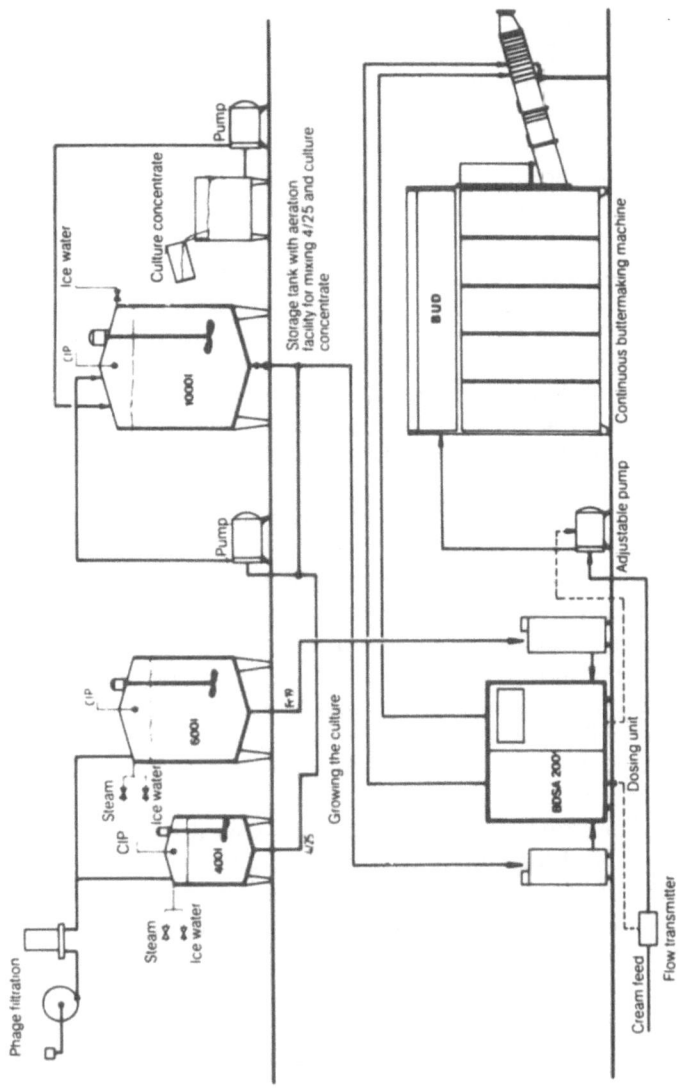

Fig. 8. Process flow diagram for the NIZO process (courtesy Westfalia Separator, UK).

whey is then passed through an ultrafiltration plant and the permeate subsequently concentrated to at least 11 per cent lactic acid. Excessive concentration can lead to lactose crystallisation on storage. The addition of 0.7 per cent of this concentrate to the butter together with the 2 per cent of starters is sufficient to reduce the pH to less than pH 5·3. Higher levels of concentrate may be used to produce a more acid butter e.g. pH 4·6. A flow diagram for the NIZO process is given in Fig. 8.

An improved process was claimed by Wiles (1978), avoiding the reliance on the starter organisms for flavour development. The characteristic flavour is achived by using a flavouring based on diacetyl, preferably a distillate from a specially cultured skim-milk (e.g. Hansen's 15X starter distillate). Acidity is increased by addition of lactic acid. Salt, which would normally inhibit the starter culture, may be added to the mixture of distillate and lactic acid to help flavour the butter to suit the UK market. Cultures may be added separately to bring the level of starter organisms close to that for traditional product (about $5000\,\text{cfu}\,\text{g}^{-1}$). The exact level of starter addition and its metabolic activity are not critical in this application so the starter may be stored for longer if necessary.

Control of Buttermaking

Traditional techniques for chemical analysis of butter are unacceptably slow for use with modern buttermakers where the production rate may exceed $80\,\text{kg}\,\text{min}^{-1}$. Economically the most important parameters to be controlled are the moisture and salt levels. Process parameters on the buttermaker must be set to maximise moisture and salt levels within the specifications, while producing an acceptable product in terms of moisture distribution, flavour and texture. Microcomputers are now increasingly used to aid process optimisation.

The level of moisture in the butter may be sensed by its effect on the dielectric constant of the butter. This information can be fed back to the control system of the buttermaker. Checks are required to maintain the calibration of the system. With salted butter a correction must be made for the effect of the increased electrolyte on the dielectric constant. An accuracy of $\pm 0\cdot 1$ per cent is claimed for systems based on dielectric measurements with unsalted butter.

In the presence of salt, a more sophisticated approach is required. One system used an array of sensors in the discharge from the buttermaker to measure the levels of moisture, density, temperature, solids-not-fat and salt. A major innovation was the measurement of salt by backscatter of gamma radiation emitted from an americium-241 source. Accuracy better than ± 0.03 per cent in measuring the salt content was claimed. Williams (1984) claimed an increased buttermaking yield when the system was introduced into a creamery. The mean moisture content of the butter increased by 0·16 per cent to 15·84 per cent and the mean salt content by 0·12 per cent to 1·77 per cent against respective maxima of 16 per cent and 2 per cent to the production requirements. However this system has similar calibration requirements to the simpler systems and with the development of rapid analytical techniques such as infrared reflectance, the potential benefit of in-line testing has been much reduced.

Microbiological aspects of cream and butter have been reviewed by Davis and Wilbey (1990) and by Murphy (1990). Bodyfelt *et al.* (1988) comprehensively describe sensory aspects with emphasis on US market requirements. More general aspects of butter analysis are covered by Davis (1986) and Wilbey (1991).

Butter Handling and Packing

Butter from the batch churn may be dropped onto butter trolleys, wheeled to the packer then either tipped or shovelled into the feed hopper. This method may still be acceptable for small-scale production but is labour intensive and potentially unhygienic. Continuous buttermakers (Fig. 9) may be set up to discharge directly into a bulk handling system. The simplest approach is to discharge directly into the feed hopper of a bulk packer, but any holdup on the bulk packer will then result in the shutdown of the buttermaker with consequent waste, both in terms of buttermaking capacity and fat in the buttermilk. Shutdown of the buttermaker may be avoided by introducing a bulk butter silo. A butter pump should be incorporated into the buttermaker where the silo is pressurised or cannot be located at the discharge from the buttermaker. Butter pumps use large interlocking rotors operating at low revolutions and discharging into wide bore pipes (typically 100 mm diameter) with large radius bends (> 500 mm radius). Butters from two buttermakers may be blended and an in-line mixer used to ensure uniform appearance and consistency.

Two types of butter silo are used. The sealed system is constructed as a vertical cylinder with a telescopic feed pipe mounted on a piston. The butter in the silo is kept at constant pressure by the action of air cylinders on the piston. Butter is discharged from the base of the silo into butter

Fig. 9. A 6 tonne h^{-1} continuous buttermaker discharging into a 10 tonne butter silo (courtesy Alfa-Laval).

pumps. This silo system has the advantage of minimising the risks of contamination, but has an operating capacity of less than 2 tonne.

The open butter silos are larger, with capacities up to 10 tonne. The sides of the silo are steeply inclined or vertical to avoid bridging. A pair of large, slowly contrarotating augers in the base of the silo discharge the butter to the butter pumps.

Care is needed in the handling of the butter to control the level of shear forces acting on the product. During the handling procedures droplets of aqueous phase can come into contact with one another and coalesce, giving rise to free moisture in the butter. Handling should either be very gentle or else designed to promote re-emulsification, for instance a high shear buttermixer may be inserted into the line to ensure a uniform product.

The use of butter production to absorb surplus milk fat results in large seasonal variations of butter production in western Europe. The majority of the butter is initially packed in 25 kg cases. Within the EC, packing is normally carried out to standards of the national intervention boards.

While manual bulk packing systems are still suitable for the smaller buttermaking lines, automated bulk packaging systems (Fig. 10) are now available to handle the output of the larger buttermakers. These lines erect and weigh the empty cartons, which may be lined with either

Fig. 10.   Automated bulk packing line (courtesy Alfa-Laval).

parchment or plastic film fed from reels. The filling operation is in two stages. The first fill is to give approximately 100 g shortweight. The carton is then reweighed, the tare for that carton deducted and a second fill made to bring the weight to within 10 g of the target. The case may then be sealed, labelled and palletised. Integration of the control systems with those of the buttermaker and butter silo minimise the manning levels required. For short periods the process could be supervised by a single operator.

The most efficient method of packing consumer portions, typically 250 g size, is by direct feed from the buttermaker via a silo to the fillers. A closed system using a pressure compensator at the filler is the most hygienic system and minimises fill variation, giving an average variation of ± 0.5 g. Filling from an open hopper will increase fill variation. Butter fed direct from the buttermaker to the fillers is relatively soft during the filling operation, so support must be provided to the packaging during this process.

The rigidity of cold stored butter and the high cost of plastic packaging, coupled to the very cost sensitive market conditions, has largely precluded the use of preformed plastic containers for butter. Most consumer butter packs use a film wrap, either vegetable parchment or a parchment-lined aluminium foil. Vegetable parchment is the cheapest of the materials used and is used extensively in the UK. The major disadvantage of parchment is its permeability. Allowance must be made for loss of water from the packs on storage. Ultraviolet light can penetrate the packaging and promote the development of oxidative rancidity, though this may be reduced by incorporating opaque pigments such as titanium dioxide on the outside of the parchment. On high temperature storage the packs may also become greasy. Laminates containing aluminium foil have much lower weight loss factors and are impermeable both to fat and ultraviolet light, so that the added cost may be balanced against the potential for a longer shelf-life and improved product quality when the butter reaches the consumer.

A small proportion of butter is packed in transparent films, e.g. cellophane. Although the packaging is attractive, it has good ultraviolet transmission properties so the butter is susceptible to off-flavours due to oxidative rancidity. The fluorescent lighting in refrigerated display cabinets is a common source of ultraviolet radiation.

The only significant market for butter that uses plastic packaging is in individual portions for catering and institutional use. In this market parchment packs are in the minority; most packs are either miniature foil bricks or plastic containers. These plastic containers are normally produced from a PVC sheet using form-fill-seal techniques where the containers are sealed by foil or aluminised films. These small portions, typically 9–14 g net weight, warm up very quickly on removal from refrigeration and rigid plastic containers help to maintain their quality of presentation at ambient temperatures.

Repackaging of Bulk Butter

Bulk packed butter is a relatively stable commodity at low temperatures. Murphy (1990) has reviewed the bacteriological aspects of butter production. Storage for six months at $-10\,°C$ has no significant effect on the quality of salted fresh cream butter. Current intervention board standards require storage at temperatures colder than $-15\,°C$. Many public cold stores however operate at lower temperatures, down to $-30\,°C$, where other more perishable commodities may be stored in the same chamber as the butter. Under these conditions the butter remains satisfactory for more than one year. Cream may also be stored deep frozen for subsequent processing into butter (or anhydrous milkfat).

Repackaging of bulk butter into consumer portions reconciles consumer demand with variations in butter production. Before repackaging, frozen butter must be allowed to warm up to an optimum temperature of 6-8 °C. During this tempering stage the humidity of the chambers should be controlled to avoid excessive condensation on the butter. Heat for the tempering process may be provided economically by carrying out the first stage of the tempering in the packet butter store, or by reusing low grade heat recovered from other processes. During tempering, the batches of butter should be checked for compositional and organoleptic quality. Up to several days, may be required for tempering large batches of frozen butter. This in turn requires large areas of specialised storage space. An alternative method is to use microwave tunnel heaters to bring the butter up to the desired temperature, thus reducing the attemperation time from days to hours. The use of microwave attemperation can also save a significant area within the butterpacking plant.

The butter must be comminuted and reblended before repacking. To achieve plasticity, the solid fat must be in a fine dispersion so that the butter mass is held together by relatively weak cohesive forces, rather than by the mechanically stronger intermeshing of large fat crystals. This reblending stage provides an opportunity to increase the salt and moisture levels to the maxima permitted (by legal or other specifications), thus improving the conversion yield and the economics of the operation. Any problems with free moisture in the butter may be overcome at this stage.

In small-scale repacking operations the blocks of tempered butter are comminuted by dropping through a shiver. The comminuted butter is then blended into a plastic mass using a batch blender with heavy-duty contrarotating 'Z' shaped blades. These blenders should preferably be driven through a variable speed gearbox, though a two speed drive is acceptable. Passage of the bulk butter through a metal detector before shiving is desirable to minimise the risks of both product contamination and damage to the process plant.

Continuous butter blenders (Fig. 11) are now availabe for efficient handling of large quantities of butter. The continuous blender includes a chopper (or shiver) for the bulk butter and provision for in-line addition of salt, water and cultures. A vacuum stage for controlling the air content may also be included. Construction of the blending and working sections is similar to that of continuous buttermakers. The capacity of the blender depends on the nature and temperature of the butter e.g. $10 \, \text{tonne h}^{-1}$ as an additional working stage for freshly produced butter, dropping to $7 \, \text{tonne h}^{-1}$ with stored butter at $7-8 \, °C$ and less than $5 \, \text{tonne h}^{-1}$ with thawed butter at $0.5-2 \, °C$. The reblended butter may be discharged into a butter silo and pumped to packaging machines in a similar way to that for freshly produced butter (Fig. 12).

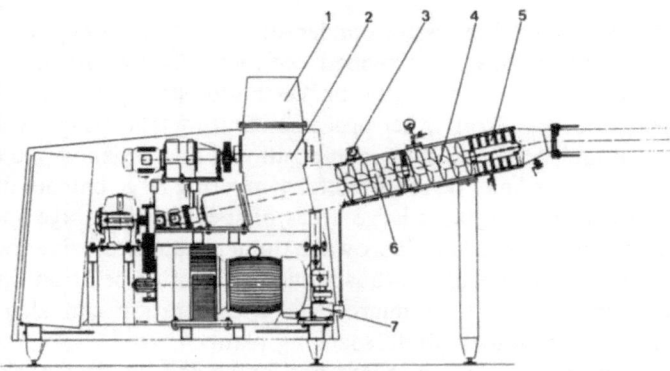

Fig. 11. Sectional drawing of continuous butter mixer/blender: (1) feed hopper (2) chopping device (3) dry-salt dosing connection (4) vacuum chamber (5) blending section (6) transporting augers (7) metering pump. (Courtesy Westfalia Separator, UK.)

Fig. 12. Pumped distribution of butter to packing machines (courtesy Alfa-Laval).

RECOMBINED BUTTER

In many countries there is now a demand for dairy products but there is not the milk production to support its production. Recombination methods for a range of dairy products, including butter have been covered in IDF publications (Jebson, 1979; Spieler, 1982; IDF, 1990). Rather than importing butter, with its strict storage requirements, local manufacture from imported ingredients has been widely adopted. The principal dairy ingredients are anhydrous milk fat (AMF) and a source of milk solids-not-fat (MSNF), usually skim-milk powder (SMP).

The production and handling of AMF has been reviewed by Rajah and Burgess (1991), Wilbey (1991) and by Munro *et al.* (1992). The AMF must be of good quality, free of any off-flavours as these would be carried over into the product. The production methods for milkfats are summarised in Fig. 13. For international trade, AMF is commonly transported in steel drums containing approximately 190 kg of fat. The fat is normally melted in the drums then pumped out into the processing plant. Manufacturing techniques for recombined butter are now largely based on scraped-surface cooling technology, as used in the magarine industry.

The aqueous phase is made up by dispersing the SMP in potable water, adding salt if a salted butter is required. Alternatively, the skim-milk may be pasteurised and fermented to provide a 'cultured' butter. Other sources of milk solids such as buttermilk powder may be used but powders containing fat are susceptible to oxidation, which limits their shelf-life. Kisza *et al.* (1990) suggested that ultrafiltered (UF) protein concentrate was superior to SMP, whereas sodium caseinate gave a poorer flavour and weaker water-binding properties.

The lipid phase is primarily milkfat, but emulsifiers such as lecithin and/or mono-diglycerides should normally be added to minimise spattering if the butter is used in frying and to avoid moisture leakage on storage. In tropical countries there may be a benefit from including some high melting milkfat fraction with the milkfat. Flavours, mainly diacetyl-based, may also be added. The emulsion may be formed on a batch or continuous basis, adding the aqueous phase to the lipid phase. If the phases have not been heat treated individually then the emulsion should be pasteurised before cooling. The emulsion may also be formed by continuous blending after partial cooling (precrystallisation) of the lipid phase.

Hot emulsion may be cooled by tubular or plate heat exchangers but cooling below about 40 °C, when the milkfat will commence crystallisation, should be carried out in scraped-surface heat exchangers (SSHES). These heat exchangers (Fig. 14) have a relatively small surface area (approximately $0.5\,m^2$ per metre tubelength) so large temperature

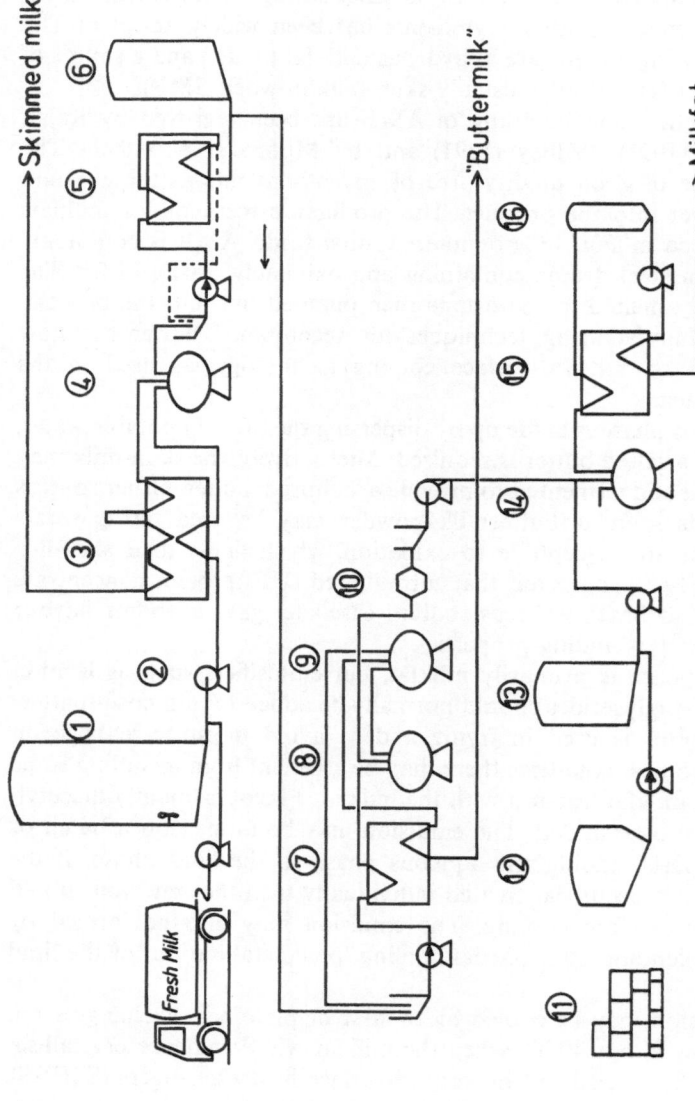

Fig. 13. Flowline for production of anhydrous milkfat (1) raw milk silo (2) pump (3) milk preheater and skim pasteuriser (4) milk separator (5) cream pasteuriser (6) cooled cream storage (7) cream heater and buttermilk cooler (8) second stage cream separator (9) buttermilk separator (10) emulsion breaker (11) bulk butter (12) butter melter (13) melted butter tank (14) milkfat separator (15) milkfat heater and cooler (16) vacuum drier. (From Wilbey (1991), courtesy Elsevier Applied Science.)

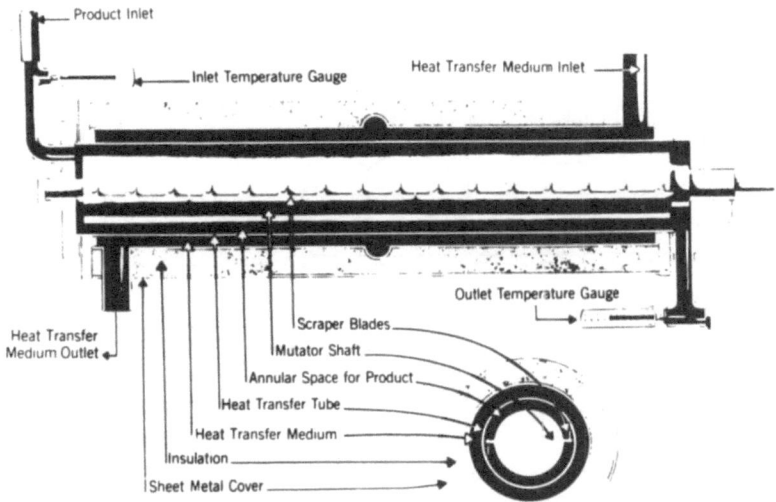

Fig. 14. Section through an SSHE (courtesy Chemtech, UK).

differences are needed to achieve adequate heat transfer. Evaporating ammonia at $-10°$ to $-20°C$ is normally used as the refrigerant for industrial plant; freon is normally used for pilot plant. Twin-bladed rotors are normally used, scraping cooled fat off the surface and blending it back into the mass of the emulsion.

The relatively short residence time of the surface film, 300–60 ms for rotation speeds of 100–500 rpm, ensures that a large number of small crystals are formed. These small crystals act as a seed for secondary crystal development in the cooled mass. Flow through the heat exchanger is relatively slow (ignoring the turbulent effect from the blades), with average velocities of approximately $0.1\,\mathrm{m\,s^{-1}}$ along the tube. Exit temperatures of 5–15 °C may be achieved at the cooler exit, depending on the plant configuration and the consistency required for the packaging operation.

The development of the fat crystals is controlled by the disruptive effect of the shear forces in the SSHE; further fat crystal development is promoted by the use of working units and, especially for packet product, a holding or crystallising tube. Figure 15 illustrates a flowline for a recombination process, the number of SSHE units employed depends on the length of unit available and on the throughput. It is common practice to place at least one worker unit between SSHE units. This provides an opportunity for further crystallisation to take place under mild shear

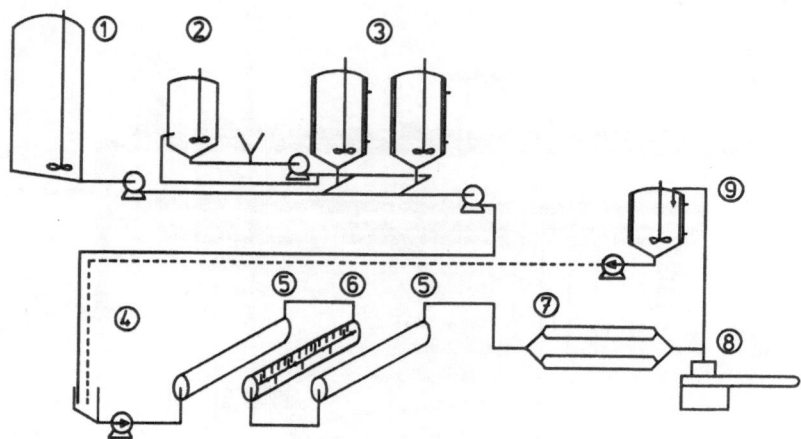

Fig. 15. Flowline for a butter recombination process: (1) melted milkfat (2) reconstitution tank and hopper for aqueous phase (3) emulsion tanks (4) high pressure pump (5) SSHE (6) worker unit (7) resting tubes (8) packet filling machine (9) rework melting and return.

conditions to create a plastic mass. The worker units normally have a larger diameter and annular cross-section than the SSHE tubes; shear is generated by the intermeshing of static and rotating pins. The cooled, worked butter may then be held in resting tubes to allow further crystallisation to give a product that is sufficiently firm to pack. Crystallisation is accompanied by release of latent heat, causing the product temperature to rise by about 5°C; the actual rise depends upon the process.

The structure of recombined butter should be similar to normal butter, with a fine crystalline network and small droplets (ideally $< 10\,\mu m$) of aqueous phase. But recombined butter does lack the globular fat component in the aqueous phase.

Butter recombination processing, as well as being based on margarine technology, may also be linked with less used alternative continuous butter processes (Alfa, Alfa-Laval, Cherry Burrell, Creamery Package Gold'n Flow, Meleshin, New Way). In these processes high fat emulsions with over 80 per cent milkfat are prepared from cream by various combinations of centrifugation and phase inversion (Wiechers and DeGegoede, 1950; McDowall, 1953). These high fat emulsions are then cooled by scraped-surface heat exchangers with varying degrees of work.

A more recent process, developed in New Zealand, provides a further link between these alternative buttermaking processes and those of recombination. This 'Ammix' process uses a blend of AMF, cream and brine to form the emulsion (Lane, 1992). The emulsion is then cooled by scraped-surface coolers and worked to produce an appropriate, plastic, texture for packaging. Butter made by this process will retain a proportion of fat globules in the structure and thus will be similar to normal butter.

SPREADABLE BUTTERS

For butter to be spreadable, it needs to exhibit plastic flow when subjected to shear at the temperature of use. Butter consists of a continuous phase of liquid fat that contains solid fat crystals, fat globules and droplets of the aqueous phase. The particle sizes should be sufficiently small for surface effects to be the major adhesive force, rather than mechanical interference, so that a relatively low yield stress is needed to bring about deformation.

Physical methods to control variation in butter hardness and to impart a softer texture have had limited success. The softening achieved may be largely confounded by temperature variations in distribution and the home. Most butters have too high a level of solid fat for easy spreading when cold. Butter is normally sufficiently plastic to spread at 15 °C, when less than 40 per cent of the fat is solid. Easy spreading characteristics are achieved at solid fat levels of 20–30 per cent.

The obvious solution in temperate climates may be for the consumer to leave sufficient butter for immediate use at room temperature, as was done before refrigerators became a normal household item. However, this approach is inconvenient and is not acceptable in hot weather or warmer climates when the butter tends to melt and deteriorate rapidly.

Substantial modification of the properties of the milkfat is needed to give a spreadable butter at domestic refrigeration temperatures (i.e. 5–10 °C). Methods of achieving this modification without the inclusion of additional ingredients have been investigated. The principal methods used have been modification of the diet of the cow and the recombination of fractionated milkfats to form butter.

Changes in the cow's diet may be reflected in changes in both the quantity and the composition of the milkfat. Storry (1981) reviewed work in this area. Where a single saturated fatty acid was fed to cows, the levels of that fatty acid plus its monoene increased in the milk. This may in turn produce a lower melting milkfat. Greater changes may be brought about by feeding protected supplements. These protected supplements were produced by encapsulation of the feedstuff by a protein–aldehyde condensation product. This modification of the protein, typically by reaction with formaldehyde, makes the proteinaceous outer layer of the supplement resistant to the rumen microflora. The particles of the protected supplement should then pass unaffected through the rumen to be broken down and digested in the intestine. Bypassing the rumen microflora increases the availability of the supplements to the cow and enables a higher level of energy to be fed to the cow without reducing its appetite. This gives an increased milk yield. The protected fat bypasses the natural degradation and hydrogenation systems of the rumen microflora. This results in the fat being absorbed as fatty acids and glycerides typical of the feedstuff rather than as short chain, saturated acids.

Change in the feed is reflected in the nature as well as the quantity of the produced milkfat (Storry et al., 1974). The degree of change depends on the type of feed and the efficiency of the protection of the fat. In work with protected rapeseed oil (Sporns et al., 1984) there was a significant change in the butter composition, though not as large as was expected.

Production of a softer butter by this means requires a major change in the feeding practices in a large area. A less saturated milkfat suitable for manufacture of a soft butter also causes problems if used in other dairy products, e.g. whipping cream. At this stage there are still substantial technological problems to be overcome in producing an effective, economic and acceptable protected feed supplement. Oldham (1990) concluded that 'by far the most, most potent means to manipulate milk composition is in the processing plant'. In the near future, Oldham predicts that large changes in the composition of product presented for processing are unlikely to be made.

The alternative to changing the nature of the milkfat at source is to select fractions of the milkfat for use in a spread. A double fractionation of the milkfat is needed to obtain high and low melting fractions that are suitable for blending. The fraction with an intermediate melting range requires a separate market.

A spreadable butter product was developed at the New Zealand Dairy Research Institute (1977) using a low melting milkfat fraction, substantially liquid at 0 °C. This was blended with a high melting fraction, water, salt and MSNF to produce a spread with acceptable consistency at 5–22 °C. The technology required for such products is similar to that for margarine, with the use of batch or continuous mixing followed by processing through swept-surface heat exchangers and a worker unit. The capital cost for this plant in addition to the anhydrous milkfat production and fractionation plants would require a premium price for the product over that of normal butter. A spreadable butter has recently been launched in the UK by Anchor Foods.

Whipped butters have been produced, particularly in the USA, using air or nitrogen to increase the volume of the product. Increases in volume of over 100 per cent are possible. Softened butter may be whipped by passing it through a high shear mixer with gas injection, or molten butter may be cooled and whipped in an SSHE. The high shear in the whipping can produce some work softening which, providing it is not compromised by temperature cycling, can soften the texture. Increasing the volume reduces the specific and latent heats per unit volume so that portions of the product may heat up more readily and become spreadable. Presence of gas will also weaken the structure. At refrigerator temperatures, the whipped butter still retains too much crystalline fat so the structure is brittle rather than plastic.

DAIRY SPREADS

The best way to produce a plastic, spreadable product is to reduce the amount of solid fat in the continuous phase. In the absence of fractionation or indirect manipulation of the milkfat, some benefit may be obtained by reducing the level of milkfat in the product. The lower level of fat may not satisfy the definition of butter and may cause problems in describing the product. The corresponding increase in the quantity of dispersed aqueous phase will also add to the potential for spoilage, particularly if the aqueous phase is not maintained as a fine dispersion. Formation of a fine emulsion may be aided by including surface-active ingredients e.g. mono-diglycerides or phospholipids in the lipid phase.

The alternative to modifications of the milkfat itself is to blend the milk fat with other fats that are liquid at 0–5 °C. This method can help reduce the cost of the product so that it may compete more readily with other spreads in the market. The mixing of milk and other fats has not been permitted in some countries, e.g. Germany and the Netherlands.

In some countries, up to 10 per cent butter may be included in margarines, e.g. UK and Switzerland. Blends of 25 per cent butter and 75 per cent margarine have been marketed in the USA. These latter products may be regarded as upgraded margarines rather than true dairy spreads.

A good example of a dairy spread is provided by 'Bregott', developed and marketed by the Swedish Dairies Association (SMR). Bregott was originally developed using ripened cream (about 35 per cent fat, pH 4·6–4·7) to which refined, deodorised soy bean oil was added in the batch butterchurn. The soy bean oil was added at approximately 20 per cent of the total fat, giving a soy bean oil content of 16 per cent in the final product. The level of soy bean oil may be varied seasonally to compensate for variations in the milkfat, and hence standardise the spreadability of the product.

The mixture of cream and soy bean oil should be churned at a lower temperature than for ripened cream alone, since the proportion of solid fat in the mixture is lower than for normal cream. The buttermilk is then drawn off, and the grains washed with cold water. A small quantity of salt may be added before working the mixture in the churn. Fat losses during churning were found to be higher for Bregott (average 0·86 per cent) than for normal buttermaking (average 0·59 per cent) (Joost, 1977). A batch of Bregott (1500–3000 kg) may be produced by this process in 2 h. The product is then packed into tubs, as the traditional foil pack would not provide sufficient support for such a soft product.

The introduction of Bregott was intended to defend the decreasing level of butter sales in Sweden by increasing the proportion of milkfat in the total yellow fats market. The success of this is demonstrated in Fig. 17, for not only did it halt the decline in milkfat sales, there was a recovery in the sales of milkfat.

This success may be compared to sales in butter and dairy spreads in the UK, where spreads production has made less use of milkfat (Fig. 16). Both countries had a similar per capita butter consumption in 1965. By 1990, the average consumption of butter in Sweden was lower than in the UK, the total milkfat consumed in dairy based spreads was higher. Total consumption of spreads in 1990 (including butter and margarine) was higher in Sweden than the UK with per capita consumptions of 13·7 and 10·3 kg respectively.

Manufacture of Bregott type products is not limited to batch churning. A continuous method was subsequently developed (Johansson, 1980) using a PSM buttermaker. Ripened cream, with a fat content of 36–38 per cent and pH 4·6 was cooled to 3–4 °C by pumping it through a plate heat exchanger, and holding it for a minimum of 3 h, during which time the temperature may have risen to 5 °C as a result of latent heat released during further crystallisation of the milkfat. The aged cream was then pumped from the ageing tank. Soy bean oil (also at 5 °C) was continuously injected into the stream of cream by a dosing pump to give

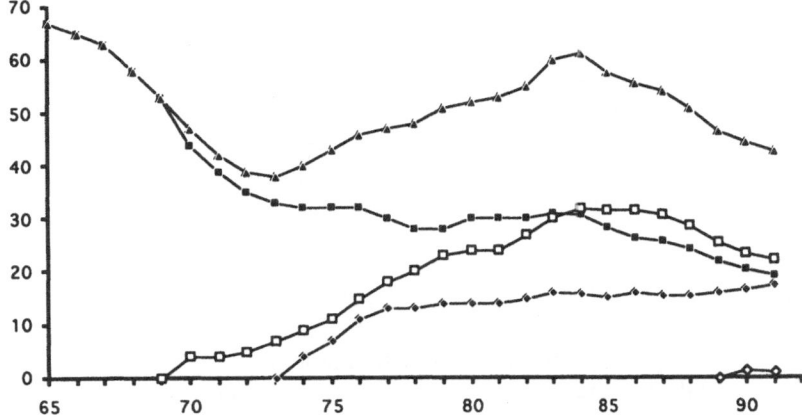

Fig. 16. Consumption of dairy based yellow fats (in thousand tonnes per annum) in the Swedish market 1965–91. ▲: total butter equivalent; ■: Butter; □: Bregott; ◆: L & L; ◇: Klick. (Based on IDF, 1984; personal communication, Svenska Mejeriernas Riksförening (Swedish Dairies Association)).

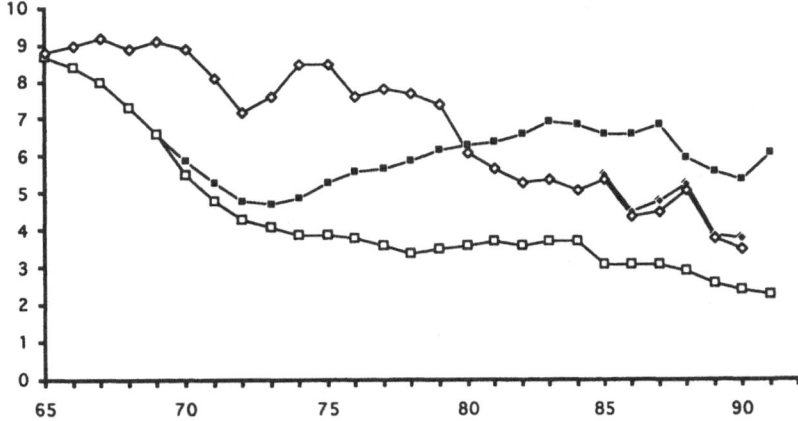

Fig. 17. Comparison of consumption patterns (kg per capita per annum) in Sweden and the UK, 1965–90. ■: Sweden, total butter equivalent; □: Sweden, butter; ◆: UK, total butter equivalent; ◇: UK, butter. (Adapted from IDF 1975–92; SMR, 1992.)

a total fat content of over 40 per cent, preferably 43–45 per cent, and the two were blended together, using an in-line static mixer, before entering the buttermaker. Though the oil must be well dispersed in the cream, it is essential that very fine droplets are not formed, since these would result in higher fat losses during churning. The cream–soy bean oil blend may be churned at 5–7 °C, and the finished product extruded from the butter-maker at approximately 10 °C; the lower temperature partially compensates for the lower solid fat content in the product. The churning of the cream results in a buttermilk containing traces of vegetable fat. This may be viewed as 'contamination' by the purist but, since most of the butter-milk produced was lactic and going for animal feed, the problem was not of major importance.

Vegetable oils have been added at other stages of the process: to the milk before separation, to the cream before pasteurisation and to the butterstream, both in the continuous buttermaker and on reblending. In all cases, satisfactory products could be made, but with injection of the vegetable oil into the butter, higher shear rates were needed to ensure adequate mixing (Dixon *et al.*, 1980). However, product flow in the dairy is eased, and the risk of accidental contamination of other dairy products is reduced if the vegetable oil addition is not made any earlier than during the feed to the buttermaker.

A more recent process for Bregott production has overcome the contamination problem and offered other processing economies (Andersson, 1989). In this 'Alfablend' process (Fig. 18) the pasteurised cream is reheated to 60 °C and reseparated to give a standardised concentrated cream with 75 per cent milkfat. The concentrated cream is cooled to approximately 20 °C and held in a precrystalliser, preferably overnight, before bending with vegetable oil, dry salt and aroma compounds. The vegetable oil may be added at 25–35 per cent for full fat spreads (> 80 per cent fat). At this stage the emulsion is still o/w. Scraped-surface heat exchangers are then used to cool the emulsion and bring about inversion to the w/o form. Worker units (or pin rotors) may be included during cooling to modify the crystallisation and ensure that the desired plastic texture is achieved.

The Alfablend process replaces buttermilk by-product with skim-mik, easier to handle with most factory systems, and reduces fat losses during production. The higher fat substitution levels (10 per cent or more) claimed for this process can also be an economic benefit.

Similar products to Bregott have now been produced in other countries (IDF, 1984). In the UK, the original concept based on continuous buttermaking, was taken a stage further by Dairy Crest Foods (the manufacturing subsidiary of the MMB of England and Wales) with their product 'Clover'. The presence of a subsidy on packet butter in the UK market reduced the price differential between butter and margarine to aid

143

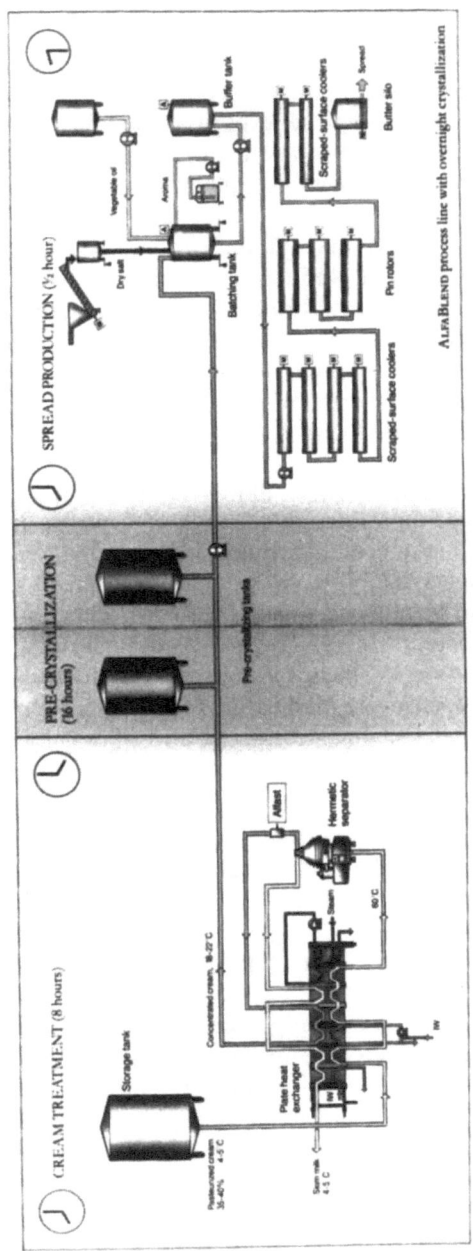

Fig. 18. Outline of Alfablend™ process line with overnight crytallisation (courtesy Alfa-Laval, UK).

butter sales. The subsidy was not available for hybrid products, so the degree of substitution had to be increased to compensate for the higher cost of the milkfat. Simply increasing the level of a vegetable oil, such as soy bean oil, would make the poduct too soft, so a third, more saturated, fat component was introduced. Under UK regulations, both butter and magarine should contain at least 80 per cent fat and have a maximum moisture content of 16 per cent; the milkfat content of margarine is limited to a maximum of 8 per cent. Reduction of the fat content avoided possible conflicts with the margarine regulations, but the addition of a monoglyceride emulsifier was needed to ensure adequate dispersion of the now larger proportion of aqueous phase.

Processing of Clover has been carried out with continuous butter-makers, and also using margarine technology. Continuous buttermakers are claimed to be more appropriate. The method adopted for the consumer trials and the subsequent launch of Clover in 1982 was to blend the non-dairy fats (soy bean oil, a more saturated fat and a monoglycer-ide emulsifier, such as Dimodan S) with the pasteurised sweet cream. The mixture may then be allowed to age overnight. The non-dairy fat should have a solid fat content of 15–35 per cent at 5 °C and 15–25 per cent at 20 °C, e.g. a blend of soy bean oil and partially hydrogenated soy bean oil. Fat addition levels were made at less than 50 per cent of the total fat, to maintain the dairy claim. A typical blend of dairy and vegetable fats would have a solid fat content of 42 per cent at 5 °C and 12 per cent at 20 °C (Wiles and Lane, 1983). Churning of the aged mixture at 7 °C may be preceded by air injection, as with butter production. The product extruded from the buttermaker may then be pumped to the filling machines, and filled into rectagular plastic tubes with a 250 g fill. The packed product is cooled before despatch.

Reduced-fat Dairy Spreads

The discussion of spreads has so far only included minor increases in moisture content to avoid restrictions imposed by composition regula-tions. If the objective is to achieve a significant reduction in energy content then a larger proportion of fat must be replaced by ingredients that provide less energy; the cheapest of these is water. There is a concensus that the description "reduced-fat spread" should be limited to products with less than 60 per cent fat.

Fat levels of 55–60 per cent on a mass basis give volume fractions of ≈60 per cent in the emulsion so that w/o emulsions may be easily achieved. In sweet cream based products, the increasing levels of water dilute the salt in the aqueous phase. The common UK salt addition rate of 2 per cent approximates to an 11 per cent solution in the aqueous phase of butter but in a product with 40 per cent moisture drops to less than 5 per cent, with a corresponding rise in the water activity a_w from 0·94 to over 0·98. At these higher a_w values, it is even more important that microbial growth be inhibited by maintaining the particles of dispersed aqueous phase at less than 10 μm. In an ideal reduced-fat spread there would then be approximately 10^9 particles of dispersed aqueous phase per ml; so that a product with an initial count of 1000 cfu g^{-1} would have organisms in approximately one in 10^6 particles, effectively limiting the extent of microbial growth and spoilage. However, in practice the sensory properties are usually improved by using a coarser dispersion that will break down more readily in the palate, so a compromise must be struck between the sensory and microbial targets.

Conventional continuous buttermaking techniques are limited in the quantity of moisture that may be retained in the product, but the addition of dosing and high shear mixing equipment may ensure a homogeneous product. An example of such a plant, capable of producing dairy spreads of 35–60 per cent fat, is illustrated in Fig. 19. A scraped-surface cooler may also be included in the plant, particularly if other fats are to be dosed into the feedstock to produce a dairy blend.

The sensory properties of reduced-fat products benefit from enrichment of the aqueous phase, particularly from increasing the level of milk proteins. Since milk proteins promote o/w emulsions, they aid breakdown of the emulsion on the palate. Milk proteins work in the opposite way to mono-diglycerides, which promote the formation and stabilisation of the w/o emulsion. Protein enhancement may be achieved by use of protein powders such as caseinates, or by treatment of buttermilk or skim-milk by ultrafiltration. An example of this approach was provided by 'Dane Lite', marketed in the UK in 1986, which contained 55 per cent milkfat, 4·1 per cent buttermilk protein, 1·6 per cent carbohydrate and 1·4 per cent salt.

Protein-enriched mixtures for aqueous phases should preferably be pasteurised then cooled to a suitable temperature for emulsification. Salt may be included in the aqueous phase; or culture or fermentation products may also be dosed into the blend.

Though the properties of the continuous phase predominate in the characteristics of an emulsion, increasing the level of the dispersed aqueous phase from 18 to 40 per cent will reduce the mechanical strength of the

Fig. 19. Low-fat butter plant with process control Mikromaster 2000S (courtesy APV Pasilac AS, Denmark).

product. However, a dairy spread based on 55–60 per cent unmodified milkfat will still be too firm for spreading at about 5 °C. Further reduction in the level of solid fat is needed, for instance by adding a lower melting fraction. The reduction of solid fat in the continuous phase must be accompanied by control of the fat crystal size by appropriate processing. A large number of small fat crystals are needed to maintain a high surface area, providing interaction of the solid and liquid fat to maintain plasticity and avoid oiling-off defects. Rapid cooling and high shear mixing are common features in achieving a well emulsified plastic structure.

Low-fat Spreads

Low-fat spreads, with approximately half the fat and energy content of butter or margarines, have been the subject of intense commercial interest. In general, the lower the fat content then the lower the inherent stability of the w/o emulsion. For low-fat spreads, with 40 per cent or less fat, the dispersed aqueous phase accounts for more than half the volume of the product. The process technology for these products has largely been adapted from margarine technology, using SSHEs and worker units in the cooling of the emulsion. Developments of the buttermaking process with in-line high shear mixing as described in the section on reduced-fat spreads, have also been used (Berntsen, 1990).

When using scraped-surface technology, the simplest method to produce the emulsion is to disperse the aqueous phase into the lipid phase on a batch basis, relying on the scraped-surface treatment to achieve the desired emulsion characteristics. Continuous blending techniques may also be used, and the emulsion may also be produced as an o/w feedstock which is then inverted during the processing to give the desired w/o product (Anon, 1989; Moran, 1990), though this approach is more applicable to low viscosity aqueous phases.

The relatively simple emulsion of a saline, aqueous phase in milkfat may be produced easily at ≈ 40 per cent fat, but has poor organoleptic qualities. A simple milkfat lipid phase also gives rise to problems. The high level of solid fat combined with the weakness induced by the high level of aqueous phase tends towards a crumbly, brittle texture in the product. As with the reduced-fat spreads, the addition of milk protein, preferably the products of fermentation processes, can give a much better product. However, at lower fat levels (about 40 per cent) the tendency for milk proteins to promote o/w emulsions does become a serious problem. This may be overcome by using higher levels of milk proteins (an increasingly expensive commodity) together with heat treatments to modify their characteristics. The function of the protein is to reduce the mobility of the particles of the aqueous phase by increasing viscosity or by gelation at refrigerator temperatures. Satisfactory organoleptic properties can be achieved at lower protein addition rates, but careful selection of both emulsifiers and hydrocolloid additives is needed in producing similar viscoelastic systems to maintain the emulsion stability. This area is the subject of many patent claims.

One of the earliest successful low-fat spreads 'Latt och Lagom' (L & L) was developed by the Swedish dairy cooperative, Mjolkcentralen Arla following the development and launch of Bregott. The objective was to produce a spreadable product with half the fat of butter or margarine, while using dairy products as the principal ingredients. Swedish regulations required a fat content of 39 ± 1 per cent for low-fat spreads, a standard that has been adopted by many countries now. In several countries, a market for low-fat spreads had already been identified by the margarine industry, and a large number of patents had been filed. Products on the market at the time were relatively simple emulsions of a saline aqueous phase in a blend of fats and emulsifier, giving relatively poor organoleptic qualities.

The approach used for L & L was to prepare an aqueous phase derived from concentrated buttermilk and containing more than 13 per cent milk protein. This aqueous phase was then dispersed in a lipid phase blended from milkfat, soy bean oil and monoglycerides. Using buttermilk in the L & L, together with milkfat, linked the product closely to both butter and anhydrous milkfat production. In Sweden, most of the butter produced has a pH of 4·6–4·7, and its production creates equivalent volumes of lactic buttermilk. Retail sales of cultured buttermilk products were relatively low, so the main outlet was disposal for animal feed. Recovery of the buttermilk protein, so that only the whey went for animal feed, would give a better return.

The method of production was patented (Strinning and Thurell, 1973) and demonstrated a combination of the technologies associated with the dairy and margarine industries. Preparation of L & L commenced with pasteurisation of the lactic buttermilk (pH $\approx 4·6$) to kill the starter bacteria; inactivation of enzymes and some denaturation of whey proteins would also take place. The pasteurised buttermilk was then passed through a nozzle bowl separator, where most of the protein was removed from the whey and ejected from the separator as a viscous white paste of buttermilk protein concentrate. This concentrate, with a protein content of 13–20 per cent, may then be cooled and stored till needed. The buttermilk concentrate contains relatively high levels of fat globule membrane, but must be converted into a suitable form for the aqueous phase by changing the casein into a soluble form. The addition of sodium citrate and sodium phosphate achieves this change by raising the pH and sequestering calcium from the casein, so that the micelles dissociate and the casein hydrates further. This is accompanied by a visual change from a white paste to a creamy viscous liquid. For a product of 7·5–8 per cent protein content, a typical protein level in the concentrate would be 13 per cent. Sodium chloride may be added at this stage, but at a lower level (e.g.

1 per cent) than for an equivalent salt flavour in butter; the presence of the other salts and the protein contribute to the saltiness of the product. Standardisation of the pH to $\approx 6\cdot4$ was achieved by adding alkali, e.g. sodium hydroxide. Preservatives, such as potassium sorbate, may also be added at this stage. The temperature of the completed aqueous phase should be at a standard level, e.g. 45 °C.

The lipid phase was made by blending the vegetable oil, typically soy bean oil, with anhydrous milkfat and minor quantities of monoglyceride emulsifier (e.g. $0\cdot3$ per cent), β-carotene for colouring, and vitamins A and D. These minor ingredients may be predissolved in a part of the milkfat. Milkfat may be used as the principal fat (Strinning and Thurell, 1980), but the product, though softer than butter, is less easily used direct from the refrigerator. The level of monoglyceride is kept low as it plays only a secondary role in the formation and stabilisation of the emulsion, with the buttermilk protein concentrate as the principal emulsification agent. Unlike the earlier protein-free, low-fat spreads, a more saturated monoglyceride may be used (Madsen, 1976). Higher levels of monoglyceride cause destabilisation of the w/o emulsion. The emulsion should be formed by adding the aqueous phase to the lipid phase. The lipid phase is normally at a higher temperature, e.g. 50 °C. Addition of the lipid phase to the viscous aqueous phase results in the formation of a stable o/w emulsion that cannot readily be reversed to the desired w/o emulsion on subsequent processing. This emulsification is carried out as a batch operation, with careful control of the addition rate and the level of agitation needed to minimise the risk of phase reversal. The presence of cultured buttermilk concentrate and milkfat in the emulsion give an acceptable flavour to the product, but this may be enhanced by addition of aroma compounds to the emulsion.

The completed emulsion may then be pasteurised and cooled using scraped-surface heat exchangers as illustrated in Figs. 15 and 20. During the cooling of the emulsion, the viscosity remains relatively low until the temperature is reduced to below 35 °C, when the viscosity increases significantly. The viscosity increase is due to crystallisation of the higher melting, saturated fats and monoglycerides, so the rate at which the viscosity increases will depend on the properties of the lipid phase.

Cooling to 35–40 °C may be achieved using water as the refrigerant, followed by further cooling using a direct expansion refrigerant, such as liquid ammonia. On a commercial scale, at least two barrels are needed for the latter stages of cooling. Shear forces developed within the SSHE continue the emulsification process, creating a wide range of particle sizes

Fig. 20. Scraped-surface pasteuriser and cooler (courtesy St Ivel, UK).

for the dispersed aqueous phase. The emulsion is stabilised at this stage by the virtual gelation of the aqueous phase, and by the reduced mobility of the now plastic, continuous lipid phase. The texture of the product may be further modified and improved by incorporating a worker unit, but the exact plant configuration depends on the product formulation and the relationship between the SSHE size and throughput.

After cooling to 8–12 °C, the pasteurised product should be piped direct to the filler for dosing into plastic tubs. The product is sealed under an aluminium foil, with the foil then covered by a plastic reclosable lid (Fig. 21). The closed system with direct feed to the filler is desirable, not only to produce a consistent product (as crystallisation of the fat continues through the filling process), but to minimise the risks of post-processing contamination.

In low-fat spreads, the aqueous phase makes up a much larger proportion of the product than in butter (approximately 60 per cent compared to less than 20 per cent). The droplets of disperse phase may also be larger than in a well-worked butter, and the level of salt in the aqueous phase considerably less than in a salted butter (2 per cent as compared to 10 per cent). In the case of the protein-enriched spreads, there are also ample nutrients, including protein and lactose, present in

Fig. 21. Multilane filler with immediate sealing to minimise contamination risks (courtesy St Ivel, UK).

the aqueous phase. Pasteurisation is sufficient to reduce the microbial population to a safe level, provided that the subsequent storage and distribution is under chilled conditions, as for most other dairy products.

The major commercial spoilage risk wih low-fat spreads is mould spoilage, particularly by those species of *Penicillium* and *Cladosporium*, which can grow under cold storage conditions. For this reason care is needed not only in the hygiene of the packaging operation but also in the procurement and handling of the packaging materials. The addition of potassium sorbate to the aqueous phase, though less effective than at pH values below 6, provides a significant measure of protection to the product.

As with Bregott, considerable interest was shown in the L & L product and the process was licensed to other producers. The first licence taken for the process was by Unigate in the UK, where different technical and marketing constituents applied. The large-scale fermentation of buttermilk was required since most butter in the UK is made from sweet cream. Use of potassium sorbate in low-fat spreads was not permitted at that time, and the presence of potentially conflicting patents provided some formulation constraints. The UK product was launched as 'St Ivel Gold' in 1977, and as with Clover, the operation of the consumer butter subsidy was a major commercial factor which, in this case, resulted in the eventual replacement of the milkfat. This was a significant change from the initial L & L concept, where defence of milkfat sales was a major consideration. The milkfat was replaced by blends of vegetable oils and hydrogenated vegetable oils, following established principles of margarine manufacture (e.g. Wiederman, 1978; Erickson *et al.*, 1980) where fat crystallisation is directed towards the β' state, as occurs with milkfat. The preservation of the flavour characteristics may be attributed to the buttermilk protein concentrate which still formed the major ingredient of the product.

The use of buttermilk as a base for low-fat spreads is a constraint on the location and extent of their manufacture, particularly where butter production is seasonal in character. This constraint has been overcome in a number of ways. In Japan, a variant of L & L has been produced using casein as the protein source. Processes employing skim-milk as an alternative to butter have also been developed (Wallgren and Nilsson, 1978).

To use skim-milk as an alternative starting material, it should first be pasteurised, then incubated with starter organisms and preferably with rennet action at about 0·001 per cent. The starter organisims should be homofermentive, since the production of carbon dioxide leads to foaming and processing problems (whey proteins tend to denature at the gas–liquid interface). Fermentation temperatures depend on the starter organisms selected, and are in the range 20–30 °C for a mesophilic cottage cheese starter culture. As with quarg, the fermentation should continue to the isoelectric point of the milk; deviation from this leads to less efficient separation of the protein.

The coagulum must then be broken up, and the suspension heated to 52–55 °C by passing it through a heat exchanger. Small temperature differentials within the heat exchanger are essential to avoid damage to the protein, as grains of overheated protein may easily block the heat exchanger, and separator, and would not be readily redissolved at later stages in the process.

Once heated, the milk is held for 15–60 min, and during this holding time, calcium diffuses from the micelle into the serum. Without the hold, the protein would not be stable to subsequent processing, and would also have a less efficient emulsifying action. (It is during this hold that any carbon dioxide present in the fermented milk comes out of solution, forming a foam in open tanks, causing operational problems and some loss of protein.) Following the hold, the skim-milk should be subjected to a further heat treatment, bringing the temperature up to 65–75 °C, depending on the stability of the fermented skim-milk. This heating must be carried out rapidly, e.g. using steam injection, with the hot skim-milk fed directly into the nozzle-bowl separator. The function of the second heating is to aid separation by reducing the viscosity of the skim-milk, to further denature enzymes, and inactivate any surviving starter bacteria. Rapid cooling of the hot concentrate is essential to prevent the development of granularity, and a reduction in the emulsifying properties of the concentrate. Similarities should be noted between the production of protein concentrates from skim-milk and the production of quarg. The use of skim-milk under these conditions produces a blander product than that based on buttermilk, thus placing greater emphasis on the role of either milkfat or flavourings to achieve a satisfactory flavour.

These processes starting with liquid buttermilk or skim-milk illustrate the complexity of the manufacturing process. Other processes are available to avoid problems with patent claims or availability of dairy ingredients, particularly fresh buttermilk. Wilton *et al.* (1978) claimed the creation of an analogue to buttermilk concentrate in the aqueous phase by dispersing fat and phosphatides in the aqueous phase. Sodium caseinate, with or without other hydrocolloids, has been widely used for manufacture of the aqueous phase of low-fat spreads. Examples of the variety of low-fat spreads are illustrated in Table I. The calcium level in sodium caseinate is already reduced, so that sequestrants such as sodium citrate and sodium phosphate may not be needed. Examples of formulations for low-fat spreads are now widely available, e.g. Hansen (1989), Madsen (1989). Caseinate is a relatively expensive ingredient and other hydrocolloids and thickening agents, such as modified starch, may be used as alternatives.

Casein-based aqueous phases suffer from the disadvantage of being sensitive to pH. Below $\approx$ pH 6 the water binding properties of the caseinate will begin to deteriorate and below $\approx$ pH 5·3 there is an increasing problem with precipitation. Lower pH values are desirable for improved microbiological stability. Whey proteins with far greater acid tolerance have been suggested, but there can be problems with inversion of the emulsion; this may be overcome by inclusion of polysaccharides (Jost *et al.*, 1990). Small particles of denatured whey protein may also mimic the sensory properties of fats on the palate. In some systems the milk protein has been omitted completely.

TABLE I

Examples of low-fat spreads sold in France and the UK

Brand	Manufacturer	Main ingredients in aqueous phase	Product composition (per cent)		
			Fat	Protein	Carbohydrate
Allégé	Bridel	Buttermilk, caseinate	41	7	1
Elle et Fine	Union Laiterie Normande	Buttermilk, carrageenan	41	0·8	2
Legertine	Evan	Concentrated buttermilk	41	8	2·5
St Florigny	Astra-Calvé[a]	Skim-milk, starch, gelatin	41	2·5	2·5
41	St Hubert	Buttermilk concentrate	41	7	1
Sylphide	Fromageries Bel	(Cream) Caseinate, gelatin	38	7	1
Tartine	Lesieur	Buttermilk concentrate	41	7	1
Anchor Low Fat Spread	Anchor Foods	Buttermilk, gelatin	40	2·8	0·1
Anchor Half Fat	Anchor Foods	Caseinate	40	6·5	0·4
Clover Extra Light	Dairy Crest Foods[a]	(Cream) Caseinate, gelatin	39·6	7·8	0·5
Gold	St Ivel[b]	Buttermilk, caseinates,	39	6·5	2[a]
Gold	St Ivel[c]	Reconstituted buttermilk, caseinates, modified starch	39	6	3
Kerrygold Light	Adams Foods[a]	Caseinate, gelatin	39	6	0·5
Shape	St Ivel	Skim-milk, buttermilk, gelatin	39·7	7·6	2·2

[a]Imported from another source.
[b]Manufactured 1985.
[c]Manufactured 1992.

Very-low-fat Spreads

The drive for marketing advantage by claims for greater energy savings has lead to the development of very-low-fat spreads, retaining a w/o emulsion with less than 25 per cent fat. Production technology is based on SSHEs.

Two examples in the UK, 'Gold Lowest' and 'Delight Extra Low' have typical fat contents of 25 per cent and 20 per cent respectively. The aqueous phase in the Gold Lowest is based on skim-milk, buttermilk and modified starch. The system is covered by a patent (Platt and Gupta, 1988) for spreads with 18–35 per cent fat, claiming the use of at least 8 per cent of one or more milk proteins plus 0·1–1·2 per cent of a modified starch, e.g. acetylated distarch phosphate and hydroxypropyl distarch phosphate. A similar application covers the use of an aqueous phase comprising at least 6 per cent milk proteins plus 1·3–4 per cent of pregelatinised starch (Gupta and Platt, 1989). The aqueous phase of the Delight Extra Low is based on maltodextrin and gelatin; the product does not contain any dairy ingredients.

The development of the very-low-fat spreads has questioned the limits of fat-continuous systems. Glaeser (1990) used simple theoretical models assuming particles of similar size in the disperse phase. Arrangement in a cubic lattice only permitted fat contents of 43 per cent fat. By adopting two particle sizes and using a cubic inner-centered lattice the minimum fat content was reduced to 34 per cent. In reality the structure of low-fat spreads is not so homogeneous as there is a range of particle sizes and, particularly in the case of the spreads with a high level of milk protein, a range of particle shapes too (Madsen, 1989). It is also possibe to create structures where the phases are bicontinuous, essentially interwoven threads of the two phases (Moran, 1990). With greater fractions of the disperse phase the particles may be distorted so that ultimately the continuous phase becomes resricted to an insulating film between the densely packed particles of disperse phase.

Developments in emulsion technology, as reported in reviews such as that by Mann (1992), have taken the limits well beyond that believed possible when low-fat dairy spreads first became available on the market. Madsen (1991) has described formulations for very-low-fat spreads with 8–10 per cent fat using maltodextrin, dried whey or milk powder, and gelatin in the aqueous phase, plus interesterified ricinoleic acid polyglycerol ester and distilled monoglycerides as emulsifiers in the lipid phase. Matsuda (1991) described the use of polyglycerol fatty acid esters based on erucic acid as emulsifiers in w/o emulsions down to 10 per cent fat. Jones and Norton (1990) describe o/w spreads down to 5 per cent fat containing more than 25 per cent of the lipid phase based on milkfat. The aqueous phase contained less than 200 ppm of protein, being based on gelling hydrocolloid(s) and with a weighted mean droplet size of less than 25μm.

CONCLUDING REMARKS

Butter will continue as a premium product in the foreseeable future, but consumption is declining and production remains in excess of its commercial market needs, at least within western Europe. Competition from less expensive spreads of marine, animal and vegetable origin (together with the exploitation of real and speculative health risks) is unlikely to diminish. A large sector of the total yellow fat market is likely, however, to be taken by spreads combining the superior organoleptic qualities of dairy ingredients with the convenience and economic advantages obtained by blending with other ingredients, especially vegetable oils. Further improvements in flavour technology will continue to intensify the competition between dairy and non-dairy products. The greater interest in diet, especially fat and energy intake, increases the opportunity for reduced-fat, hybrid products. With the development of these hybrid products, one can expect a continuing transfer of technologies across what were traditional divisions in the food industry.

REFERENCES

Anon (1989). Low calorie butter made from dairy cream. Research Disclosure No. 303, p. 516.
Andersson, K. (1989). PCT International Patent Application WO 89/10700 A1.
Berntsen, S. (1990). *Scandinavian Dairy Information*, **4** (3), 68.
Bøgh-Sørensen (1992). *IDF Bulletin*, **271**, 32.
Boutonnier, J. L. and Dunant, C. (1985). In: *Laits et produits laitiers*, Volume 2 (Ed. F. M. Luquet and Y. Bonjean-Linczowski), Lavoisier, Paris, pp. 443–504.
BS 998: 1969. *British Standard Specification for Vacuum Salt for Butter and Cheese Making and other Food Uses*. British Standards Institution, London.
Bodyfelt, F. W., Tobias, J. and Trout, G. M. (1988). *The Sensory Evaluation of Dairy Products*, Van Nostrand Reinhold, New York.
Davis, J. G. (1986). In: *Quality Control in the Food Industry*, Volume II (Ed. S. M. Herschdoerfer), Academic Press, London, pp. 178–86.

Davis, J. G. and Wilbey, R. A. (1990). In: *Dairy Microbiology*, Volume II (Ed. R. K. Robinson), 2nd edn, Elsevier Applied Science, London, pp. 40–108.

Dixon, B. D. (1970). *Australian Journal of Dairy Technology*, **25** (2), 82.

Dixon, B. D., Cracknell, R. H. and Tomlinson, N. (1980). *Australian Journal of Dairy Technology*, **35** (2), 43.

Erickson, D. R., Pryde, E. H., Brekke, O. L., Mounts, T. L. and Falb, R. A. (Eds) (1980). *Handbook of Soy Oil Processing and Utilisation*, American Soy Association and AOCS, St Louis.

Forman, L. (1990). In: *Proceedings of the XXIII International Dairy Congress*, Volume II, IDF, Brussels, p. 1490.

Frede, E., Precht, D. and Peters, K. H. (1983). *Milchwissenschaft*, **38** (12), 711.

Glaeser, H. (1990). *Dairy Industries International*, **55** (9), 29.

Gupta, B. B. and Platt, B. L. (1989). UK Patent Application 2 215 343 A.

Hansen, R. (1989). *North European Food and Dairy Journal*, **55** (6–7), 125.

IDF (1975–92). *Consumption Statistics for Milk and Milk Products*, Annual Bulletins of the IDF, Brussels.

IDF (1984). *The World Market for Butter*, Document 170, IDF, Brussels.

IDF (1990). *Recombination of Milk and Milk Products*, Special Issue No. 9001, IDF, Brussels.

Jebson, R. S. (1979). *IDF Bulletin*, **116**, 30.

Johansson, M. S. J. (1980). UK Patent 1 582 806; US Patent 4 209 546.

Jones, M. G. and Norton, I. T. (1990). European Patent Application EP 0 372 625 A2.

Joost, K. (1977). *Svenska Mejeritindningen*, **69** (3), 13.

Jost, R., Dannenberg, F. and Gumy, D. (1990). In: *Proceedings of the XXIII International Dairy Congress*, Volume III, IDF, Brussels, p. 1490.

Kessler, H. G. (1980). *Deutsche Molkerei-Zeitung*, 1820–8.

King, N. (1964). *Dairy Science Abstracts*, **26** (4), 151.

Kisza, J., Hadal, A., Laszek, R. and Staniewski, B. (1990). *Brief Communications and Abstracts of Posters of the XXIII International Dairy Congress*, Volume II, IDF, Brussels, p. 308.

Lane, R. (1992). In: *The Technology of Dairy Products*, (Ed. R. Early), Blackie, Glasgow, pp. 86–116.

Madsen, J. (1976). *Food Product Development*, April, pp. 72–80.

Madsen, J. (1989). *Low-calorie Spread and Melange Production in Europe*, TP107-1e, Grindsted Products, Brabrand, Denmark.

Madsen, S. (1991). 8–10 per cent Fat Spread. Reseach Disclosure No. 332, pp. 921–2.

Mann, E. J. (1992). *Dairy Industries International*, **57** (2), 17, **57** (3), 15.

Matsuda, K. (1991). European Patent Application 0 430 180 A2.

McDowall, F. H. (1953). *The Buttermakers Manual*, New Zealand University Press.

Mick, S., Mick, W. and Schreier, P. (1982). *Milchwissenschaft*, **37** (11), 661.

Mortensen, B. K. (1983). In: *Developments in Dairy Chemistry*, Volume II (Ed. P. F. Fox), Applied Science Publishers, London.

Moran, D. P. J. (1990). *Dairy Industries International*, **55** (5), 41.

Mulder, H. and Walstra, P. (1974). *The Milk Fat Globule*, C. A. B. & Pudoc.

Munro, D. S., Cant, P. A. E., Macgibbon, A. K. H., Illingworth, D., Kennett, A. and Main, A. J. (1992). In: *The Technology of Dairy Products* (Ed. R. Early), Blackie, Glasgow, pp. 117–45.

158   R. A. Wilbey

Murphy, M. F. (1990). In: *Dairy Microbiology*, Volume II (Ed. R. K. Robinson), 2nd edn, Elsevier Applied Science, London, pp. 109–30.

New Zealand Dairy Research Institute (1977). UK Patent 1 478 707.

NIZO (1976). Netherlands Dairy Research Institute, British Patent 1 516 576.

Oldham, J. D. (1990). In: *Proceedings of the XXIII International Dairy Congress*, Volume I, IDF, Brussels, 714–25.

Spieler, R. (1982). *IDF Bulletin* **142**, 132.

Platt, B. L. and Gupta, B. B. (1988). UK Patent Application 2 193 221 A.

Samuelsson, E. and Pettersson, K. I. (1937). Årsskrift för Alnarps lantbruks-mejeriog trädgårdsinstitut, cited by Mortensen, B. K. (1983) Physical properties and modification of milk fat, in *Developments in Dairy Chemistry*, Volume II (Ed. P. F. Fox), Applied Science Publishers, London.

Sporns, P., Rebolledo, J., Cadden, A. M, and Jelen, P. (1984). *Milchwissenschaft*, **39** (6), 330.

Strinning, O. B. S. and Thurell, K. E. (1973). UK Patent 1 455 146; US Patent 3 922 376.

Strinning, O. B. S. and Thurell, K. E. (1980). UK Patent Application 2 066 837 A.

Storry, J. C., Brumby, P. E., Hill, A. J. and Johnston, V. W. (1974). *Journal of Dairy Research*. **41**, 165.

Storry, J. C. (1981). In: *Recent Advances in Animal Nutrition* (Ed. W. Haresign), Butterworths, London, pp. 3–33.

UK Intervention Board for Agricultural Produce (1990). *LP13: Intervention Purchase of Butter*, Intervention Board, Reading.

Wallgren, K. and Nilsson, T. (1978). UK Patent Application 2011 942 A.

Walstra, P. and Jenness, R. (1984). *Dairy Chemistry & Physics*, Wiley, New York.

Wiechers, S. G. and DeGoede, B. (1950). *Continuous Butter Making*, North-Holland, Amsterdam.

Wiederman, L. H. (1978). *Journal of the American Oil Chemists Society*, **55** (11), 823.

Wilbey, R. A. (1991). In: *Analysis of Oilseeds, Fats and Fatty Foods* (Ed. J. B. Rossell and J. L. R. Pritchard), Elsevier Applied Science, London.

Wiles, R. (1978). UK Patent 1 597 068.

Wiles, R. and Lane, R. (1983). European Patent Application 0 106 620.

Williams, G. R. (1984). *Journal of Society of Dairy Technology*, **37** (2), 66.

Wilton, I. E. M., Envall, L. O. G., Sunderstroem, K. L. and Moran, D. P. J., (1978). US Patent 4 071 634.

Chapter 4

Drying of Milk and Milk Products

The late M. E. Knipschildt and G. G. Andersen

APV Anhydro AS, Søborg-Copenhagen, Denmark

Milk is extremely perishable, yet for a number of reasons it is desirable to preserve it for later consumption. Today, drying is the most important method of preservation. The advantage is that using modern techniques, it is possible to convert the milk to powder without any loss in nutritive value, i.e. milk made from powder has the same food value as fresh milk. The disadvantage is that drying consumes a lot of energy; in fact, no other process in the dairy industry has a higher energy demand per tonne of finished product. This is due to the fact that approximately 90 per cent of milk is water, and all the water has to be removed by evaporation using heat. Modern membrane techniques make it possible to remove some of the water mechanically, without heat, but this method has limitations so it is not widely used in practice. Nevertheless, the sharp increase in the cost of energy has caused significant developments in process and equipment, making it possible today to convert milk into powder with an energy consumption per tonne of finished product approximately half that required years ago.

HISTORY AND DEVELOPMENT

For a long time, butter and cheese were the only dairy products known to have extended keeping qualities, but in 1856, a new dairy product with long shelf-life was introduced – sweetened condensed milk. It was followed some 30 years later by evaporated milk. The manufacture of condensed milk grew steadily until approximately 1950 (Hall and Hedrick, 1966) but since then a constant decline has occurred.

From 1860, private and cooperative dairies started to be built in the USA, and from 1880 in Europe. In these dairies, butter and cheese were manufactured on a large scale. The skim-milk was sent back to the farmers for animal feed or dumped, and the whey was disposed of in rivers and lakes. This was the usual pattern for many years; right up till about 1930, when utilisation of the by-products slowly commenced.

Drying of milk was introduced at the beginning of this century on a modest scale. It took many years before equipment and processes were developed for commercial use. From approximately 1930, the spray drying process started to gain importance, but the great developments have taken place since the Second World War.

Dairy Research Institutes throughout the world have been somewhat slow in recognising dry milk as a product deserving the same attention as the classical products: cheese, butter and liquid milk. Substantial developments have been made by equipment manufacturers and dairy factories. However, during the last 20–30 years, useful work has been carried out at several dairy research institutes with regard to the manufacture of dry milk and whey, and the importance of these products is now generally recognised.

Milk drying has become an essential part of the long chain between the farm producer and the final domestic consumer, and dry milk manufacture has grown into a large industry of considerable international importance. The diversification of the dry milk industry has been described by Robinson (1982) outlining the potential use in the food industry of the non-fat milk solids and the whey proteins.

Due to the dry milk industry, it is possible to make milk available in regions that are unfavourable for dairying, or where the milk production is insufficient. Milk produced in Australia, New Zealand, or in Europe is consumed in the Far East, in the Middle East, or other countries where milk powder is converted into liquid milk or UHT milk in recombining dairies. In most countries where milk production has become highly organised, the industry has to face great seasonal variations in the milk supply. The most important way of dealing with the surplus milk in the flush season is to dry it. During the lean season, it is often necessary to 'stretch' the milk by adding powder (to tone the milk). In this way, for instance, it is possible to keep cheese production at a constant level all the year round to meet the demand for cheese, and to obtain full utilisation of equipment and manpower.

The drying process plays a decisive part in achieving two objectives which today are topics in the developed part of the world. One is to meet the increasing shortage of proteins in the developing countries, as milk powder is one of the food products most widely used in relief programmes. In this context it should be mentioned that several years ago UNICEF started to fight starvation by means of milk proteins in dry form, and UNICEF has contributed to the erection of milk powder plants in several countries all over the world. The other objective is to fight pollution, and in achieving this, the drying process is of great importance. It would not be possible to build the large cheese factories we have today without a drying plant to take care of the large quantities of whey. A drying plant is the most efficient disposal plant for the whey.

The removal of water from the milk takes place in two stages. The first stage is concentration by vacuum evaporation, and the second stage is drying. Ninety per cent of the water in the milk is removed in the evaporator and only ten per cent in the spray dryer. However, the energy required per kg water evaporated in the spray dryer is 16–20 times the energy required per kg water removed in the evaporator.

In absolute terms, the energy consumption of the dryer is approximately twice that of the evaporator if a six-effect evaporator and a two-stage spray dryer is used. If a four-effect evaporator is used, the energy consumption of the evaporator and the dryer will be approximately the same. The above illustrates that although only 10 per cent of the water is removed in the dryer, this should not lead to the assumption that the efficiency of the dryer is of minor importance.

EVAPORATION

Concentration of milk at low temperature by vacuum evaporation is based on the physical law that the boiling point of a liquid is lowered when the liquid is exposed to a pressure below atmospheric pressure. The milk is brought to boil at low temperature in an apparatus with negative pressure. The maximum boiling temperature accepted in a modern evaporator is approximately 70 °C corresponding to an absolute pressure of 230 mm (9·05 inch Hg).

The first person to use vacuum evaporation for the concentration of milk was Gail Borden, who started a condensing factory in Connecticut, USA, in 1856. Concentration alone does not improve the keeping quality of milk and, therefore, sugar (sucrose) was added to inhibit bacterial growth in the concentrate. That was the start of sweetened condensed milk.

As often happens when a new product is introduced, the consumers were slow to accept it, and the factory had to close a few years later. However, Gail Borden believed in his product and started a new factory a few years later. He died in the town of Borden, Texas in 1874, but his company, The Borden Condensed Milk Company, grew steadily and is today one of the largest in the industry.

The next important development was the introduction of concentrated milk with a long shelf-life, but without the use of sugar as a preservative. It took place in Switzerland, where The Anglo-Swiss Condensed Milk Company started to manufacture sweetened, condensed milk in 1866 in the first condensed milk factory in Europe built in Cham, Canton Zug. The operator of this factory, John Meyenberg, conceived the idea of giving the concentrated milk a heat treatment severe enough to destroy the enzymes and to reduce the bacterial count. The revolutionary idea in his progress was that of avoiding recontamination after heat treatment. He suggested putting the evaporated milk in cans, sealing the cans, then sterilising the sealed cans. He designed a revolving steriliser in which the cans were heated to 115 °C. This innovation became the basis of a new industry, the manufacture of unsweetened condensed milk to be known as evaporated milk. In 1884, Meyenberg was granted a patent for his process of preserving milk without addition of a preservative, and in the following years he built several factories in the USA to produce evaporated milk. In 1902 the Anglo-Swiss Condensed Milk Company sold its entire American interests to The Borden Condensed Milk Company, and the factory in Switzerland merged with Henry Nestlé of Vevey, Switzerland.

Since the Second World War, the manufacture of sweetened condensed milk and evaporated milk has been in steady decline in favour of milk powder. However, they are still sold in some of their traditional markets in the Far East and South America. As the countries where condensed milks are favoured have sufficient milk production, milk powder is often used for local manufacture of sweetened condensed milk and evaporated milk. This calls for powder manufactured especially for that purpose.

Principle of Evaporation

It was several years before vacuum evaporators were used to concentrate milk before drying. Most of the development took place after the Second World War, and the rise in energy costs since 1973 has led to a new generation of evaporators. The first evaporators to be used in conjunction with dryers were installed around 1930; they were made of copper or aluminium. Since the Second World War only stainless steel has been used; its great advantage is that it permits chemical cleaning. Whereas all parts of an evaporator made of copper or aluminium had to be accessible for manual cleaning, this is no longer necessary when chemical cleaning can be used, and it presents new possibilities in design.

Over the past thirty years, the tubular falling film evaporator has been universal in the dairy industry; its only competitor of importance has been the plate evaporator, introduced in 1957. In 1923 Richard Seligman, founder of the APV company in England, invented the plate heat exchanger, and ever since it has been of paramount importance to the dairy industry. It was, therefore, natural for APV to design an evaporator where the heating bodies consist of plate heat exchangers. Known as the plate evaporator, it has been very successful and offers the advantage of low building height; this makes it possible to install it in existing buildings. A further advantage is that the heating surface can be varied easily by adding or removing some of the plates. This facility makes the plant very versatile and capable of handling a wide range of products. The plate evaporator held a strong position in the dairy industry for approximately 10 years. Since then plate evaporators have been used only in other industries, as the capacities called for in the dairy industry have been outside its range.

To fully comprehend the advantages of the tubular falling film evaporator, the principle of the tubular natural circulation evaporator, used in the dairy industry for many years, will first be explained.

Natural Circulation Evaporator

The natural circulation evaporator is also known as the climbing or rising film evaporator (Fig. 1). The milk is brought to the boil in tubular heat exchangers (calandrias) consisting of a large number of tubes through which the product rises. The tubes are heated externally by a steam chest. The milk enters the bottom of the calandria and boiling takes place near the base of the tubes. The vapour bubbles expand rapidly; this draws a thin film of product at high velocity up the walls of the tubes. From the top of the calandria, the product is discharged into a separator vessel where the vapour is removed and used as the heating medium in the following effect. The vapour from the last effect is led to a condenser. The product is piped back to the base of the calandria (recirculated) until the desired concentration is reached. Natural

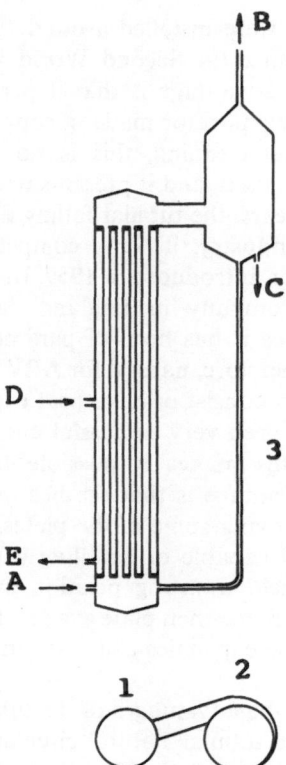

Fig. 1. Circulation evaporator. (A) product (B) vapour (C) concentrate (D) boiler steam (E) condensate (1) calandria (2) separator (3) recirculation tube.

circulation evaporators are usually built as double-effect plants. The tube length is normally 4 m, and the diameter 25 mm.

Falling Film Evaporator

In the years after the Second World War, tank collection of milk was rapidly adopted in countries with high levels of milk production. In consequence, dairy factories grew much larger and they continue to grow as increased throughput reduces fabrication costs per tonne of finished product.

Milk powder factories demanded evaporators with much larger capacity than known before, and at the same time higher concentration, better steam economy, and improved product quality were called for. These factors led to the design of the falling film evaporator, first introduced in 1953 by Wiegand in Germany. The falling film principle offers essential advantages and was soon adopted by all the leading evaporator manufacturers in Europe.

For many years, the tubular falling film evaporator has been the only evaporator installed in the dairy industry. High steam economy is obtained by recompression of the vapour created by the boiling of the milk. This is done by adding energy to the vapour, either by a steam jet or by a mechanical compressor. The two methods of recompression have given their names to the two types of evaporators used today; TVR stands for thermal vapour recompression, and MVR for mechanical vapour recompression.

TVR Evaporators

Calandrias are used as heat exchangers as in natural circulation evaporators, but the tubes are up to 15 m long and 50 mm (2 inches) in diameter (see Fig. 2). Milk is introduced at the top of the calandria. Gravity moves it down the inside surface of the tubes as a thin film, aided by the vapour which also moves downwards. The vapour is separated from the milk in a separator placed at the base of the calandria. The vapour is used as the heating medium in the following effect. From the last effect the vapour goes to the condenser.

Part of the vapour from the first effect's separator is used together with boiler steam as the heating medium in the first effect's calandria. The pressure – and consequently the temperature – of the vapour from the separator is increased by use of a thermocompressor (Fig. 3). Boiler steam (usually at a pressure of 6–10 bar) is introduced through a nozzle to create a steam jet in the mixing chamber. The speed of the mixture of boiler steam and vapour is reduced in the diffuser, consequently the pressure and the temperature are increased. This makes the mixture suitable as a heating medium in the first effect's calandria.

The amount of vapour sucked into the recompression unit will often be equal to the amount of boiler steam. The design of a thermal vapour recompressor is critical, but the unit is inexpensive, and the saving it provides in steam consumption is very considerable. It has no moving parts and requires no maintenance. The thermocompressor must be effectively sound insulated. The performance of a thermocompressor depends on the pressure of the boiler steam and on the temperature

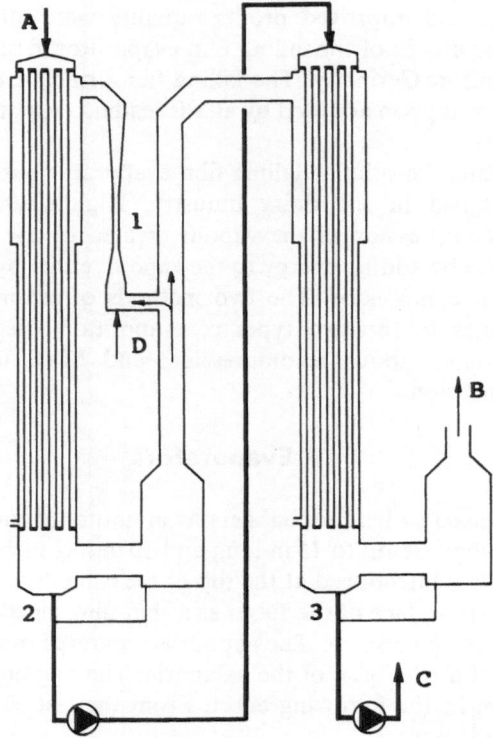

Fig. 2.   TVR falling film evaporator. (A) product inlet (B) vapour (C) concentrate
(D) boiler steam (1) thermocompressor (2) first effect (3) second effect.

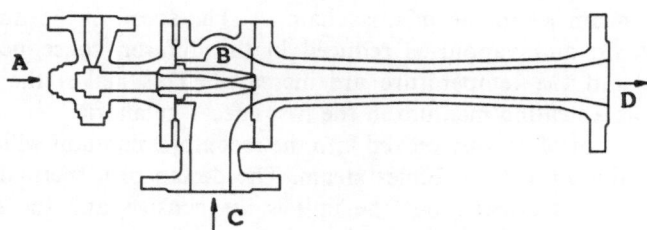

Fig. 3.   Thermocompressor. (A) boiler steam (B) steam nozzle (C) vapour from
separator (D) recompressed vapour.

difference (Δt) between the jacket temperature in the first effect (temperature of the mixture of boiler steam and vapour) and the temperature of the vapour from the separator. The suction ratio (amount of vapour to amount of steam) increases with the steam pressure and increases with decreasing Δt.

1 kg steam of 9 bar will suck

1·0 kg vapour at Δt 13 °C
1·4 kg vapour at Δt 10 °C
2·3 kg vapour at Δt 6 °C

Uniform distribution of the milk over all the tubes is essential, as burning-on, i.e. deposits or scale in the tubes, will occur quickly if the tubes are 'starved'. A distribution device is placed at the top of the calandrias; different manufacturers use different designs of the distribution head. The milk is supplied to the top of each calandria at a temperature slightly higher than the boiling temperature in the calandria, so that the liquid flashes on entry. The flash vapour assists in the uniform distribution of the liquid over the tubes.

The purity of the condensate is governed by the efficiency of the separators. To keep entrainment loss (i.e. loss of milk solids in the vapour) at a minimum, it is essential to ensure high vapour velocity in the separators with tangential inlet to give centrifugal motion. The adoption of the cyclone principle in the design of the separators has removed the need for the non-sanitary deflecting plates used earlier in the separators.

The most important advantages of falling film evaporators over circulation evaporators are as follows.

(1) A condition for the operation of a climbing film evaporator is a fairly high temperature difference (Δt) between jacket temperature and boiling temperature. Below a certain Δt, the evaporator will not operate at all, and a temperature difference of 15–20 °C is not unusual. The large temperature difference between the tube wall and the product promotes scaling of the tubes.

In a falling film evaporator, the energy needed to move the product through the tubes is no longer provided (as in climbing film evaporators) by the difference in temperature outside and inside the tube. This means that the falling film evaporator can operate at a much lower Δt, which gives a number of advantages. The lower the Δt, the smaller the risk of scale formation in the tubes, i.e. the operating time before cleaning is required can be extended. As deposits in the tubes are usually small, even after 22 h operation, cleaning can be done by CIP, whereas the tubes in a circulation evaporator often require manual cleaning.

The low Δt opens the possibility of using six or seven effects. The temperature drop available over the effects is the difference between the boiling temperature in the first effect and the boiling temperature in the last effect (condenser temperature). To avoid protein denaturation the boiling temperature in the first effect should not exceed 68–70 °C, and it is hardly feasible to have a lower temperature in the last effect than 43 °C. This gives 25 °C to be divided over six effects, i.e. a temperature difference of only 5 °C between the effects or between heating medium and product. A specific steam consumption (kg steam per kg water evaporated) as low as 0·09 is obtained with a multiple-effect evaporator, whereas a double-effect plant would use more than four times as much. Normally there is almost no scaling in the tubes even after long operation times; this is due to the small Δt. Of course the investment in a multiple-effect evaporator is substantially higher, but with today's high energy costs, the payback time for the additional effects is only a few years.

(2) The falling film evaporator offers gentle heat treatment due to single-pass operation and short holding time. Undesirable bacteriological, physical and chemical changes in the milk, as a result of excessive heat treatment, are thus avoided. Whereas the circulation evaporator holds a large milk volume, the milk quantity in a single-pass evaporator is small. From start, a circulation evaporator can take up to 1 h or more before the desired concentration is reached, but in a falling film evaporator the concentrated milk is drawn off within a few minutes from the start of the plant.

(3) The low Δt possible in falling film evaporators is favourable to vapour recompression, consequently the energy consumption is low. In a natural circulation evaporator it is necessary to operate with a high Δt in order to obtain a high concentration; this is due to the increased viscosity of the product. The low Δt used in a falling film evaporator makes it possible to reach high concentrations without scaling of the tubes.

(4) The tendency to foam formation in the separators of circulation evaporators, which has an adverse effect on the entrainment loss, is eliminated in single-pass evaporators.

Preheaters

Extensive utilisation of the heat in the vapour for preheating the feed to the evaporator is essential in order to obtain a high thermal efficiency. The milk is normally drawn from milk silos at 4 °C, and heated in spiral heaters placed in the condenser and in the calandrias (Figs. 4 and 5). The preheater in the condenser also serves the purpose of condensing some of the vapour, thereby saving cooling water, but provision must be made in the design for adequate condensing facilities in case an elevated feed temperature is used. The spiral preheaters have the disadvantage of not being accessible for mechanical cleaning and inspection, but cost less than external straight-tube preheaters placed adjacent to the calandrias. The spiral preheaters should not be used for high temperatures, and special preheaters are required for the manufacture of heat classified powders.

Care should also be taken when using the spiral preheaters for partly concentrated milk. In the case of 'backwards' product flow in order to avoid a high concentration in the effect where the temperature is lowest, the product will, for instance, pass from the last effect to the effect before last. The product temperature is thus below the boiling temperature in the calandria to which it is pumped, and the milk with high total solids must, therefore, be heated. For this purpose, an external straight-tube preheater is best. In some cases, where a spiral preheater has been used, it has become blocked-up as a consequence of power failure or other disturbance. If that happens, it is necessary to bypass the preheater and seal it off, or try to force out the congealed concentrate with a high pressure pump (homogeniser), assuming the preheater is designed for the high pressure.

Today preheaters and condensers, for instance in connection with MVR evaporators, can be made as an external plate heat exchanger.

Condenser

A shell and tube condenser should be used where the vapour moves through the tubes and cooling water is passed over the tubes. This type of condenser is called a surface condenser, and it ensures that no carry-over from one medium to another takes place. In a jet condenser, the cooling water is sprayed into the vapour so that the cooling water may be contaminated, and in the case of incorrect operation, some cooling water may be sucked into the last effect, contaminating the concentrate.

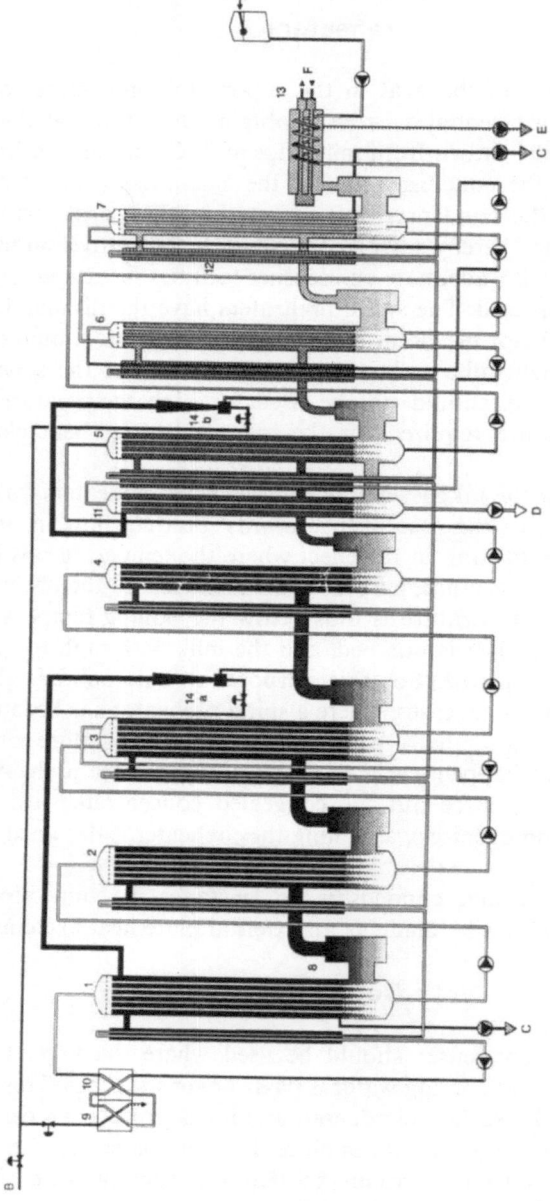

Fig. 4. Seven-effect TVR evaporator with finisher. (A) feed product (B) boiler steam (C) condensate outlet (D) product outlet (E) vacuum (F) cooling water (1) first effect (2) second effect (3) third effect (4) fourth effect (5) fifth effect (6) sixth effect (7) seventh effect (8) vapour separator (9) pasteurising unit (10) heat exchanger (11) finisher (12) preheater (13) condenser (14a,b) thermocompressors. (Courtesy APV Anhydro AS, Denmark.)

Fig. 5.   Calandria (courtesy APV Anhydro AS, Denmark).

Vacuum Producing Equipment

The purpose of the vacuum equipment is to produce and maintain a vacuum in the plant by removing the air and non-condensable gas. Normally, a water ring pump is used, but the vacuum can also be produced by steam jet ejectors, which have low maintenance costs and therefore may be preferred if the costs of the steam for the ejectors are lower than the costs of the power required for the pump.

Finisher

Where high total solids content in the concentrate is required, it is usual to provide the evaporator with a finisher, designed especially to handle viscous products. It is advantageous that the finisher is a small, separate evaporator where the boiling temperature can be chosen to suit the product. Further, where automatic control of the total solids in the concentrate leaving the evaporator is used (automatic density control), a quick response is obtained when the density controller governs the steam supply to the finisher, rather than the steam supply to the main evaporator or the milk intake. Variations in these parameters will give a slow response in a multiple-effect evaporator.

Condensate

The condensate from the first effect is usually led back to the boiler. To protect the boiler against contaminated condensate, the purity of the condensate can be measured by a conductivity meter which activates a bypass valve in case the conductivity exceeds a certain predetermined value. The condensate from the other effects may be used for preheating the air to the spray dryer and, after that, for cleaning.

To ensure maximum heat transfer in the calandrias from the steam (vapour) side to the product, provision must be made for efficient drainage of condensate, and for venting of the steam side to remove all air.

Pumps

A large evaporator requires many circulation pumps. Cavitation in the pumps causing objectionable noise can be eliminated by placing adjustment valves on the pressure side of all the pumps.

Wetting Rate

The wetting rate is the proportion between product and heating surface and is a key design figure for a falling film evaporator. It is equally critical whether the heating surface is too large or too small for a specified evaporation duty. Every tube requires a certain minimum quantity of product to cover the entire surface. Therefore, in a single-pass evaporator, the number of tubes per calandria will decrease from effect to effect because the milk volume is constantly reduced as the concentration is increased. The first effect is always the largest, and in large evaporators, the calandria will be divided into two stages in order to overcome the problem of distribution of the product over a very large number of tubes, and to obtain the correct wetting rate. When the first effect is in two stages, the milk flow will be in series, but the steam flow will be parallel. Partitions are sometimes placed in the calandrias, and the product is pumped from one section to the other so that the correct wetting rate is maintained over a large heating surface.

When comparing different proposals for an evaporator, one should compare the heating surface offered. Some suppliers specify the inside surface of the tubes, others the outside surface, without making it clear which area they refer to. The difference between the inside and outside tube areas can be approximately 5 per cent. Figure 6 shows a six-effect TVR evaporator.

Fig. 6. Six-effect TVR evaporator (courtesy APV Anhydro AS, Denmark).

Matching Evaporator and Dryer

If the duty is only one product, for instance skim-milk, it is obviously no problem, but if different products should be evaporated and spray dried, it becomes necessary to use a feed rate other than the design figure. The spray dryer will operate equally well at a reduced feed rate, as it is simply a question of reducing the air inlet temperature. This, of course, will reduce the thermal efficiency of the dryer, but otherwise, it will not cause any problem.

As far as the evaporator is concerned, it is possible to reduce the evaporation capacity by reducing the supply of steam to the evaporator. However, the evaporator will not operate satisfactorily when the milk intake is reduced so much that the wetting rate becomes too low. The range within which the capacity of the evaporator can be varied with no adverse effect is referred to as the turn-down ratio. The desired turn-down ratio should be specified initially as it is a design parameter. The following example illustrates the problem. Let us assume that an evaporator and a spray dryer are designed to handle 10,000 kg h^{-1} skim-milk of 9 per cent solids, and as a secondary duty, whole milk (full cream milk) of 12 per cent solids is to be dried. The performance data are as in Table I. If the intake of whole milk exceeds 7382 kg h^{-1}, the dryer will not be able to handle all the concentrate.

TABLE I

Product/total solids	skim-milk (9%)	whole milk
Intake, kg h^{-1}	10 000	7 382
Evaporation in evaporator, kg h^{-1}	8 125	5 537
Concentrate, 48 per cent solids, kg h^{-1}	1 875	1 845
Evaporation in spray dryer, kg h^{-1}	937	937
Powder, kg h^{-1}, moisture content	938/4%	908/2·5%

Turn-down ratio of evaporator:

in terms of product: $\dfrac{7382 \times 100}{10,000} = 74$ per cent

in terms of evaporation: $\dfrac{5537 \times 100}{8125} = 68$ per cent

Normally, an evaporator can be made with the above turn-down ratio, if it is specified at the design stage. Two thermocompressors will be supplied – one for the full duty and one for the reduced duty – as a thermocompressor has a narrow operation range.

Of course, it is possible to operate the evaporator at its full capacity on whole milk, but then the excess of concentrate has to be stored. That is seldom the right solution, because the provision of storage facilities and the energy for cooling and reheating the concentrate bring nothing but extra costs.

The total solids of skim-milk may vary, therefore a powder manufacturer sometimes specifies that a plant is to be designed for only 8 per cent solids in the milk 'to be on the safe side'. When, in actual fact, the milk contains 9 per cent solids, the powder manufacturer finds to his surprise that the dryer cannot cope with the rated intake of milk. This is hard for him to understand as the milk contains less water than assumed. However, at the rated milk intake, the dryer must have 12·5 per cent higher capacity if the milk contains 9 per cent solids instead of 8 per cent. The evaporator has, of course, spare capacity. Exactly the opposite is the case if the milk has a lower total solids content than the plant is designed for, so that if the solids in the milk vary, the evaporator must be designed for the minimum solids, the dryer for the maximum solids content in the intake milk.

During recent years, the size of the tubular, falling film evaporator has increased and it is now available with a capacity up to 60 tonne h^{-1} water evaporation.

MVR Evaporators

Mechanical vapour recompression has been known for many years, and was used in milk evaporators in Switzerland during the Second World War due to lack of fuel for raising steam. Electrically driven turbocompressors were installed for recompression of the steam, but due to the high cost of this machine, MVR evaporators were not further developed in the dairy industry for many years.

Nuclear power and the steep rise in fuel costs have changed the relative costs of steam and power, and during the past 10 years, MVR evaporators have found increasing use in the dairy industry. At the same time, advances have been made in recompression techniques, making an MVR evaporator as reliable as a TVR evaporator. Today, the choice between the two systems is an economic exercise. Previously, the position was that where the running costs of MVR were lower, the annual utilisation of the plant had to be high to justify the higher investment, but today there is not a great difference in the costs of the different evaporators. Therefore, the difference in direct running costs will normally be the decisive factor. It is not always a question of comparing the costs of steam and electric power. With natural gas being available in many countries, it can be economical to drive the compressor by gas. Either the compressor can be driven by a gas motor, or electric power to drive the compressor can be generated by a gas motor.

If a dairy factory has insufficient steam to supply a TVR evaporator, there may be a good case for an MVR plant, eliminating the need for a new boiler. However, the available power supply should also be looked into, as an MVR evaporator requires an electric motor of a size not normally used in dairies; as an example, an MVR evaporator with a capacity of 30 tonne h^{-1} water evaporation requires a 400 horse power motor for the compressor.

Whereas only part of the vapour is recompressed in a TVR evaporator, all the vapour is recompressed in an MVR evaporator. Normally, the evaporator consists of only one effect, and the boiling temperature can be chosen as the most suitable for the product. The calandria is divided into sections, and the product is fed by pumps through the sections in series. The number of tubes in the sections is reduced, ensuring the correct wetting rate as the product volume is reduced with increasing concentration. For capacities exceeding approximately 15 tonne h^{-1} water evaporation, two or more calandrias may be used in parallel, i.e. the boiling temperature is the same in the two calandrias.

Apart from the steam used for start-up, an MVR evaporator requires no steam and no cooling water. However, if high total solids is to be achieved, it is advantageous to use a steam-heated finisher for the last concentration. In this way it will be easier to control the evaporator. If skim-milk has to be concentrated to 50 per cent solids, it can, for instance, be concentrated to 42 per cent solids in the MVR effect, and further up to 50 per cent in the steam-heated finisher (Fig. 7). In this case, 96 per cent of the evaporation will take place in the MVR effect, and only 4 per cent in the finisher. Thus the amount of steam and cooling water required by the finisher is insignificant. Extensive preheating of the feed takes place in plate heat exchangers utilising the heat in the condensate.

Suitable mechanical vapour compressors made of stainless steel are available from various manufacturers. They are expensive items, therefore, it is costly to keep a spare compressor. It is often more favourable to design the evaporator in such a way that it can also be used as a TVR evaporator, i.e. a thermal vapour recompressor is installed as a 'spare' instead of a spare mechanical compressor. If the mechanical compressor breaks down, the evaporator can be operated on steam during the hours when steam is available. This seems to be the right solution if prices for fuel and electric power are likely to change later on, thus making a steam-heated plant more favourable than an electrically driven plant. In

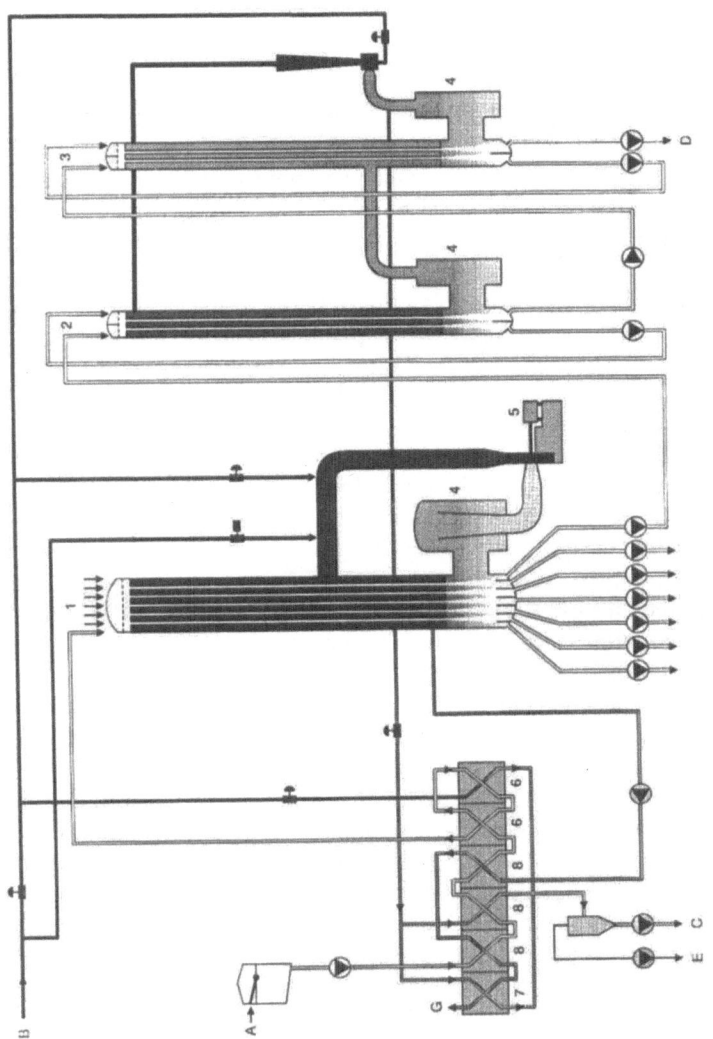

Fig. 7. MVR evaporator. (A) feed product (B) steam (C) secondary condensate outlet (D) product outlet (E) vacuum (G) primary condensate outlet (1) MVR effect (2) first finisher effect (3) second finisher effect (4) vapour separator (5) mechanical compressor/high pressure fan (6) pasteurising unit (7) condenser (8) preheater. (Courtesy APV Anhydro AS, Denmark.)

addition, having a dual system is a good insurance against changing costs of fuel and power, which could lead to an incorrect investment in the initially chosen system. Likewise, if a TVR evaporator is chosen initially, it would be wise to design the evaporator in such a way that it can be converted into an MVR later, and to reserve space in the layout for a mechanical compressor.

With a temperature rise of 10 °C of the vapour from the suction side of the compressor to the pressure side, the power consumption of the compressor will be 0·042 kW per kg water evaporated. The power consumption will be reduced with decreasing Δt, but, of course, the heating surface has to be increased at unchanged capacity. To optimise an evaporator is a complex task, and it is tedious to do without the aid of a computer.

Instead of a turbocompressor it is now possible to use a high pressure, heavy duty fan, specially designed for the purpose. The fan is a far simpler machine than a compressor, and significantly less costly. Several evaporators have been fan operated for a number of years with very good results. The fan-operated evaporator represents a new generation, and it will, no doubt, contribute considerably to make mechanical vapour recompression more widely used than before. The fan must be designed to operate under vacuum with a casing to be vacuum-tight at a pressure below 0·1 bar. The revolutions of the fan are adjusted continuously by means of an AC motor with frequency converter.

Some entrainment from the separators cannot be avoided and, therefore, the recompression unit must be cleaned at intervals. It is not easy to clean a compressor, whereas a fan can be cleaned by water supplied from an injection system permanently placed in the fan. This means that the fan can be included in the daily CIP cleaning of the evaporator. The correct operation of the fan can be checked from the control panel by electric temperature sensors measuring the temperature of the bearings as well as a device for vibration measurements. The fan should be provided with soundproof insulation or should be placed in a sound-absorbing room and should be connected to the ducts by compensators preventing vibrations from being transmitted to the rest of the plant and allowing the ducts to expand when heated. An impeller can be kept as a spare, but is normally not required.

A temperature rise of up to 5 °C from the suction side of the fan to the pressure side can be obtained, and a higher Δt can be achieved by using two fans in series. A fan-operated evaporator is often designed for a Δt of only 4 °C, which ensures long runs without scaling of the tubes. The corresponding power consumption of the fan is 0·012 kW kg^{-1} water evaporated. Figure 8 shows a heavy-duty fan used in milk evaporators.

Fig. 8. Heavy-duty radial fan for MVR evaporating plants (courtesy Piller, Germany).

DRYING

Milk and milk products are dried by either the roller process or the spray process; the spray process dominates today.

Roller Drying

Roller drying was the first method used to dry milk. It originates from the USA where it was invented in 1902 by John A. Just. In the USA the process is also referred to as drum drying. James Hatmaker of London, England, purchased the Just patent and improved the process. The resulting modification is known as the Just-Hatmaker process, and was widely used for many years all over the world. In several countries a roller dryer was referred to as a 'Hatmaker'.

The principle of the roller drying process is that the milk is applied in a thin film upon the smooth surface of a continuously rotating steam-heated metal drum. The film of dried milk is continuously scraped off by a stationary knife located opposite the point of application of the milk.

The dryer may consist of a single drum or a pair of drums. The milk may be preconcentrated before it is applied to the drum, and the degree of concentration varies with the design of the dryer. The rule is that a single-drum dryer can handle milk of higher concentration than a double-drum dryer. The dried milk film is milled to break up the film. The particles of a roller dried powder are solid, flat, irregular flakes containing no air. The absence of occluded air retards oxidation, resulting in good keeping quality.

The first Just-Hatmaker roller dryer was installed in 1903 in the creamery at Sherborne, Dorset, England, by the company Cow & Gate. This machine was so successful that the company installed two more roller dryers a year later in Somerset and Dorset to produce 'dried full cream milk, dried half-cream milk, and dried separated milk from pure English milk' as the advertisement read. The new product was marketed as 'germ-free and water-free milk'. The claim for a 'germ-free' product was no exaggeration, because it is characteristic of the roller process that the milk is exposed to a high temperature heat treatment which safeguards the bacteriological quality of the dried product. For this reason, roller drying became a standard process for manufacture of infant foods for many years.

However, the excessive heat treatment results in a high degree of protein denaturation. The solubility of the dried product is poor, and the reconstituted milk has a cooked flavour. These features render it unacceptable as a substitute for liquid milk. The chocolate industry prefers roller dried milk, claimed to give a desirable flavour to the chocolate. When manufacturers of spray powder are requested by the chocolate industry to produce a powder with properties closely resembling roller powder, it can be done by giving the milk an intense heat treatment before drying.

The maintenance costs of roller dryers are high, as resurfacing of the drum must be done at frequent intervals to keep a smooth surface, and the knife to remove the product from the drum requires frequent sharpening. The capacity per unit is low, and with the centralisation of dairy factories it became impossible to handle large milk quantities by the roller process. Another factor which today is a decided disadvantage is the pollution from roller dryers. Above the drum is installed a hood which collects and carries the vapour that arises from the milk drying. The hood exhausts to the outside through a stack, and as the entrainment of fine milk flakes into the vapour can be considerable, the exhaust from roller dryers presents a pollution problem unless the vapour is led to a wet scrubber.

Spray drying was without practical importance in the dairy industry until the First World War. Between the First and the Second World War roller drying and spray drying developed in parallel, but after the Second World War, most dryers installed in the dairy industry have been spray dryers. The roller dryer does not meet today's requirements with regard to powder quality, high capacity, high preconcentration, and low operating and maintenance costs.

Manufacturers of milk powder within the European Community (EC) have the advantage of a guaranteed outlet at minimum price from the intervention system, but this arrangement only applies to spray powder. The fact that roller powder is not covered by the intervention system has been a further incitement for powder manufacturers within the EC to shift from roller drying to spray drying.

Spray Drying

Spray drying can be described as the instantaneous removal of moisture from a liquid. To achieve this, the liquid is converted into a fog-like mist ('atomised'), whereby it is given a large surface. The atomised liquid is exposed to a flow of hot air in a drying chamber. The air has the function of supplying heat for the evaporation and, in addition, it acts as carrier for the vapour and the powder. When the atomised product is in contact with the hot air, the moisture evaporates quickly and the solids are recovered as a powder consisting of fine hollow spherical particles with some occluded air.

Spray drying is a gentle drying method. The material to be dried is suspended in the air, and the drying time is very short. Air inlet temperatures up to 215 °C are used for the drying of milk, but due to evaporation, the temperature drops immediately in the drying chamber, and the milk solids will not reach a temperature approaching the air inlet temperature. The temperature in the drying chamber is equal to the air outlet temperature, which is approximately 95 °C when single-stage drying is used, and lower with multiple-stage drying. The product temperature will be 20–30 °C below the air outlet temperature provided a co-current dryer is used, i.e. a dryer where product and air move in the same direction.

The evaporating capacity of the dryer is determined by the difference between the air inlet temperature and the air outlet temperature. Also the heat loss from the dryer affects the capacity, and efficient insulation of the hot air ducts to the dryer and the drying chamber is essential. Further, the humidity of the intake (ambient) air affects the capacity. The humidity of the ambient air varies within a wide range and can have a considerable adverse effect on the capacity in a warm and humid climate. The low air temperature in the winter gives a low humidity in the air, hence it increases the capacity of the dryer. The powder is cooled immediately as it comes out of the dryer, either in a pneumatic powder cooling system or in a fluid bed. Figure 9 shows two different types of single-stage dryer. In (A) the drying chamber has a flat bottom from which the powder is removed pneumatically by means of a rotating powder collector which sucks the powder from the chamber. The powder is separated from the air in the cyclones. Under each cyclone is placed a rotary valve (air-lock) through which the powder is discharged into the pneumatic powder cooling system. In (B) the drying chamber has a conical bottom from which the greater portion of the powder is removed by gravity and discharged into the fluid bed powder cooler. Only the fine powder particles are carried with the air to the cyclones. In the following paragraph a description of various components of a spray dryer is given.

Components of a Spray Dryer

Air filtration

Efficient air filters for the drying air and the cooling air are essential to prevent the air from contaminating the powder. Only dry filters are used today – either cleanable or throw-away filters. The use of filters where the pressure drop over the filter is automatically kept constant ensures a constant amount of air in the dryer, and this prevents the capacity of the dryer from being reduced as a result of reduced air intake due to dirty filters. The filters are designed in such a way that clean filtering material is supplied from a roll activated by the pressure drop over the filter. The used filter material is simultaneously collected on another roll. The saving in labour for supervision and cleaning of air filters contributes to the costs of new filter rolls.

The filter for the drying air is not required to destroy any airborne bacteria as it is unlikely that bacteria will survive when the air is heated. However, as the cooling air is not subjected to this 'pasteurisation', it is advisable to use bacteria-absorbing filters for the cooling air in places where airborne bacteria are likely to be found. If an air filter inadvertently gets wet, it must be thoroughly cleaned or replaced immediately as there will be milk solids collected in the filter which provide an excellent growth medium for bacteria when moisture is present.

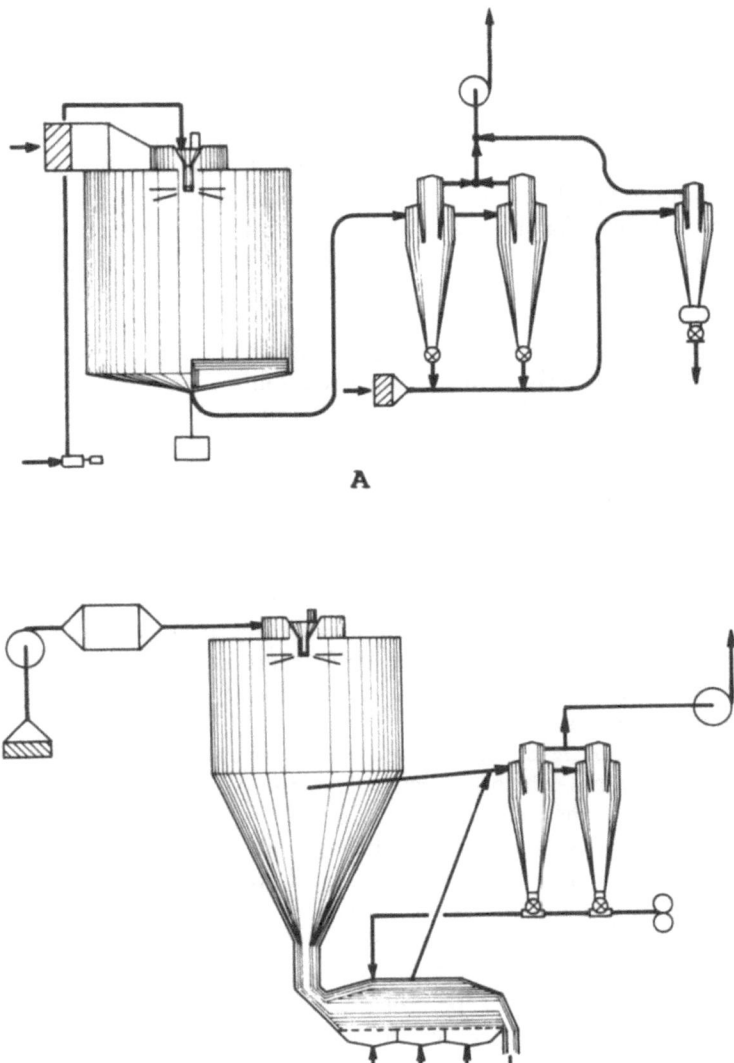

Fig. 9. Single-stage spray dryer. (A) drying chamber with flat bottom (B) drying chamber with conical bottom and fluid bed. (Courtesy APV Anhydro AS, Denmark.)

Today it is quite common (as explained later) to use fabric filters at the end of the dryer to collect the fine material escaping from the cyclones. If the fabric filter is designed in such a way that the powder collected in the filter is suitable for mixing into the powder from the dryer, special attention must be paid to the filter for the intake air. If the filter for the intake air is less efficient than the fabric filter for the exhaust air, there will obviously be an accumulation of impurities in the fabric filter. Therefore, the filter for the intake air must be more efficient or at least as efficient as the fabric filter for the discharged air.

Feed pump

The feed pump for a rotating atomiser must be a sanitary, volumetric pump with a variable speed drive. A nozzle atomiser requires a feed pump to give a pressure up to 300–400 bar, and it must be a three-piston pump to avoid pulsations.

Atomiser

Atomisation of the concentrated milk is the principle of the spray drying process. By atomisation the concentrate is converted into droplets of size 10–200 μm, with the greatest portion in the range 40–80 μm. Two systems are used, stationary pressure nozzles or rotating atomisers. The use of pressure nozzles originates from the USA where they are still used to a great extent. The rotating atomiser was developed in Europe where it is still the preferred system (Fig. 10).

The pressure nozzle is a simple and inexpensive item, but it requires a high pressure pump which is costly. The pump must have a variable speed drive unless the feed rate is controlled by a bypass valve after the pump. This latter system is not desirable as it involves recirculation of the concentrate at a temperature where bacterial growth can take place and further, it can cause the formation of foam.

The principle of the rotating atomiser involves feeding the concentrate onto a disc rotating at high speed. The product is fed by a liquid distributor into a chamber at the centre and moves to the periphery of the disc through a number of vanes. As the product leaves the disc it is broken up into droplets. The peripheral speed of the disc is usually 130–150 m s^{-1}.

The rotating atomiser is a high speed, precision piece of equipment and is therefore expensive, but the feed pump is considerably less costly than the high pressure pump required for nozzle atomisation. The high rotating speed of the disc can be obtained by using a direct coupled, high

Fig. 10. Rotating atomiser (left) and nozzle atomiser (right) (courtesy APV Anhydro AS, Denmark).

speed electromotor with a frequency converter, or by using a standard motor with gear or belt transmission. The belt transmission is simple and reliable provided the right kind of belt is used. It is easy to change the rotating speed of the disc by fitting another size pulley to the spindle. The centrifugal force available in a rotating atomiser to break up the liquid is much higher than the force from the pressure in a nozzle atomiser, therefore the rotating atomiser can handle more viscous products (i.e. concentrate with high total solids) than a nozzle atomiser.

A nozzle is exposed to considerable wear and therefore the orifice must be of a hard metal. The control of a dryer with a nozzle atomiser is more difficult than a dryer with a rotating atomiser, because the feed rate cannot be varied without changing the pressure unless another orifice is fitted. This change cannot be made during operation. When the pressure is changed, the atomisation is also changed which is undesirable. When a rotating atomiser is used, a variation in the feed rate will not affect the atomisation as it is governed only by the rotating speed of the disc.

To take up variations in the total solids of the feed to the atomiser, it is necessary to vary the feed rate to the atomiser so as to maintain a constant temperature of the outlet air from the dryer and thereby a constant moisture content of the powder. This is done automatically by means of a temperature sensor in the duct carrying the outlet air. The temperature sensor governs the speed of the feed pump which will supply more or less concentrate to the atomiser if the outlet temperature is above or below the set point. Great variations in the total solids of the feed cannot be tolerated when nozzle atomisation is used, as variations in the feed rate give a significant change in the atomisation.

The question of whether one atomisation system is better than the other cannot be answered in a simple way. For some products one system is better than the other and therefore, if the aim is to have a versatile dryer, the type of spray dryer should be installed where both atomisation systems can be used. It is not necessary to have two feed pumps because the high pressure pump can be used without pressure with the rotating atomiser.

Nozzle powder has less occluded air than powder from a rotating atomiser and therefore a higher bulk density. A bulk density of $0.7\,\mathrm{g\,cm^{-3}}$ is normal for skim-milk powder produced by nozzle atomisation, whereas approximately $0.6\,\mathrm{g\,cm^{-3}}$ is normally achieved when a rotating atomiser is used.

Whole milk powder and fat-filled powder produced by nozzle atomisation are claimed to contain less free fat than powder produced by means of a rotating atomiser. It is obvious that the high pressure pump and passage through the nozzle have a homogenising effect which will reduce the amount of free fat. The same result can be obtained by homogenising the concentrate before feeding it to a rotating atomiser. The rotating atomiser can handle concentrate of higher total solids. A nozzle atomiser is not suitable for whey concentrate containing lactose crystals, as the crystals wear the nozzles.

If powder from different products or a powder with certain characteristics is to be produced, the dryer must have both atomisation systems.

Air heating system

Steam has traditionally been used for heating the drying air, but in a modern milk dryer, the air inlet temperature may be as high as 215–

220 °C, and steam is seldom available at the pressure required to produce such high temperatures. When steam is used, the air temperature is usually boosted by another heating system.

Saturated steam of 10–12 bar gives an air temperature of 170–180 °C as the heating surface is normally rated to give an air temperature of 8–10 °C below the temperature of the steam. Superheated steam is not suitable as it requires an air to air heat exchanger which is much more costly than a saturated steam to air heat exchanger. If only superheated steam is available, it can be saturated by injection of water before it enters the air heater.

The corrosive nature of steam and condensate makes it necessary to use stainless steel tubes welded into stainless steel inlet and outlet headers. Aluminium fins are tension wound onto the tubes to give good heat transfer characteristics. The condensate stage is connected to the steam stage by means of manifolds incorporating a stainless steel orifice plate. This system has the advantage that no steam traps are required. Care must be taken that the steam–air heater is manufactured according to the code accepted by the authorities. The code may be national or international, such as TÜV, Veritas, or Lloyd's.

Air cooling has to be used for the powder cooling system when the ambient air temperature is above 10–15 °C. The air cooler must be connected to a drain to take away the moisture condensed from the air, and it must be provided with an efficient droplet eliminator. It must be designed to cool the air below the required temperature, as the air has to be reheated a few degrees to bring it above saturation temperature. Condensate or hot water can be used for the reheat section.

Liquid phase heating (hot oil system) is widely used to heat the drying air, either to do all the heating or to boost the temperature from a steam–air heater. Figure 11 shows a diagram of a liquid phase heating system. An oil- or gas-fired heater is used to heat a heat transfer fluid to 330–350 °C. The hot fluid is pumped to an air heater battery similar to that employed when steam is used. A high thermal efficiency is obtained by preheating the return fluid in the stack of the unit used for heating the heat transfer fluid.

The economiser built into the stack is designed to cool the flue gases from an oil burner from about 500 °C to 180 °C by means of the return fluid (at approximately 160 °C) from the air heater. The stack temperature must, of course, be kept above the condensation temperature for the flue gases. Where a gas burner is used, it is possible to cool the flue gases further by the return oil. The thermal efficiency of a liquid phase heating

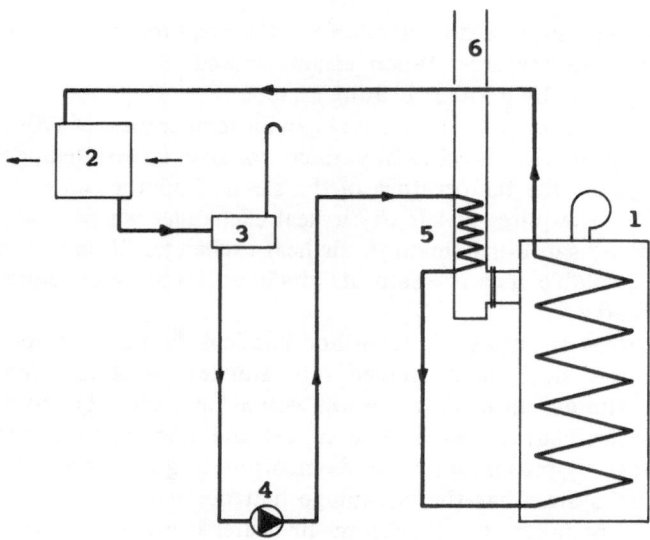

Fig. 11. Liquid phase heating system. (1) burner (2) air heater (3) expansion tank (4) circulation pump (5) economiser (6) stack for flue gases.

system with economiser is claimed to be close to 90 per cent. It is a simple system operating in closed circuit at atmospheric pressure, and it is well suited to heat the drying air to the desired high temperature.

Another system widely used for heating the air is indirect oil or gas heating. In this system, the oil or gas is burnt in a cylindrical chamber and the air is drawn over the external surface thus gaining heat. Before the combustion gases are exhausted to the atmosphere they are cooled by the intake air, but the thermal efficiency will not exceed 85 per cent. The material used for a heat exchanger of this kind should be heat resistant stainless steel. The heater will be damaged if it is not constantly interlocked with the fans; this makes it impossible to run the burner without the fans in operation. Electric air heaters can be used, but are normally not competitive on running costs. They are, however, sometimes used as booster systems.

Direct gas-fired air heaters have been in use for several years in milk dryers in North America. During recent years, availability of natural gas and high fuel prices have created a considerable interest in this system. Figure 12 shows a direct gas-fired air heater. No heat exchanger is required. The mixture of the combustion gases from the burner and ambient filtered air is led to the drying chamber as the drying medium.

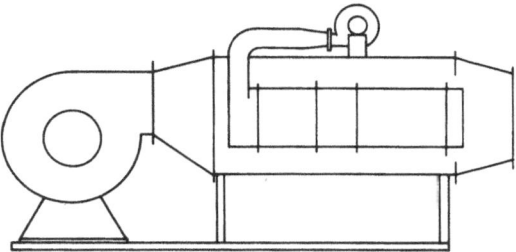

Fig. 12. Direct gas-fired air heater.

The absence of a heat exchanger brings the thermal efficiency to practically 100 per cent, i.e. a saving in fuel consumption of 10–20 per cent is obtained compared with indirect heating. Further, the capital investment in a direct system is substantially less than for an indirect system. Maintenance costs are lower, and the quick response and flexibility of a direct system is also an advantage.

In spite of the great attraction offered by direct gas firing, it is not yet widely used outside North America. The reason is recognition during recent years of the danger of contaminating the milk powder product by direct gas firing. The contamination is produced by oxides of nitrogen contained in the combustion gases; it is therefore imperative to minimise the quantity of nitrogen oxides in the combustion gases. When nitrogen oxides get in contact with food products, they can react with amines associated with the proteins to produce nitrosamines (Challis *et al.*, 1982), which are classified as cancer promoting when assimilated. Nitrosamines can also be formed when amines react with nitrite present in food products. Nitrite is sometimes added to food products, or it is formed from nitrate by bacteriological conversion.

Nitrites and nitrates are therefore undesirable products in food commodities as they can lead to formation of nitrosamines which constitute a health hazard. Research work has shown that an increased content of nitrite and nitrate is found in milk powder prepared in direct gas-fired dryers, as much as double that found when indirect heating was used (Harding and Gregson, 1978). It was found in Australia (Rothery, 1968) that when direct gas firing is used, the nitrite content of the spray-dried sodium caseinate is higher than in skim-milk powder. The interesting discovery was made by Rothery that the amount of nitrite could be significantly reduced by an injection of steam into the flame, whereby the flame temperature was reduced, hence smaller amounts of nitrogen oxides were produced. In this way the nitrite content of spray-dried caseinate was reduced from about 28 ppm to about 3 ppm, and the nitrite content of skim-milk powder was reduced from 1 ppm to less than 0·1 ppm. The work done by Rothery clearly demonstrates the requirement for minimising the amount of nitrogen oxides created by the combustion. This recognition has recently led to the development of a new generation of gas burners to be used in the food industry.

Oxides of nitrogen are formed as part of all combustion processes in which air is used as the oxidant. When nitrogen and oxygen are heated, nitrogen oxides are formed by thermal fixation of the nitrogen. Another source is the conversion of chemically bound nitrogen in the fuel. The nitrogen oxides found in combustion gases are mainly NO and NO_2, but also small quantities of higher oxides are formed. NO_x is conveniently used as a common designation for all the nitrogen oxides in the combustion gases, both those created from the air by the thermal process and those originating from nitrogen in the fuel. Only during the past ten years has it been recognised that special burners, producing low NO_x in the combustion gases, are required in the food industry in order to prevent the formation of nitrosamines in food products exposed to the combustion gases.

Fuel-bound nitrogen is present in oil and coal where it contributes significantly to the amount of NO_x in the combustion gases. Fuel-bound nitrogen can account for over 50 per cent of the amount of NO_x in the combustion gases from oil, and approximately 80 per cent when coal firing is used. When gas is burned the whole amount of NO_x in the combustion gases is generated from the air as a result of the thermal process. This is because there is practically no fuel-bound nitrogen in gas. Some consider it should be perfectly safe to use methane, propane or butane from refineries in direct-fired milk dryers, because these gases contain no impurities. This is a fundamental misunderstanding. Natural gas also contains very few impurities, for it undergoes a cleaning process so that possible impurities are stripped off before the gas is supplied to the consumers, and the combustion is practically complete. It is not impurities in the gas that give cause for concern – it is the NO_x formed by the combustion. This process takes place regardless of the kind of gas used.

The rate of thermal formation of NO_x from oxygen and nitrogen is strongly temperature dependent. The amount of NO_x generated by combustion is governed by the flame temperature and the duration of the top temperature phase. At flame temperatures in the range 1500–1600 °C, the formation of NO_x starts to be significant, therefore the flame temperature should be kept below this level but high enough to ensure complete combustion. The duration of the top temperature phase should be short, but combustion must not be inhibited too early by dilution with the air to be heated as this will produce undesirable intermediate combustion products in the air. A low flame temperature can be obtained, as mentioned earlier, by injection of steam into the flame. However, this is not a practical method in connection with spray drying, as it will increase the amount of moisture in the drying air and the outlet temperature of the air will need to be higher to maintain the specified moisture content of the powder. In low NO_x burners, a flame temperature of 1200–1300 °C is obtained by using over-stoichiometric combustion with an excess air level of up to 90 per cent. Uniformity of the air–gas mixture is essential in ensuring not only controlled combustion but also the absence of packets of gas-rich mixtures, the combustion of which would result in localised high temperatures and high NO_x levels.

Figure 13 shows a low NO_x burner. The venturi gas–air mixer is designed to give a uniform mixture of the gas and air before it enters the flame stabiliser. Increased stability of the flame is achieved by using a refractory-lined fire tube. The NO_x level ex-burner (i.e. before dilution with air) is guaranteed to be below 1 ppm by volume. Conventional gas burners will produce an NO_x level ex-burner of between 50 ppm and several hundred ppm (typically about 150 ppm). The NO_x level of the drying medium getting in contact with the atomised milk concentrate will be substantially lower than the NO_x level ex-burner due to dilution by the air to be heated. At an NO_x level of 1 ppm ex-burner, the NO_x level of the drying medium will be 0·12 ppm if a temperature of 200 °C of the mixture of combustion gases and air has to be achieved, and lower in the case of a lower drying temperature. The figure 0·12 ppm should be understood as the increase in NO_x level (i.e. NO_x added from the combustion), as ambient air can contain variable concentrations of NO_x produced by industry, power stations and the exhaust gas from motor cars.

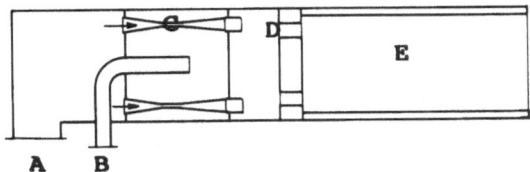

Fig. 13. CAX low NO_x gas burner. (A) combustion air entry (B) gas entry (C) gas–air mixer (D) premixer burner stabilising nozzles (E) combustion chamber. (Courtesy Urquhart-Engineering Co., UK.)

Food products are, of course, in contact with the combustion gases from bakeries, from the gas ovens used in households, and from grills open to the combustion gases from coal. However, as the surface of such food products is small compared with the volume, it has been found that the increase in nitrite, nitrate and nitrosamines is so small that it is not possible to measure it with the methods available today. It is a different matter when milk is spray dried, because the product is given a very large contact surface in a spray dryer.

Direct gas firing was used in milk dryers in North America long before the problems associated with this system were recognised, but it is of note that mostly skim-milk and whey are spray dried in North America, whereas in Europe the production of whole milk powder and fat-filled powder is significant. It may be that the problems are more pronounced when powder containing fat is produced, but further research work is required to verify that.

The authorities in Europe and in other countries realise that the use of direct gas firing for spray drying of milk presents a complex problem, therefore they do not impose any restrictions. It is not allowed and it is not forbidden. In Holland the answer from health authorities to applications to use direct gas firing in spray dryers for edible milk powder has been that the heating system used is not subject to any legislation, but it is the manufacturers' responsiblity to ensure that certain limits for the content of nitrite, nitrate, and nitrosamines in the powder are not exceeded. It is specified that the content of nitrosamines must be below 1 ppb, and WHO has issued standards for the maximum content of nitrite allowed in food products. Several milk dryers in Holland are equipped with direct gas firing. The authorities have the right to demand analysis for nitrite, nitrate, and nitrosamines in samples taken at any time. In practice, samples are taken every third month to ensure the correct operation of the burners. These analyses are costly and tedious to make.

Another product formed by the combustion of gas is water. This does not constitute a health hazard, but it affects the capacity of the dryer. From the combustion of one standard cubic metre of gas approximately 1·7 kg water is formed, which is present as vapour in the drying medium. It has been shown (Muir *et al.*, 1981) that the water of combustion in a direct, gas-fired spray dryer is equivalent to approximately 20 per cent of the water to be evaporated in the dryer. The increased moisture content of the drying medium has the effect that a higher outlet temperature of the dryer must be used than when the dryer is indirectly heated. This means that a higher air quantity is required to obtain the same evaporating capacity of the dryer, i.e. the economic advantage from direct gas firing is partly offset by the loss in the efficiency of the dryer caused by the water liberated when gas is burned. In a warm and humid climate, where the ambient air has a high moisture content, the extra moisture in the drying air generated by the combustion can lead to such a high outlet temperature – when single-stage drying is used – that a serious loss in powder solubility can occur. In addition, it may result in more deposits in the chamber, as the deposits will increase with a higher water content in the outlet air. This applies especially to products with a high content of carbohydrate and hygroscopic products such as whey.

To underline the significance of the above, the following incident may be relevant. A powder manufacturer decided to change from indirect heating to direct gas firing for economic reasons. He approached the health authorities who advised him to install a low NO_x burner. He did not get in contact with the supplier of the spray dryer, but went to a company that specialised in the supply of gas burners. They made it clear that they knew nothing about spray drying, but they undertook to deliver a burner with guaranteed low NO_x formation. The burner was installed, but to the surprise of the powder manufacturer it was only possible to obtain the specified moisture content of the powder at an air outlet temperature very much higher than used before. This drastically reduced the capacity of the dryer, but worse than that, the powder was useless as the solubility was very poor. The powder manufacturer had followed the request from the authorities to install a low NO_x burner. The burner supplier had fulfilled his guarantee, and nobody was to blame for the poor result of the conversion of the air heating system. A consultant found that a burner using steam injection to obtain the low NO_x content had been installed. This extra moisture together with the water of the combustion of the gas raised the moisture content in the hot air supplied to the dryer by an amount equal to 40 per cent of the water to be evaporated in the dryer. Consequently, a very high air outlet temperature was required to obtain the specified moisture content of the powder. The direct air heating system had to be abandoned, and the indirect system was reinstalled.

Drying chamber

Two different shapes are shown in Fig. 9, the chamber with flat bottom, from which the powder is sucked from the chamber by means of a rotating powder collector, and the chamber with a conical bottom, from which the greater portion of the powder is removed by gravity.

The advantages of the flat bottom chamber are low building height, forced removal of the powder from the chamber (i.e. controlled residence time), and easy access for cleaning. It is also an advantage that hammers or vibrators on the chamber are not required.

There are several reasons for the use of a conical chamber.

(1) The incorporation of two- or three-stage drying.
(2) A requirement for the separation of the powder into a coarse fraction from the chamber, and a fine fraction from the cyclone. By recycling the fines to the atomisation zone, a dust-free, partly agglomerated powder is obtained.
(3) The manufacture of a powder with a fat content exceeding 35–40 per cent. Powder with a high fat content has a tendency to stick to the interior surface of the ducts and cyclones, therefore the greater portion of the powder must be discharged directly from the chamber, so that only the small quantity contained in the air will pass through the ducts and cyclones.

Only skim-milk powder is comparatively free-flowing; both whole milk powder and fat-filled powder have poor flow characteristics. To avoid a build-up of powder on the conical bottom it is necessary to use external hammers or vibrators on the cone. This presents two problems. One is the problem of noise, which is very objectionable and very difficult to overcome. The other is the risk of the development of cracks in the stainless steel sheet due to metal fatigue caused by the external hammers. The impact of the hammers can normally be adjusted, but it is important to equip the system with a blocking device. This prevents the development of cracks by preventing the hammers from causing an impact higher than is regarded as safe. The operator tends to use the maximum possible impact to facilitate operation and cleaning of the dryer.

It is difficult to detect fine cracks in a stainless steel sheet, but it can be done by applying a special liquid to the sheet so that the cracks become visible. There have been incidents of bacteriological contamination originating from an accumulation of moist powder in cracks in the stainless steel sheets. To avoid these problems it is essential to use a steep cone (small included angle), so that the number of external hammers and their impact is reduced to a minimum. The steep cone does increase the height of the chamber, which adds to the cost of the building, but is necessary if powder is removed at high moisture content for subsequent drying, as the moisture will reduce the flow characteristics of the powder.

The two parts of a spray dryer requiring most design skill and experience are the atomiser and the air distributor, placed on top of the drying chamber. The hot air is admitted to the chamber through the air distributor. It contains a number of vanes to give the air a rotary movement following the direction of rotation of the atomiser. This imparts a swirl pattern to the particle cloud and moves it downwards into a vortex. The intimate mixing of the drying air and the atomised product is essential to obtain the highest utilisation of the heat, i.e. to produce a low air outlet temperature. Further, the air distributor must direct the product cloud downwards to prevent moist powder impinging on the walls of the chamber, particularly at the level of the atomiser disc.

In a chamber designed for a rotating atomiser, a nozzle atomiser can be used instead. Four or more nozzles are employed, not in a vertical position but placed at an angle to the vertical. In this way, the pattern of the product cloud is similar to that produced by a rotating atomiser. The pattern of the product cloud has to be similar in order to use the same chamber configuration and the same air distributor for both atomisation systems.

The chamber ceiling will be heated to a temperature equal to the air inlet temperature at the point where the hot air enters the chamber. To avoid charring of milk solids at this point, the ceiling is provided with an annular cooling ring around the air distributor. Slightly heated air is blown through the cooling ring to remove enough of the heat to bring the surface temperature down. Unheated ambient air should not be used as it may cause condensation. Water cooling has been used in recent years; the temperature regulated by a thermostat.

The chamber volume is sized in relation to the volume of drying air, and in such a way that the number of air changes gives an air residence time in the drying chamber of 25–30 s. The drying chamber is the most space demanding part of a spray dryer, and a building to house a large drying chamber represents a considerable investment. Drying chambers can be designed to be erected in the open without a building, and Fig. 14 shows a spray dryer for milk where the drying chamber and the vertical indirect oil-fired air heater are placed externally. The feed pump, control panel, cyclones, fans, and bagging-off equipment are placed in an adjacent building.

The insulation of a chamber installed in the open is increased from 100 mm to 200 mm. The temperature inside the chamber can, for the purpose of assessing the heat loss, be assumed to be 100 °C. The outside temperature may be approximately 15 °C if the chamber is housed in a

Fig. 14. Spray dryer with drying chamber and indirect oil-fired air heater in the open (courtesy APV Anhydro AS, Denmark).

building, but could be -20 to $-30\,°C$ if the chamber is installed in the open. The temperature difference across the wall is thus increased from $85\,°C$ to say $130\,°C$, i.e. by approximately 50 per cent. By increasing the insulation by 100 per cent, ample compensation has been made for the larger Δt even when the heat loss due to convection is considered, therefore the external installation of the chamber does not cause any loss in the efficiency of the dryer. The insulation is provided by a cladding of plastic-coated, mild steel sheets. The cladding is extended downwards to cover the support of the chamber and upwards to cover a light steel structure (penthouse) on top of the chamber to accommodate the atomiser. In this way the chamber looks like the milk silos, which were moved to the outside long ago. There is no operation, cleaning or other function to be done externally. The door giving access to the chamber is placed in a wall recess in the adjacent building, and the access to the penthouse for the atomiser is also from this building. The space under the chamber is connected to the building by a short corridor, where a fluid bed can be installed.

Milk dryers with external chambers have been built in several countries. The only problem encountered in some cases has been to obtain planning permission when a building is deemed to suit the surroundings better.

Powder recovery system

Cyclones have been used for many years to separate the powder from the air. The principle of cyclone separation is based on the centrifugal force exerted on a particle similar to the principle of a centrifuge. The separation efficiency of a cyclone varies inversely with the diameter of the cyclone, and directly with the square of the tangential velocity of the air and with the mass of the particles to be separated. It can be expressed by the formula

$$\frac{mV^2}{d}$$

where m = the particle mass
V = the air velocity
d = the diameter of the cyclone.

Of course the shape of the cyclone, i.e. the ratio between the cylindrical part and the conical part, also affects the efficiency.

The loss of solids from a system of cyclones for a milk dryer is often expressed as a percentage of the total powder production, although only a smaller portion of the powder (20–30 per cent) may pass through the cyclones. The greater portion, discharged directly from a conical drying chamber into a fluid bed, does not pass the cyclones. Even if it did pass the cyclones, it would hardly contribute to the loss as it consists of large particles. The fines suspended in the air are conveyed to the cyclones, and the particles below 5–10 μm constitute the loss to the atmosphere.

For the powder manufacturer it is logical to express the efficiency of the powder recovery system as a percentage of the total powder production. To assess the loss with regard to the environment, it is necessary to express it as the amount of solids in relation to the volume of discharged air. Normally the loss is expressed in mg solids per standard cubic metre of air. In North America the loss is expressed in grains per standard cubic foot, and the relationship is 1 grain (st. ft)$^{-3}$ = 2275 mg (st. m)$^{-3}$.

As dryers have got bigger, the absolute amount of milk solids discharged to the atmosphere from the cyclones has reached levels that can no longer be tolerated for environmental reasons, particularly where the dryer is close to a residential area. In most countries, legislation has been introduced during recent years to limit the emission levels by specifying the maximum permitted amount of solids, where the dust does not constitute a health hazard, contained in the discharged medium. In Europe the permitted emission level from milk dryers varies between 50 and 100 mg (st. m)$^{-3}$ air discharged. In the USA the permitted emission level from milk dryers varies from 0·01 grain (st. ft)$^{-3}$ in California to 0·04 grain (st. ft)$^{-3}$ on the east Coast (23–91 mg (st. m)$^{-3}$). In practice, the cyclone loss varies within wide ranges depending on several technical and technological factors; Table II shows some typical cyclone losses from three dairy products.

TABLE II
Powder loss from cyclones

Product	Loss	
	Percentage of total production	*(mg/st.m)$^{-3}$*
Skim-milk	0·5	200
Whole milk	0·35	150
Whey	1·0	400

It will be seen that the loss from cyclone separators in all cases exceeds the permitted emission level. Therefore, it is not possible today to install a milk dryer where the powder recovery system consists of cyclones alone. Further cleaning of the air is necessary to bring the loss to the atmosphere within the permitted limit. The powder discharged to the atmosphere represents a significant financial loss. When the air is cleaned further, it should therefore be done in such a way that the solids recovered can be added to the powder from the dryer to increase the yield. The following three methods can be considered.

(1) The use of a wet collector or wet scrubber after the cyclones to trap the greater portion of the solids by scrubbing the air with water. The water is recirculated and periodically renewed.

The disadvantages are that the milk solids are not recovered, disposal of the scrubbing water presents a problem, and the scrubber itself is hygienically unacceptable in a dairy factory because the conditions in the scrubber are ideal for rapid growth of micro-organisms. To overcome the first two problems – recovery of the solids and disposal of the scrubbing medium – the so-called milk scrubber was introduced. Unconcentrated milk is used as the scrubbing medium instead of water to absorb the solids from the air. The milk goes from the scrubber to the evaporator, and the solids recovered are contained in the concentrate to the dryer and thus increase the yield. However, the third problem – bacteriological contamination – is not solved. On the contrary this problem becomes highly critical, because all the product passes through the scrubber, the very opposite of using water, where the scrubber is kept separate from the product.

The idea of using the milk as the scrubbing medium originated in the USA, where it was used 50 years ago but abandoned due to the bacteriological hazard. There is a wet zone in a scrubber and at the air exit there is a dry zone. The transition between wet and dry represents a moist zone where conditions for the growth of bacteria are ideal. By assessing the bacterial count in the milk before and after the scrubber, it was found that the number of bacteria increased immensely after a few hours operation. High temperature pasteurisation of the milk was necessary before it entered the evaporator. In plants where a large number of bacteria are present, there is a considerable risk of the generation of toxins. The toxins are not destroyed by the heat treatment commonly used, and it is difficult to analyse the powder for toxins, so that by employing a milk scrubber, the powder manufacturer runs the risk of producing milk powder contaminated with toxins.

It is incomprehensible that milk scrubbers have recently been put on the market again. Milk producers are encouraged to produce milk with low bacterial counts, and then an unhygienic piece of equipment, that generates bacteria at a high rate, is installed in the factory. It is against all logic, and against the continuous trend of raising product quality.

Milk scrubbers were first installed around 1976 on dryers in several countries, and they were marketed by reputable companies alleging that precautions were taken in the design to prevent bacterial build-up. Today they have practically all been taken out, or are only used where powder for animal feed is produced. A duplex or triplex installation is required for edible products as the scrubber must be cleaned after 3–4 h operation. The complicated changeover procedure must be designed to exclude the possibility of the cleaning solution getting into the product. The rinse water from cleaning five times during a 20 h run puts a considerable load on the effluent plant. The introduction of milk scrubbers into a modern dairy factory has been a step backwards and should not be repeated.

(2) Another attempt to minimise the amount of dust discharged to the atmosphere has been made by employing highly efficient cyclones used in the chemical industry. Their operating principle is that secondary air at high pressure is introduced into the cyclone to increase velocity, thereby increasing efficiency. However, these cyclones have not been successful in the dairy industry as they are unsuitable for powder containing fat. When the powder is exposed to the high speed of rotation in the cyclone, a high amount of free fat is generated and the powder sticks to the interior surface of the cyclone. As the cyclones are not sanitary, neither are they suitable for non-fat powder.

(3) The third method of collecting the dust in the air from the cyclones is to insert a textile filter (fabric filter or bag collector) before the air is discharged to the atmosphere. The filter is not used as the main collector, i.e. the cyclones are retained, and the filter is employed as a secondary collector.

A textile filter is not a piece of equipment that really belongs in a dairy factory. However, the fact that a filter is the only known solution to the problem has caused considerable development in the design of filters suitable for milk dryers. If a conventional filter is installed, the recovered product is kept separate and sold as animal feed. This makes it difficult to justify the filter from an economic point of view. By installing a filter especially designed for milk dryers, a first grade product is recovered, which can be continuously blended back into the main stream of powder from the dryer. In this way the filter serves two purposes; it protects the environment and increases the yield.

The loss of powder from the filter is extremely small. Normally less than 10 mg (st. m)$^{-3}$ air is guaranteed, which will meet even the most strict standards for environment protection. If the air from the cyclones contains milk solids in the order of 200 mg (st. m)$^{-3}$ air, the filter will recover 95 per cent. To illustrate what is involved, it is useful to look at an example. If a spray dryer produces 4 tonne h^{-1} of skim-milk powder and is in operation 4000 h per year, has a stack loss from the cyclones of 0·5 per cent and a filter with an efficiency of 95 per cent, the powder recovered from the filter amounts to 76 tonne per year. At a value of £1200 (sterling) per tonne, it represents a financial contribution of £91,200 per year.

Figure 15 shows a filter designed for milk dryers. The air from the cyclones is introduced into the bottom part of the filter from where it enters the inside of the bags. The solids are thus collected inside the bags contrary to conventional filters where the solids are collected on the outside of the bags. The air goes through the filter bags into the filter housing and then to the atmosphere.

The filter bags are provided with stainless steel rings preventing the bags from collapsing when exposed to negative pressure. The bags are continuously cleaned by a slowly rotating suction arm connected to two collection ports located at positions where the arm is held stationary for a short time by means of a timer. In this way, the powder accumulated in the bags is sucked out by a high pressure, stainless steel fan and separated from the cleaning air in a small cyclone. The small cyclone has a higher efficiency than the large cyclones employed for the drying air. The air from the small cyclone is passed back into the bottom part of the filter, i.e. the pneumatic cleaning system operates in a closed circuit. Powder that drops from the bags into the bottom part of the filter is discharged into the pneumatic cleaning system by means of a rake attached to the rotating suction arm. Compressed air normally used for cleaning of filter bags, is not required. The product does not come into contact with the baghouse which only contains the filtered air to be discharged. The absence of product in the baghouse has the following advantages.

(1) A baghouse is not sanitary, and product sticking to the interior surface can become a 'nest' of bacteria – particularly if care is not taken to avoid condensation at start-ups. Dry cleaning of a baghouse for powder would not be feasible in practice.
(2) The baghouse can be constructed of mild steel. Only the bottom and the chamber into which the dust-laden air is introduced must be made of stainless steel. The chamber is provided with cleaning doors that give easy access for dry cleaning.
(3) Due to the fact that the baghouse only contains the cleaned air, it does not constitute any hazard for dust explosion, therefore no special precautions are required.

The powder recovered from the filter is discharged from the cyclone of the pneumatic cleaning system into the powder from the

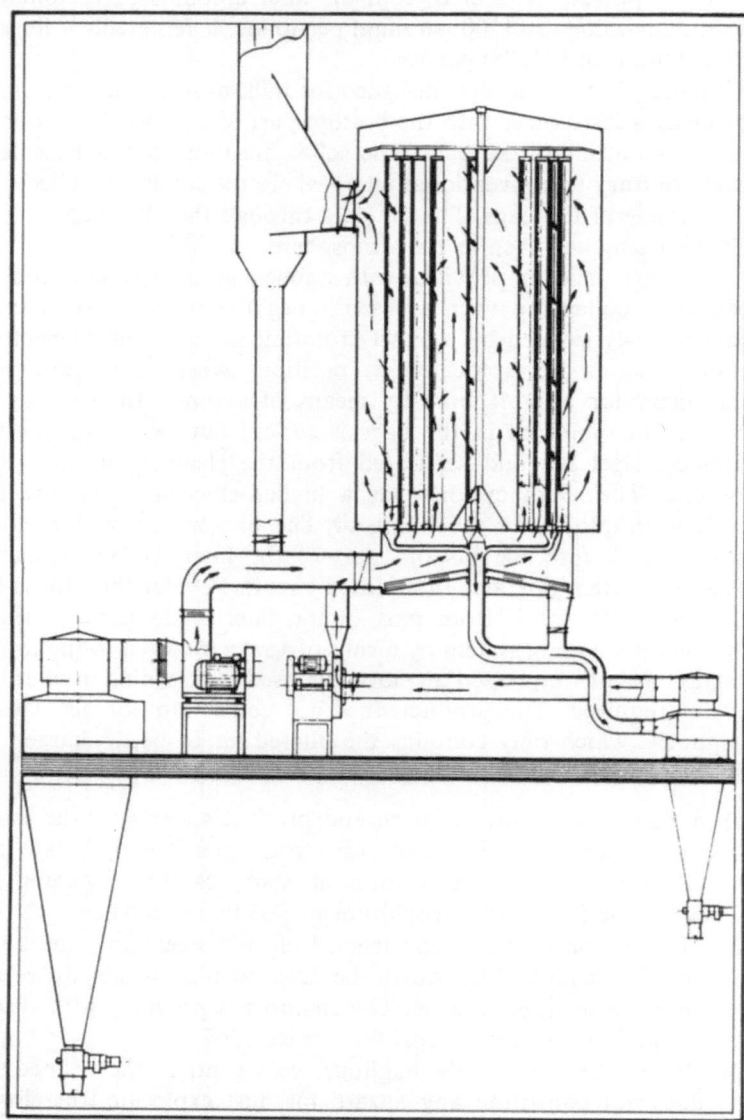

Fig. 15.   Textile filter (courtesy APV Anhydro AS, Denmark).

dryer. Where the powder from the dryer is conveyed pneumatically to silos, it is convenient to connect the cyclone outlet to the pneumatic powder conveying system. The powder from the filter consists of very small particles, i.e. the texture is different from that of the powder from the dryer, but this will not be apparent by continuous back-mixing. The filter powder does not differ from the powder from the dryer with regard to solubility and bacterial count. Sediment tests made on the filter powder can sometimes show disc B indicating that the filter needs cleaning, but that is without practical importance, because the quantity is too small to have any influence on the sediment of the powder resulting from back-mixing.

It is imperative that a bypass with airtight dampers is installed to connect the dryer directly to the atmosphere when it is wet cleaned. When the dryer is idle, the filter bags must be protected against moisture pick-up, either by supplying heated air to the filter or by flowing dehumidified air through the filter. If the main fan for the dryer is placed between the cyclones and the filter, care must be taken that the powder is not contaminated in the fan, i.e. the fan must be provided with an efficient, rust protecting coating or be made of stainless steel. The fan can also be placed after the filter and, as it can then be standard, a separate silencer must be provided.

Besides giving environmental protection and increased yield, the textile filter has another important function. It provides clean air for heat recovery, i.e. transfer of heat from the outgoing air to the inlet air. For this purpose a heat exchanger is required. If the air is used directly from the cyclones, the milk solids in the air will foul the heat exchanger. This will not happen when the air has been cleaned in the textile filter. By recovery of the heat from the outgoing air, a fuel saving of 20–25 per cent is obtained. The installation of a textile filter is often initiated by environment requirements, but can be a profitable investment that provides both solids recovery and heat recovery.

Heat recovery
During recent years several methods have been introduced to transfer heat from the outgoing air to the inlet air, either by air to air heat exchangers or by air to liquid heat exchangers.

Air to air heat exchangers can be of the plate type or tubular type. The air intake is usually far from the point where the air is discharged, therefore an air to air heat exchanger requires considerable duct work. Fouling of the surface in contact with the outlet air will occur if the heat exchanger is used without a filter after the cyclones. Fouling reduces heat transfer, therefore the heat exchanger must be cleaned frequently. Fouling does reduce the amount of powder discharged to the atmosphere, but the milk solids are lost and the frequent wet cleaning adds to the load on the effluent plant. In the case of a blocked cyclone, the heat exchanger will be clogged with powder.

A much simpler heat recovery system can be established when the air from the cyclones is cleaned in a filter. In this case, exhaust air is passed over the surface of a finned tube heat exchanger transferring heat to the liquid inside the tubes. The heated liquid, which can be water with glycol, is pumped to an air heater battery in the inlet air stream to preheat the inlet air. As it involves heat transfer twice, the resultant preheater temperature is slightly lower than when an air to air heat exchanger is used, but the system is much simpler and less costly to install. Where an indirect oil- or gas-fired air heater is used, the temperature of the water in the heat recovery system can be raised by running the water through an economiser in the flue gas stack for the air heater. In this way, the water temperature is boosted and a higher preheat temperature is achieved.

It is most rewarding to recover heat from the outgoing air in a system where the temperature of the outgoing air is high; that is the case when single-stage drying is used. Assume an air inlet temperature of 215 °C, ambient temperature 15 °C, and a preheat temperature of 70 °C, obtained by heat exchanging ingoing air with outgoing air. Then instead of heating the air by 200 °C, the air will only have to be heated by 145 °C with heat recovery, i.e. a saving of 27·5 per cent is obtained. In a three-stage dryer where the temperature of the outgoing air is lower, the saving by heat recovery is approximately 20 per cent. The payback time for a heat recovery system is short when the yearly utilisation of the dryer is high.

Thermal efficiency

The thermal efficiency of a spray dryer is the ratio between the heat utilised for evaporation and the total heat consumption. The heat required for secondary drying is normally not included in this context, therefore the thermal efficiency as a percentage is expressed by

$$\frac{T_i - T_u}{T_i - T_a} \times 100$$

where T_i = air inlet temperature
        T_u = air outlet temperature
        T_a = ambient air temperature.

It is apparent that the efficiency is increased with increasing T_i and T_a and with decreasing T_u. The maximum air inlet temperature to be used in a milk dryer is 215–220 °C, and the ambient air temperature can be raised by employing heat recovery.

The aim of increasing the efficiency by reducing the outlet temperature has led to the introduction of two- and three- stage drying. Multiple-stage drying is based on the fact that, whereas most of the water is readily evaporated from the concentrate, a more severe heat treatment – higher temperature or longer drying time – is required to remove the last few per cent of the moisture. In other words: the rate of evaporation is fast until a dry particle surface has been formed, but it takes more heat or longer time to evaporate the moisture contained in the particles.

Two- and Three-stage Drying

Figure 16 shows a two-stage dryer with a fluid bed for secondary drying. The powder is discharged from the drying chamber at 5–6 per cent moisture into the vibrating fluid bed dryer and cooler (Fig. 17). The fluid bed consists of a stainless steel perforated plate through which air is blown upwards. The powder is 'fluidised' by the air, and high turbulence occurs, which provides excellent conditions for heat transfer between air and powder.

In the first section the powder is dried to its final moisture content by air at 100–120 °C, and in the second section the powder is cooled by air at approximately 10–15 °C. The powder layer is approximately 100–200 mm high, and the fairly long residence time in the fluid bed (10–20 min) makes it possible to perform the drying at a low air temperature. The air used for drying and cooling is collected in the hood over the perforated plate from where it goes to the cyclones for separation of the fines carried with

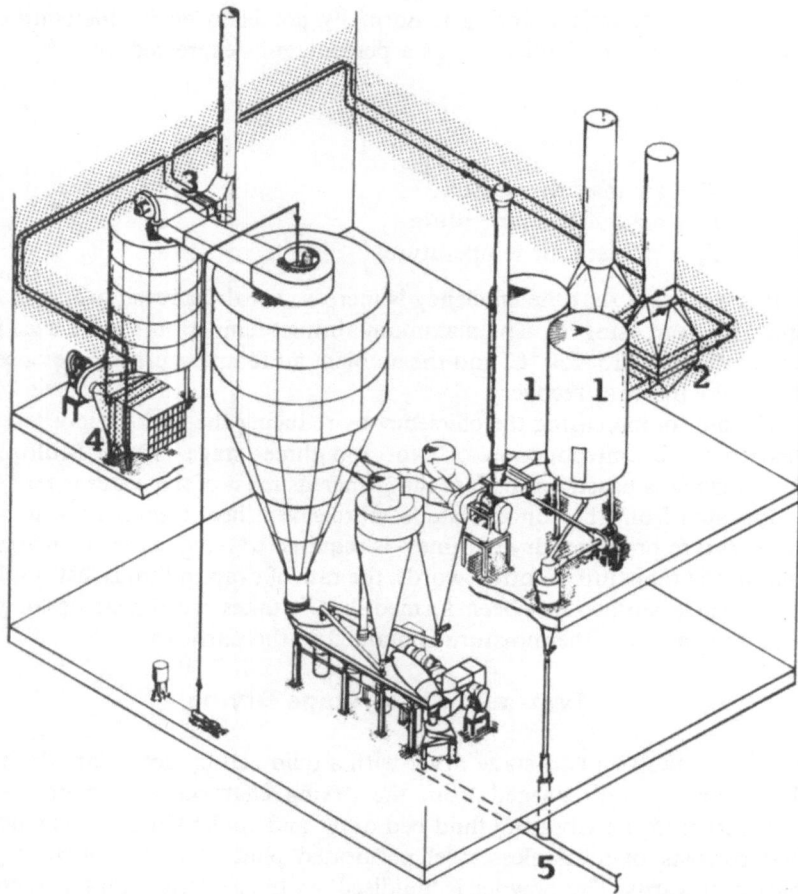

Fig. 16. Two-stage dryer with solids recovery and heat recovery. (1) textile filter (2) heat exchanger for heat recovery (3) booster in flue gas stack (4) air preheater (5) fluid lift system to powder silo. (Courtesy APV Anhydro AS, Denmark.)

the air. The high moisture in the powder from the chamber is obtained by using a low outlet temperature. The low air outlet temperature improves the thermal efficiency of the dryer, and it also improves the powder quality – especially in terms of solubility, bulk density, and free fat. The dryer shown in Fig. 16 has both solids recovery (textile filter) and heat recovery. Liquid is used to transfer the heat from the outgoing air to the

Fig. 17. Fluid bed (courtesy APV Anhydro AS, Denmark).

intake air, and the temperature of the liquid is boosted in the flue gas stack for the indirect oil-fired air heater.

Three-stage drying (Fig. 18) was introduced in order to further improve the thermal efficiency of the drying process by transferring a greater portion of the evaporation from the first stage to the second and third stages. The second stage is a fluid bed placed at the outlet of the conical part of the drying chamber. This fluid bed is static (non-vibrating), and the semi-dry powder drops over an adjustable weir into an external fluid bed for final drying and cooling. A low outlet temperature from the drying chamber can be used, as the powder is delivered from the first drying stage to the integrated fluid bed at a high moisture content.

The air leaves the drying chamber through a vertical duct placed at the base of the cone in the centre of the chamber. The off-centre discharge of the powder makes this arrangement possible, and it is an improvement as the conventional, horizontal outlet air duct in the chamber is difficult to clean inside and collects powder deposits on the outside. Further, the horizontal air duct prevents the swirl pattern of the drying air from being extended into the cone, as the duct is obstructing the rotary movement of the air. Without the horizontal air duct, it is possible to support the swirl movement by the so-called 'wall sweep arrangement', i.e. the tangential introduction of secondary air through a number of narrow vertical slits

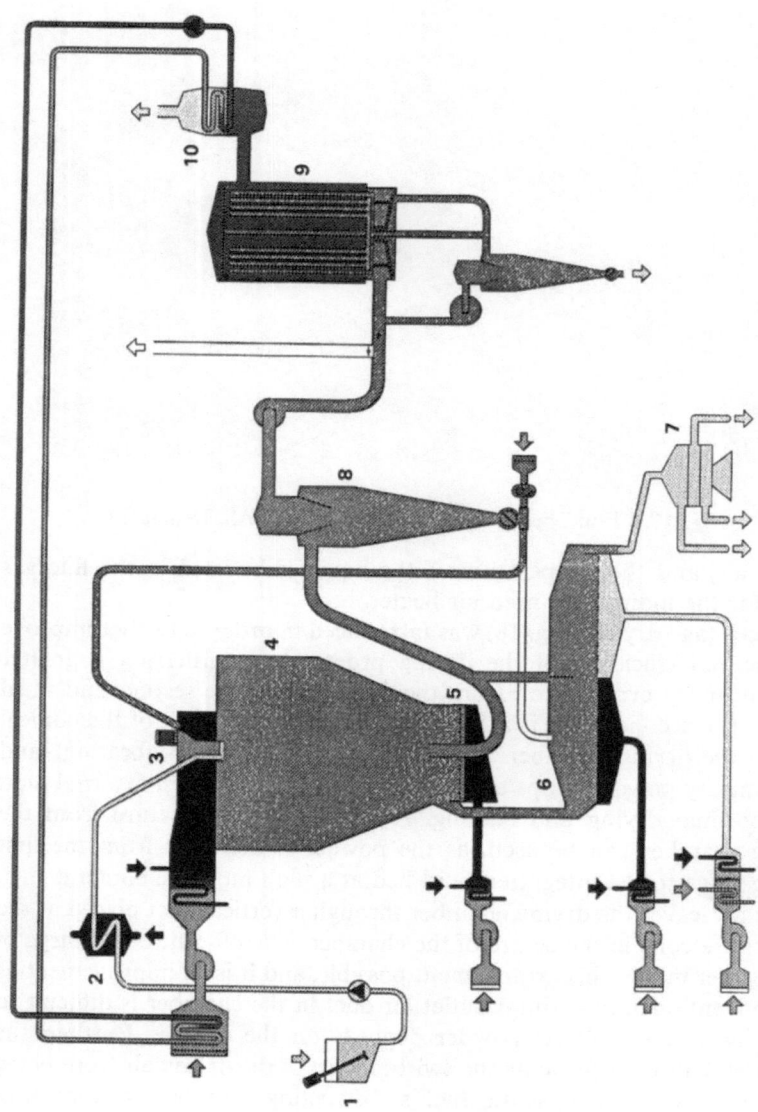

Fig. 18. Three-stage dryer with solids recovery and heat recovery (1) feed tank (2) concentrate preheater (3) atomiser (4) spray drying chamber (5) integrated fluid bed (6) external fluid bed (7) sifter (8) cyclone (9) bag filter (10) liquid-coupled heat exchanger. (Courtesy APV Anhydro AS, Denmark.)

placed in the drying chamber wall at the transition point between the cylinder and the cone. The temperature of the sweeping air is slightly lower than the temperature of the air in the chamber. The high speed air rotation prevents the moist powder from lodging in the cone, and reduces the amount of powder carried with the air to the cyclones. The combination of wall sweep and a steep cone reduces the requirement for external hammers, i.e. the number of hammers as well as their impact can be reduced.

The external fluid bed can be substituted by a pneumatic system for drying and cooling the powder if the height under the chamber is insufficient for the installation of a fluid bed. This may be the case when an existing single-stage dryer is converted to two- or three-stage drying. The two-stage dryer and the three-stage dryer can operate with either rotating atomiser or pressure nozzle atomiser, and Table III shows the heat required to evaporate 1 kg water in single-, two- and three-stage dryers. In all three cases the air inlet temperature is 215 °C, and the dryers are without heat recovery. As heat recovery reduces the heat consumption by approximately 20 per cent, the heat consumption of a three-stage dryer with heat recovery approaches half of that of a single-stage dryer without heat recovery. This clearly illustrates the progress made in energy conservation, and the saving per tonne of powder is even greater when the contribution from a low energy demanding evaporator is included in comparison with a conventional evaporator.

TABLE III
Heat required to evaporate 1 kg water.
Air inlet temperature 215 °C

Type of dryer	Heat (Kj)
Single-stage	5000
Two-stage	4100
Three-stage	3550

The advantages offered by three-stage drying are reduced heat consumption, reduced space demand and improved powder quality. The amount of air required to obtain a certain capacity is the smallest possible when three-stage drying is used, so the costs of a textile filter for cleaning the outgoing air are reduced.

CIP Cleaning

Wet cleaning of a dryer should be restricted to a minimum to reduce the risk of bacteriological contamination as moisture is favourable for growth of bacteria. However, to maintain disc A when the test for scorched particles is made, it is necessary to wet clean a spray dryer at certain intervals. In fact, this test is often used as an indicator of when wet cleaning is required. If the dryer is being used for skim-milk, wet cleaning should only be performed about every three weeks, but often the requirement for cleaning is dictated by a change of product.

For many years the operators have had to dismantle the ducting and other plant items for manual cleaning, but as the dryers are getting bigger and more complex, automatic cleaning in the shortest possible time is required. Cleaning in place is well established in the dairy industry, but only during recent years has it been applied to spray dryers. This is done by fitting stationary spray balls into the plant so the spray covers the entire interior surface. They are permanently placed in ducts, cyclones and fluid beds (above and below the perforated plate) as shown in Fig. 19. In a large dryer, 60–80 spray balls are required. It is possible to leave the spray balls in ducts that carry powder, cyclones and fluid beds because when the plant is in operation, heated purging air is blown through the entire CIP system and this prevents the product from penetrating into the spray balls.

The spray balls are connected in sections; it would require too much water to clean the whole plant at the same time. The cleaning cycle can be programmed, as experience will dictate the time required for each section to be cleaned. Normally the aim is to clean the dryer in 2–3 h, including subsequent drying with hot air, but a longer time is required if the dryer has been used for fat-filled milk which tends to form hard deposits. And fat-filled milk is not the only factor to lengthen cleaning time. It is important that all surfaces in the dryer are completely dry before operation commences; air at 90–100 °C should be drawn through the plant after wet cleaning to sanitise the dryer.

Only the drying chamber is not cleaned by a built-in cleaning device (the sprinklers shown in Fig. 19 in the chamber ceiling are for fire protection). The chamber is cleaned by removing the atomiser and lowering a rotating jet tank cleaner down into the chamber. The rotating jet cleaner is very effective as the water hits the surface at high pressure and produces a mechanical action for the removal of wall deposits.

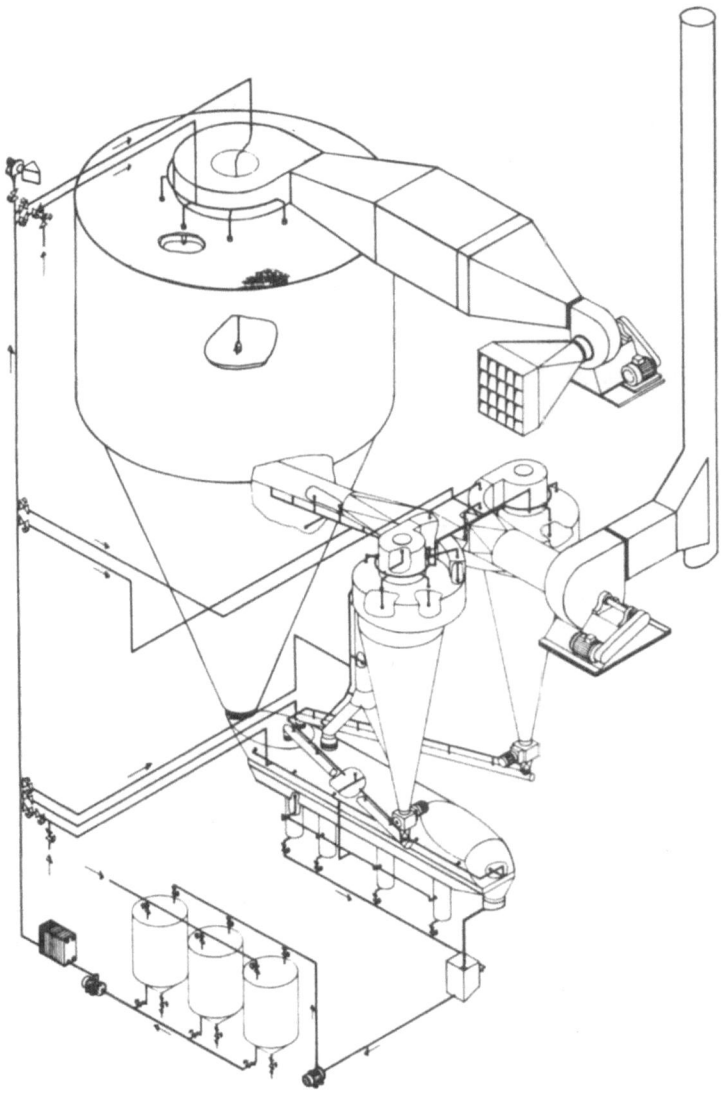

Fig. 19. Spray dryer with in-place cleaning system (courtesy APV Anhydro AS, Denmark).

All cleaning water leaves the plant through the fluid bed and can be collected in tanks. As the cleaning water from the first rinse contains some milk solids it may be used for animal feed. Cleaning solution is normally collected in a tank for reuse. If the plant has a textile filter for the outlet air, it is advisable to have interlock switches on the CIP pump and the damper for the filter bypass system. This ensures the pump cannot be started unless the filter is bypassed. Access door, inspection doors, and pressure relief panels are normally opened after cleaning, and frames and gaskets are cleaned manually. Special attention must be paid to design all doors, panels and joints in such a way that they are completely leakproof.

A condition for the success of the CIP system is that the dryer is cleanable in place, i.e. that the plant items are of good hygienic design with no product traps or dead space, and that ducts are at the correct slope for the removal of water. If they are not properly designed, fittings for temperature probes can provide uncleanable crevices. It is difficult to provide an existing dryer with a CIP system as the dryer may not be leakproof and the plant items may not be of hygienic design.

Tall Form Dryer

The tall form spray dryer (Fig. 20) has pressure nozzle atomisation by means of one or more vertical nozzles. Vertical nozzles require a tall drying chamber of small diameter. Its height is approximately five times its diameter; this rules out use of a rotating atomiser.

The absence of a swirl movement of the air in the chamber is claimed to result in a true co-current flow of air and product, the so-called plug flow. This gives a short residence time, the advantage of which is somewhat counteracted by the higher outlet temperature required for single-stage dryers to achieve the desired moisture content of the powder. Installation of an external fluid bed after the spray tower makes a lower outlet temperature possible. This gives better product properties, as the heat treatment of the powder is reduced.

Spray Bed Dryer

The spray bed dryer (Fig. 21) has been developed to meet demands for various milk powder products with specific properties. It can be designed as a two-stage or three-stage dryer. In both cases the product is led to the integrated fluid bed with a high moisture content; then it can be dried to

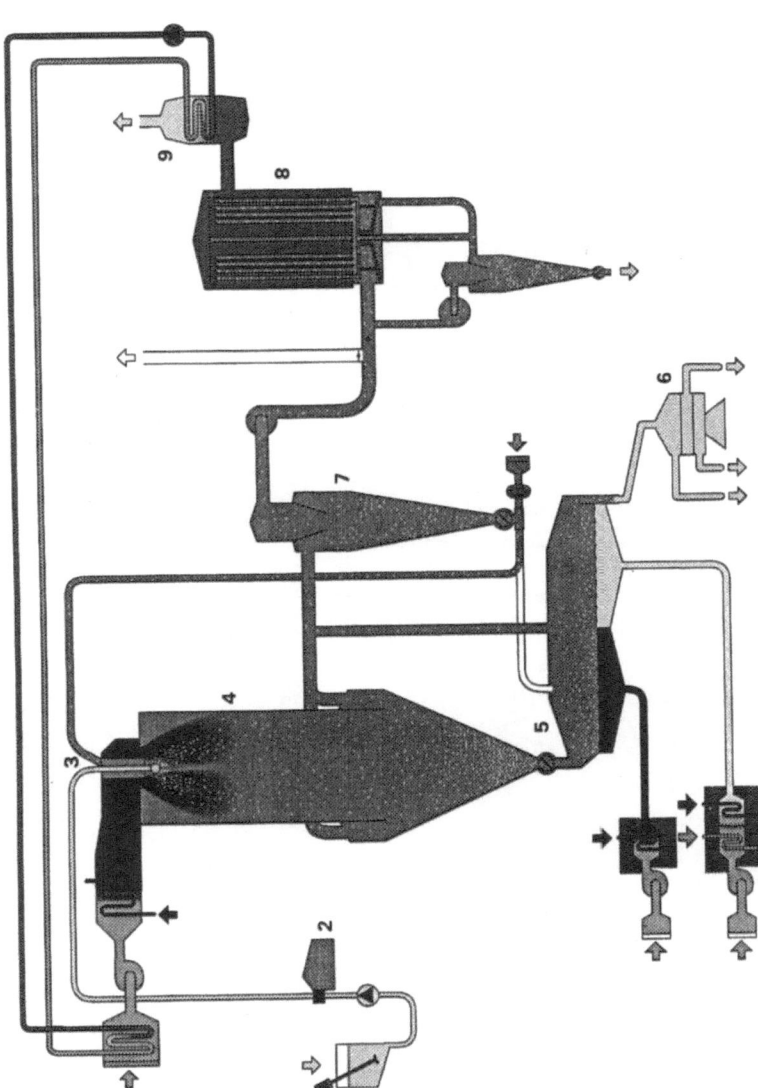

Fig. 20. Tall form spray dryer with solids recovery and heat recovery. (1) feed tank (2) high pressure pump (3) nozzle atomiser (4) drying chamber (5) external fluid bed (6) sifter (7) cyclone (8) bag filter (9) liquid-coupled heat exchanger. (Courtesy APV Anhydro AS, Denmark.)

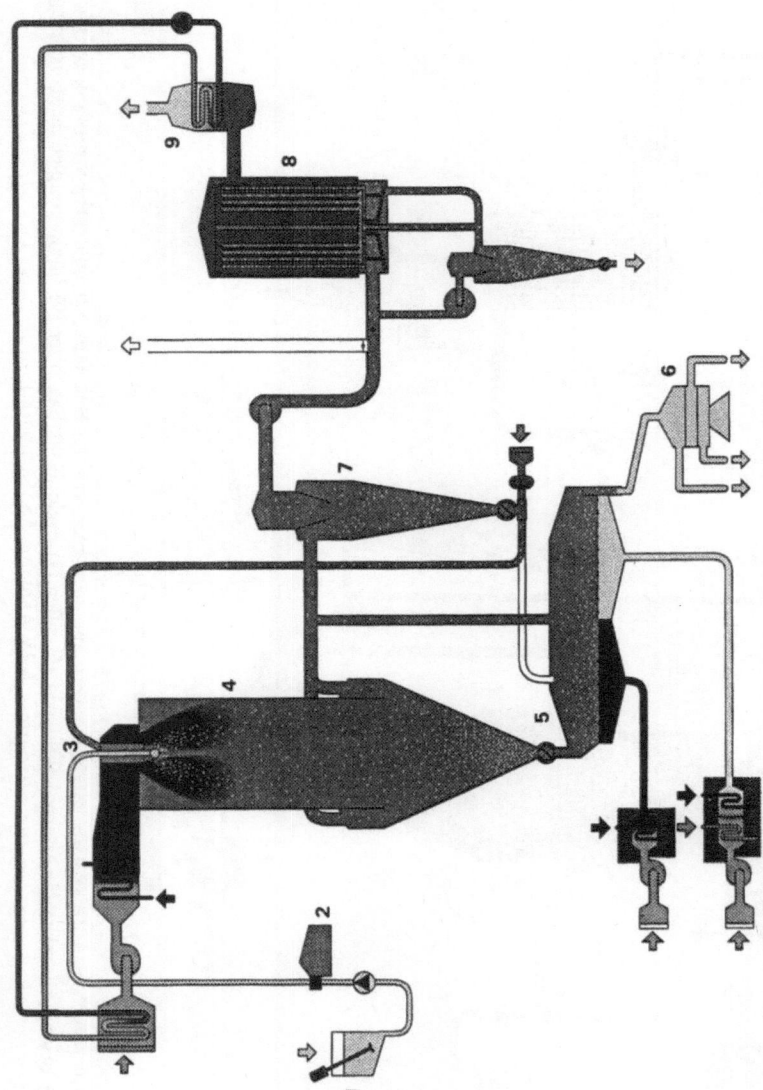

Fig. 21. Spray bed dryer with solids recovery and heat recovery. (1) feed tank (2) high pressure pump (3) concentrate preheater (4) atomiser (5) spray drying chamber (6) integrated fluid bed (7) external fluid bed (8) sifter (9) air outlet (10) cyclone (11) bag filter (12) liquid-coupled heat exchanger. (Courtesy APV Anhydro AS, Denmark.)

the required moisture content in the integrated fluid bed or led to an external fluid bed for final drying and cooling. Both nozzle and centrifugal atomisation can be applied.

The spray bed dryer is characterised by air outlets at the top of the drying chamber. The spent drying air travels along the walls of the drying chamber and helps to prevent deposits on the walls. The fines carried up from the bed by the fluid bed drying air agglomerate with the moist particles sprayed down by the atomiser. The small fraction of fines that is not caught in this way travels with the exhaust air to a powder collector. Here it is separated from the air, and can either be taken back to the integrated fluid bed or to the atomisation zone for agglomeration. An extra nozzle atomiser can be inbuilt above the integrated fluid bed to improve agglomeration or to add liquid.

Pillsbury Dryer

The University of Minnesota and Pillsbury Co., Minneapolis have jointly developed a dryer, known as the Filtermat® Dryer (Fig. 22). The drying chamber is rectangular (box type) with a number of vertical pressure nozzles in the ceiling. As the height of the chamber is insufficient for

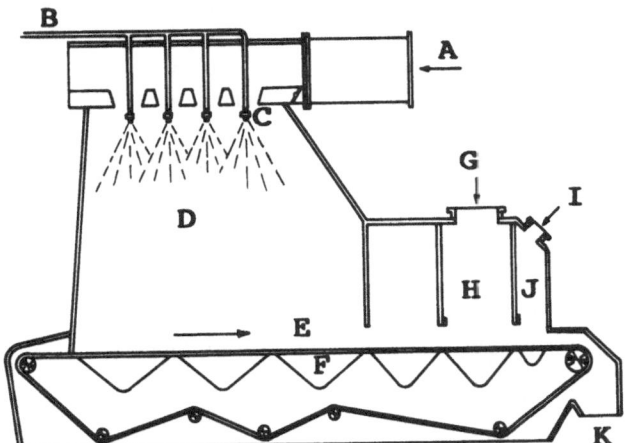

Fig. 22. Filtermat® (Pillsbury) Dryer. (A) hot primary air supply (B) product feed line (C) high pressure atomising nozzles (D) spray drying primary stage (E) moving belt (F) exhaust air (G) warm secondary air supply (H) bed drying secondary stage (I) cooling air supply (J) cooling stage (K) product discharge.

drying to the final moisture content, the moist product builds up into a mat on a porous moving belt. The drying air goes through this layer of powder, which acts as a filter for the exhaust air. The moist product is moved on the belt into following sections for further drying and cooling.

Compared with a fluid bed, the air moves in the opposite direction in the Filtermat Dryer. This results in a compact cake of product which breaks off at the end of the moving belt. Milling of the product from the belt is necessary to achieve a powder. The belt is cleaned by brushes or by water before it is again exposed to product.

The belt dryer was originally developed for the drying of food products like tomatoes, fruit and vegetables, but is also used in the dairy industry for the manufacture of special products with a high content of fat or sugar. The design is not suitable for large capacities as there is a limit to the size of the moving belt. It is not possible to use a rotating atomiser in this type of dryer.

TECHNOLOGY

Concentration

In an effort to improve the overall efficiency of the process, the tendency has been to concentrate the milk as much as possible before drying. In modern plants, skim-milk and whole milk are concentrated to 48–50 per cent solids, whey to 55–60 per cent and permeate to 60–70 per cent.

It has been suggested that evaporators and dryers should be developed to handle milk of higher total solids. However, the use of higher concentrations is not an engineering problem related to the design of evaporators and dryers, it is a technological problem. Experience shows that if milk is concentrated beyond the above mentioned concentrations, the quality of the powder suffers, especially the solubility.

The stability of the casein is decisive for the solubility of the powder; poor solubility is usually due to coagulation of the casein. Casein stability is extremely sensitive to variations in pH and salt balance, and both are essentially affected by concentration of the milk. The titratable acidity increases very nearly in direct proportion to the concentration. The hydrogen ion concentration also increases; whereas the pH of fresh milk is approximately 6·7, the pH of 48–50 per cent concentrate is approximately 6·1. Only a very slight increase in the acidity of the original milk is enough to decrease the heat stability of the casein, as a change in pH of the milk is magnified by concentration. Although far from the isoelectric point, a pH may be reached where fine casein precipitation starts taking place, which results in poor solubility of the powder. The interaction of such factors as protein content of the fresh milk, concentration, temperature, acidity and salt balance determines the degrees of casein stability.

There are indications that the concentrations used until now are not far from the optimal conditions. When these concentrations are used, seasonal variations in the composition of the milk are often sufficient to cause destabilisation of the proteins. The stability is restored by decreasing the concentration – a reduction of only 1–2 per cent in total solids is usually sufficient.

It is well known that lactation period affects milk composition. Milk composition also changes when cows are moved from stable to pasture; the protein content rises during the lactation period, which markedly increases the viscosity of the concentrate. The viscosity increases further on severe heat treatment.

It is desirable to have sufficient latitude between the evaporator and the dryer to maintain the capacity when the level of concentration has to be varied over the year. When whole milk has to be dried, it is sensible to design the evaporator for 50 per cent solids and the spray dryer for 48 per cent. During the time when the composition of the milk permits the maximum concentration to be used, it is easy to reduce the capacity of the dryer by reducing the air inlet temperature.

The chemical and physical properties of a concentrate with high total solids affect the properties of the resultant powder. A closer study of factors that affect the properties of a concentrate as total solids increase is necessary before attempting to use higher concentrations.

Concentrate Heating

The concentrate is usually discharged from the evaporator at a temperature of about 45 °C, unless backwards flow or a separate finisher is used, in which case the concentrate temperature may be higher. The viscosity of concentrated milk is significantly affected by temperature (Fig. 23). It is desirable to feed the concentrate to the atomiser at the lowest possible viscosity, and to achieve this, the concentrate has to be heated to 70–75 °C before atomisation, whereby the following advantages are obtained.

(1) A saving of energy, as it is more efficient to heat the concentrate by means of steam or hot water than to use heat from the air supplied to

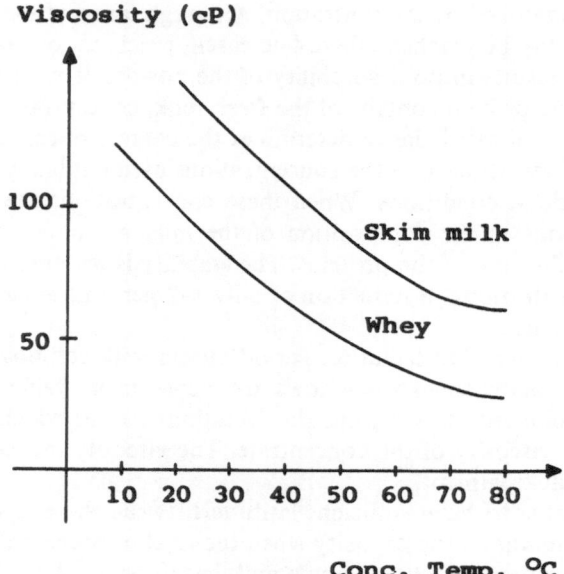

Fig. 23.   The effect of temperature on the viscosity of concentrated skim-milk and whey.

the dryer. The evaporation capacity of the dryer is increased by approximately 5 per cent when the concentrate is supplied to the atomiser at 70 °C instead of 50 °C.

(2) The average size of the droplets into which the concentrate is atomised is reduced due to the lower viscosity. This facilitates drying, making it possible to reduce the air outlet temperature and still maintain the specified moisture content of the powder. The lower outlet temperature results in increased capacity (i.e. lower heat consumption per tonne of powder) and more gentle drying, which results in improved powder quality, especially improved solubility. This is of particular importance when concentrate with a high total solids is dried in a single-stage spray dryer, as the outlet temperature can get near the critical temperature.

(3) As the temperature conditions in the last stages of the evaporator and in the feed tank for the dryer can favour the growth of microorganisms, heat treatment of the concentrate immediately before drying will help to safeguard the bacteriological quality of the powder.

Heating of the concentrate must be done in such a way that no holding takes place, i.e. the concentrate heater must be placed immediately before the atomiser. This is necessary in order to avoid the so-called 'age thickening' which occurs very rapidly at temperatures above 60 °C. The time-dependent thickening of hot concentrate is irreversible and results in a very high viscosity.

If the concentrate heater is placed at floor level in the dryer room, the holding time in the feed line to the atomiser can be sufficient to cause thickening of the concentrate. The concentrate heater should be inserted in the feed line between feed pump and atomiser, and it should be placed at the top of the drying chamber next to the atomiser. A tubular heat exchanger of the coil type, where the heating medium is vacuum-steam or hot water, can give satisfactory service when the temperature difference between product and heating medium is small, and the product has a high velocity. A scraped-surface heat exchanger may also be used, but if the concentrate heater is to be suitable for nozzle atomisation, a tubular heat exchanger can more easily be designed for the high pressure.

Powder Manufacture to a Certain Heat Classification

Heat classification refers to skim-milk powder. It is not a grading requirement, but is of practical importance in describing the suitability of powders for various specific uses. The powder is classified according to the heat treatment to which the raw milk has been subjected before converting it to a powder. The heat classification is expressed indirectly by the level of undenatured whey protein in the powder. To fully appreciate the heat classification, it is useful to review the composition of milk proteins and their specific properties.

When unheated skim-milk is acidified to pH 4·6, approximately 80 per cent of the proteins precipitate. These proteins are the caseins, and pH 4·6 is called the isoelectric point. It is well known that precipitation of the caseins also takes place by the action of rennet or by saturation with sodium chloride.

The proteins remaining in suspension after precipitation of the caseins are called the whey proteins, and they represent approximately 20 per cent of the total protein content. The two main groups of whey proteins are the lactalbumins and the lactoglobulins. Unlike the caseins, they do not precipitate with acid unless they have been exposed to heat treatment to such an extent that denaturation takes place. The term 'denaturation' indicates that the proteins are no longer in their natural configuration; in this context, it is used to describe the change in the molecular configuration of the proteins induced by heat. When cooled they do not again assume their original configurations, i.e. the change is irreversible. There are various stages of denaturation according to the heat treatment; excessive heat treatment causes a more extended change in the configuration of the proteins molecule. Denatured whey proteins do not precipitate from the milk, nor do they precipitate when milk powder is reconstituted, unless the heat treatment has been excessive. Denaturation of the whey proteins changes their properties significantly. They lose biological activity and change functional properties, and they precipitate when the milk or whey is acidified to pH 4·6 or saturated with sodium chloride.

The level of undenatured whey proteins in milk powder is indicated by the 'whey protein nitrogen (WPN) index'. This is the amount of undenatured whey protein nitrogen in non-fat milk solids measured in mg g^{-1} of powder. The method for the determination of the WPN index has been given by the American Dry Milk Institute (ADMI, 1971). The principle is that a sample of milk powder is reconstituted in distilled water. The caseins and the denatured whey proteins are precipitated with sodium chloride then filtered off. The level of undenatured whey proteins in the filtrate is assessed by the turbidity method using a spectrophotometer or colorimeter. The heat treatment classification for non-fat dry milk is given in Table IV. The heat classification does not indicate the level of denatured whey proteins, because the exact content of whey proteins in the milk is usually unknown. Any estimate of the extent of denaturation depends on an assumed average value of whey protein nitrogen in the raw milk, normally 8–9 mg g^{-1} non-fat milk solids. In milk with 8 mg g^{-1} solids, not more than 25 per cent of the whey proteins can be denatured if a low heat powder is to be produced, and at least 81 per cent (approximately) must be denatured if the powder is to be classified as 'high heat'. The content of whey protein nitrogen in the raw milk varies with the origin of the milk (geographical and breed) and is subject to seasonal variation as well.

TABLE IV

Class	WPN Index (mg g^{-1} powder)
Low heat	not less than 6
Medium heat	above 1–5, but below 6
High heat	not more than 1–5

WPN can range from 6 to 13 mg g^{-1} solids. At certain times of the year, it is impossible to produce powder with a WPN index above 6 mg g^{-1} when, at the same time, the milk must be pasteurised according to legislation. It does not seem reasonable that there may be cases where powder, according to ADMI's definition, is not classified as 'low heat' in spite of the fact that all the whey proteins may be undenatured – and this could be the case when the WPN index of the raw unheated milk is below 6. It would be more correct to relate the heat classification to the ratio between undenatured (or denatured) whey proteins and the total amount of whey proteins in the milk.

The amount of denatured whey protein nitrogen in the powder can be assessed in the following way. The raw unheated milk is acidified to pH 4·6, whereby the caseins precipitate. Milk reconstituted from the powder is also exposed to isoelectric precipitation, whereby the denatured whey protein co-precipitates with the caseins. By comparing the two results, the amount of denatured whey protein is found. The ADMI test gives the amount of undenatured whey protein in the powder, and so, by adding the result from the two tests, the total amount of whey protein (denatured plus undenatured) in the powder is found. The total nitrogen can also be determined by the Kjeldahl method.

In the above we have dealt only with denaturation of the whey proteins, but it should also be mentioned that the caseins can be denatured by heat. Such terms as destabilisation or coagulation are suggested to describe the change in the caseins produced by heat, in order to indicate that the change is of a somewhat different nature from the heat-induced changes in whey proteins. Very drastic heat treatment is required to coagulate the caseins, and the preheat conditions sufficient to denature a substantial amount of the whey proteins normally do not affect the caseins. When milk is roller dried, the heat treatment is so severe that the caseins will coagulate; this accounts for the poor solubility of roller powder. Whey proteins that have been heat denatured to such an extent that they do not redisperse into a stable suspension, also contribute to poor solubility.

In liquid milk, only the lactose and some of the inorganic salts exhibit true molecular solubility. The proteins are present in colloidal suspension and the fat is dispersed in microscopic droplets. The 'solubility' of milk powder is primarily determined by the physical and chemical condition of the proteins, i.e. the ability of the proteins to undergo redispersion into a stable suspension. Therefore, when the term solubility is employed in describing the property of milk powder, it implies suspension stability, not true molecular solubility. The milk constituents, which are in true solution, are not important in this context because they do not become insoluble by conversion of the milk to powder, therefore they do not affect the solubility index of the powder.

Denatured whey proteins will redisperse into a stable suspension, when the powder is reconstituted, unless the heat treatment has been extreme. However, coagulated casein does not form a stable suspension when the powder is reconstituted, but appears as a sediment. The solubility index is mainly a measure of the coagulated casein. By analysis of the sediment obtained when solubility tests are made, it has been confirmed that the insoluble matter mainly consists of coagulated casein with small amounts of fat and mineral complexes; denatured whey proteins are only found in rare cases. Therefore, in the whole process of dried milk manufacture, it is of the greatest importance to avoid destabilisation of the casein.

Many years ago, Wright (1933) pointed out that the insoluble substance in milk powder is mainly coagulated casein, and it was found that the dangerous range of concentration is when the milk is at 84–90 per cent total solids. At this concentration the milk exhibits maximum heat sensitivity for insolubilisation. Wright found at 88 per cent total solids, it took only 0·4 s to coagulate 50 per cent of the casein at 100 °C. It is therefore essential to avoid slow drying when the concentration reaches the dangerous level, i.e. at the end of the drying process. At this stage the product temperature should be as low as possible, i.e. low temperature drying should be introduced at the end of the process.

It has often been suggested that the poor solubility of a milk powder is due to a high level of protein denaturation. This is not an incorrect statement but it is inaccurate. Only if reference is made to denaturation (coagulation) of the casein is it correct that protein denaturation has an adverse effect on the solubility of the powder. Denaturation of the whey proteins does not normally affect the solubility of the powder. If that were the case, it would not be possible to make high heat powder of good solubility. This is, of course, not so and it is equally untrue that low heat powder is synonymous with powder of good solubility. There is no relationship between WPN index and solubility index, and whey protein denaturation is not synonymous with insolubilisation.

In prescribing the heat treatment required to produce powder of a certain WPN index, it needs to be recognised that the heat treatment during the whole process is cumulative. The heat to which the milk is subjected in the various maufacturing stages contributes to the resultant WPN index of the powder. The effect on whey protein denaturation of pasteurisation, preheating before evaporation, evaporation, concentrate heating and drying must be considered. It is easy to measure the effect of each heating stage by determining the WPN index after each stage.

The conventional pasteurisation conditions required to ensure the bacteriological quality of the product have only a small effect, as a more severe heat treatment is required to cause any appreciable amount of whey protein denaturation. Only at times when the content of whey protein in the raw milk is very low, will it be necessary to use gentle pasteurisation conditions when low heat powder is to be manufactured. This may not be acceptable, unless the bacteriological standard of the milk is particularly high.

The time–temperature conditions in evaporators vary according to the design of the plant. If whey protein denaturation is to be avoided during evaporation, the boiling temperature must not exceed 70 °C, and recirculation, where the product temperature is high, should be avoided.

As the quantity of whey proteins and their response to heat treatment vary, it is not possible to prescribe the exact heat treatment required to produce powder of a certain WPN index. It is useful to measure how the complete process, without preheating, alters the WPN index from raw milk to powder for, with that knowledge, experience will soon be obtained as to the preheat conditions that result in powder of a certain heat classification. It is normally necessary to alter the preheat conditions over the year.

High heat powder was previously manufactured by heating the milk to 95 °C and holding it for 20–30 min, but the large throughput that is common today makes this long holding impossible, as it would require a very large holding capacity. The long holding has been substituted by ultra-high temperatures. Preheating equipment is required to heat the milk to temperatures varying from 75 to 130 °C and a holding capacity is required to effect holdings from 15 s to 3–4 min. The design of preheating equipment capable of operating at 130 °C for 20 h before cleaning is not a simple matter. Heating by direct steam injection would meet this requirement, but it is far too expensive in running costs as regeneration is not possible. In addition, it is necessary to reduce the milk temperature to the normal feed temperature, which is a few degrees above boiling temperature in the first effect. If the milk were fed in to the evaporator at 130 °C, the rate of flashing would be far too high. A further disadvantage attached to direct steam injection in this context is that, as the steam condenses, this extra water must be removed again in the evaporator. Another problem is that steam injection can cause casein coagulation, which results in powder of poor solubility.

Heat treatment of milk with plate heat exchangers has a high regeneration effect and is therefore attractive with regard to running costs. The final heating by steam to reach the desired pasteurisation temperature is in the range 3–5 °C. However, the required duplex installation will increase the initial investment. Due to the leak-safe dairy valves, the changeover valve cluster is rather simple, and computer control of the changeover procedure makes it almost bumpless.

Direct heating in a vessel with steam from a flash vessel, built into systems with one, two or three sets of vessels, is another type of heat treatment system with regeneration. The regeneration effect is not so high as that for plate heat exchangers, and the final heating by steam is normally in the range 10–15 °C. The final heating is either by direct steam injection or infusion heating. In the evaporator, excess steam can be utilised but foaming problems can give poor condensate quality. Operation times of 20 h can be achieved with systems of this type.

A heat treatment system with tubular heat exchangers, using water as regeneration medium with final heating by steam, can also be used together with an evaporator. A low temperature difference between water and milk together with high velocity ensure operation times of 16–20 h.

Only a few evaporators in Europe have sufficient preheating facilities to manufacture a heat-classified powder and, therefore, most of the powder is of undefined heat classification. The demand for powder manufactured for specific uses, i.e. powder with defined properties above the normal grading requirements, will no doubt increase in future, therefore new evaporators should have the required preheating facilities. In New Zealand and Australia, practically all powder is manufactured to a defined heat classification because the purposes for which the powder is used demand specific heat-induced properties.

The organoleptic properties of the reconstituted milk are affected by the heat treatment of the milk before drying. A cooked flavour is found in milk heated sufficiently to denature the whey proteins, as free sulphydryl compounds are liberated. Low heat powder is of importance to the recombining dairies to avoid a cooked flavour in the reconstituted milk. Low heat powder should also be used in cheese manufacture where fresh milk can be 'stretched' by addition of 30–40 per cent reconstituted milk. High heat powder is required for baking. When powder is used for the manufacture of condensed milk and evaporated milk, a very specific heat treatment is required to obtain the right viscosity and, in the case of evaporated milk, the powder must be heat stable to sustain the sterilisation. For these uses, the WPN is often not a sufficient criterion for the suitability of the powder. The right combination of temperature and holding time is found by testing the powder and varying the heat treatment until a satisfactory result is achieved.

Whole milk is normally given a rather severe heat treatment before evaporation in order to develop the antioxidants that delay oxidation of the fat. Sulphydryl compounds are released by the heat treatment and act as antioxidants. A more complete destruction of oxidising enzymes also takes place.

MILK POWDER QUALITY

Standards for dry milk products based on specific analytical procedures were defined many years ago by the ADMI. This grading system and the methods of analysis have gained worldwide acceptance and are equally useful to consumer and manufacturer. The grading system contains the basic requirements that the products must meet, but it is not designed to describe the suitability of the product for any specific use; supplementary tests may be necessary for that. Table V shows ADMI's specification for powder to be designated 'extra grade'. This specification is widely accepted as the basic requirement for edible milk powder. Powder offered for sale to the intervention board of the European Community must meet this specification or a more strict one, because the intervention board sometimes limits some of the items to increase the rejection rate. It should be mentioned that a significant quantity of powder is rejected by intervention, due not to faults in the powder but to faults in the packing.

The great value of ADMI's grading system is that detailed analytical procedures are laid down for measurement of all the specified properties. This gives an objective, consistent and uniform international

TABLE V
Quality requirements issued by ADMI for extra grade powder

	Skim-milk powder	*Whole milk powder*
Butterfat	max. 1·25%	min. 26%
Moisture	4%	2–5%
Titratable acidity	0–15%	0–15%
Bacterial estimate, per g	max. 30,000	max. 30,000
Solubility index	max. 0–5 ml	max. 0–5 ml
Scorched particles	A	A

measurement of quality to facilitate the purchase and sale of the products.

The bacteriological estimate (or standard plate count) is the estimated total number of organisms per gram of powder, and is determined as laid down by ADMI. Specifications for powder for particular purposes often call for a very low plate count. The bacteriological quality should, in all cases, be specified in more detail to verify the absence of pathogens. Special analysis are required for coli, salmonellae and staphylococci. A special microbiological specification would be a low spore count. In general, it is regarded as unnecessary to establish standards for spore counts when the bacteriological quality of the powder is under control. But for powder that in its use offers favourable conditions for spore germination and cell multiplication, it can be significant to limit both aerobic spore formers and anaerobic spore formers. Since the heat treatment during processing generally does not destroy spores, milk must come from farms having a history of low spore counts.

By the solubility test, the amount of insoluble matter is determined in the reconstituted milk; distilled water is used. The solubility index is the volume of sediment (in ml) from 50 ml of reconstituted milk. In some countries, the solubility is reported on a volume percentage basis. Thus, for a milk powder with a solubility index of 0·5 ml, the insoluble residue is 1·0 ml per 100 ml, and this would be reported as '99·5 per cent solubility'. Similarly for a powder with a solubility index of 0·25, the insoluble residue is 0·50 ml per 100 ml, which would be reported as '99 per cent solubility'. When expressing the limits, the designation 'maximum' is used in relation to ADMI's index, whereas the same limit is expressed as 'minimum' when reference is made to the volume in per cent.

The test for scorched particles (also called the sediment or purity test) is the only one where there may be room for a subjective judgement, as the sediment pad containing the impurities is graded visually against standard pads ranging from A to D.

So far we have dealt with the powder properties on which ADMI's grading system is based, but individual powder users have their own items included in the specifications.

The most common non-grading item is the bulk density of the powder. Unfortunately, there are different methods for determination of the bulk density, and as the method used has a marked effect on the results obtained, the method should be prescribed together with the desired value. According to British Standards, the density is measured after the sample has been dropped ten times from a height of 150 mm onto a soft pad (a folded duster). The density is expressed in $g\,ml^{-1}$ $(g\,cm^{-3})$. Sometimes the reciprocal value is used which is referred to as the 'specific volume' expressed in $ml\,g^{-1}$ $(cm^3\,g^{-1})$. The powder density is obviously of considerable interest from an economic point of view, as it influences the cost of storage, packing and transport.

The amount of free fat in the powder is of particular importance when instant powder is produced. The free fat is defined as the portion of the fat extracted from whole milk powder by washing with a solvent, such as light petroleum, for a defined period, normally 15 min. The free fat is expressed as a percentage of the powder, or as a percentage of the total fat. Normally a free fat content of 1 per cent (maximum) of the powder is stipulated for instant powder which equals very nearly 4 per cent of the total fat; the fat content of whole milk powder comprises nearly one quarter of the powder. It is possible to reduce the level of free fat in the powder by homogenising the concentrate immediately before atomisation.

Further powder requirements are the heat treatment classification, already dealt with, and 'instant' properties, to be covered next.

Factors Affecting Powder Properties

It is obviously important for the powder manufacturer to know which factors affect the powder properties. It may be raw milk quality; it may be process conditions. If it is process conditions, it must be known where in the process the powder properties are influenced and how variations in operation conditions affect powder quality.

Of the quality requirements for extra grade powder, according to ADMI listed in Table V, the fat and moisture content are entirely a function of processing. For whole milk powder, it is a condition that the ratio of fat to non-fat milk solids in the raw milk is such that the dried product will contain a minimum of 26 per cent fat.

The characteristics which are influenced by the raw milk quality are the acidity, the microbiological count and – indirectly through the acidity – the solubility. The bacteriological quality of the raw milk affects the quality of the powder, both directly and indirectly. In the direct sense, organisms that occur in the raw milk and survive the process will appear in the powder. In this context the presence of thermophilic and thermoduric organisms in the raw milk are of greatest importance. Whereas other microorganisms can be controlled to an acceptable level, thermo-resistant organisms will survive the process and contribute to the plate count of the powder.

The number and type of bacteria in the raw milk indirectly affect the quality of the powder. The natural level of lactic acid in milk direct from the cow is very low, with a titratable acidity of approximately 0·11 per cent. During the time interval between milking and processing, lactic acid may develop as a result of bacterial activity. One of the most numerous microorganisms in normal raw milk is *Streptococcus lactis*. It grows well at 21–35 °C and produces lactic acid using the lactose as a nutrient. The developed acidity can lead to serious defects in the finished product. Not only may the titratable acidity exceed the permissible level, but the same may apply to the solubility. A high lactic acid content will result in an increased hydrogen ion concentration, and this causes problems with protein stability during processing. The acidity problem is magnified by concentration of the milk because the titratable acidity increases very nearly in direct proportion to the concentration. The level may be reached where the heat stability of the proteins is seriously affected, which results in poor solubility of the powder.

In practice, the greatest problem attached to a high bacterial content of the milk is normally not a high plate count of the powder, but a high developed acidity, which can be reached very quickly. It is not uncommon for the total titratable acidity at the processing point to reach 0·16 per cent lactic acid and, on hot summer days, values of 0·17 per cent or even 0·18 per cent are not unknown in milk which has travelled far and has been too warm. In this condition the milk becomes highly unstable. It causes problems in processing and yields an unsatisfactory finish product. Although the strength of the acid expressed by the pH value – not the volume expressed by the titratable acidity – controls the heat stability of the milk, it is common practice to express the acidity as titratable acidity. This is the number of millilitres of sodium hydroxide solution required to neutralise 100 ml of milk using phenolphthalein as indicator. Different units are used for titratable acidity. The difference lies in the strength of the NaOH solution used and Table VI indicates the relationship between the different units and the corresponding content of lactic acid.

Soxhlet Henkel degrees are the number of millilitres of 0·25 N NaOH required to neutralise 100 ml of milk. Thörner degrees are the number of millilitres of 0·1 N NaOH required to neutralise 100 ml of milk. Dornic degrees are the number of millilitres of 0·11 N NaOH required to neutralise 100 ml of milk.

One millilitre of 0·11 N NaOH = 0·01 g lactic acid, i.e. 1 Dornic degree corresponds to 0·01 g lactic acid per 100 ml of milk or 0·01 per cent.

TABLE VI
Designation for titratable acidity of milk

Soxhlet Henkel	Thörner	Dornic	% Lactic acid
1	2·5	2·25	0·0225
2	5·0	4·5	0·045
3	7·5	6·75	0·0675
4	10·0	9·0	0·0900
5	12·5	11·25	0·1125
6	15·0	13·5	0·1350
7	17·5	15·75	0·1575
8	20·0	18·0	0·1800
9	22·5	20·25	0·2025
10	25·0	22·5	0·2250
11	27·5	24·75	0·2475
12	30·0	27·0	0·2700

Therefore, the number of Dornic degrees divided by 100 gives the content of lactic acid in per cent.

The acidity of the milk at the processing point should preferably be about seven Soxhlet Henkel degrees in order to produce a first class powder, and should be rejected as a raw material for a first class, edible product if the acidity exceeds 8 degrees. The acidity of the powder (i.e. of the milk reconstituted from the powder) cannot be derived directly from the acidity of the milk, as there is an initial drop in titratable acidity when the milk is heated, due to the release of carbon dioxide. However, raw milk is far from the only cause of quality defects in the powder. Many of the problems arise in and around the process, and efforts to improve powder quality as a whole necessitate close attention to both milk quality and the process.

The conventional approach is to blame only the dryer for process-related defects in the powder. But the fact is that only one property of the powder – the moisture content – is entirely governed by the dryer. Some of the other properties, which are a function of the process, may be determined by the dryer, by the evaporator or by other stages in the process. Some properties are governed by conditions outside the dryer. This is the case when whey is dried, as the desired non-hygroscopic nature of the powder is chiefly governed by crystallisation of the lactose between the evaporator and the dryer. To a great extent, free fat and bulk density are determined outside the dryer, and the heat classification is determined entirely outside the dryer. Too high a plate count of the powder is more often caused by poor sanitation somewhere in the process rather than by poor bacteriological quality of the raw milk. Powder solubility and scorched particles are also functions of the process.

Solubility is probably the most difficult problem in connection with milk drying; not all the factors affecting solubility are known. The air outlet temperature is critical. Assuming the acidity of the milk is within the acceptable range, the problem may be related to too high a total solids in the concentrate, which reduces the heat stability of the proteins and causes too high a viscosity. The high viscosity can also be a consequence of an unusually high protein content of the milk, or a high heat treatment of the milk. The high viscosity results in large droplets when the concentrate is atomised, which requires an increase in outlet temperature to reach the desired residual moisture. The high outlet temperature may overheat the milk solids and cause insolubilisation. The answer is to increase the rotational speed of the atomiser (or the pressure of the nozzle atomiser), to reduce the degree of concentration, or to heat the concentrate.

It can be very useful to apply the solubility test to the concentrate to find out if it contains insoluble matter. If the solubility index of the concentrate is equal to the solubility index of the powder, the fault is not in the dryer. The opposite is true if there is no insoluble matter in the concentrate. It can also be useful to analyse the sediment obtained when the solubility test is done.

Scorched particles usually originate from milk solids that have been held up somewhere within the process, and are thereby overheated or burnt. The most frequent cause of scorched particles is deposits in the drying chamber, atomiser or air distributor. The heat will darken or char this material, and particles which break off get mixed with the product to appear in the powder as dark specks. Scorched particles may be present in the concentrate if burning-on of the product occurs in the evaporator. By applying the test for scorched particles to the concentrate, it may be possible to locate where in the process the contamination has taken place. Sometimes a filter for the concentrate is inserted between the evaporator and the dryer. Another cause of scorched particles may be inadequate filtration of the drying air or poor maintenance of the filter, whereby impurities are introduced to the product by the air. The air may contain milk powder dust which becomes scorched when passing through the air heater.

The bulk density of the powder is chiefly governed by the total solids of the feed to the atomiser, as the density increases with increasing total solids. The density is very sensitive to variations in the concentration. It is therefore desirable to have an evaporator with automatic control of the total solids in the concentrate. The density of the drying air decreases with increasing air temperature, but this effect is slight. It might be expected that particle size would affect the density but, in practice, particle size variations have insignificant effects. Agglomeration is part of the instantising process, and it has a marked effect on the density. A heavily agglomerated powder is very light and its density is often used to express the degree of agglomeration.

The density of individual particles is determined by the density of the solids, i.e. the product composition, and by the amount of air entrapped within the particle. Powder produced by high pressure nozzle atomisation contains less entrapped air than powder from a rotating atomiser. But to some extent, the benefit of this on the denisity is offset by the fact that a rotating atomiser can accept concentrate of a higher total solids than a nozzle atomiser. The amount of entrapped air in the powder from a rotating atomiser can be reduced by using a steam swept atomiser. By admitting steam to the product in the atomiser disc, steam is incorporated into the particles instead of air. The steam condenses practically instantaneously, and the particles shrink and become more solid. The tapped density of skim-milk powder is normally 0.58–$0.60 \, \text{g cm}^{-3}$, but with steam atomisation it is possible to increase the density by approximately 20 per cent to 0.70–$0.72 \, \text{g cm}^{-3}$.

How Process Conditions Affect Powder Properties

Experiments have been made at the Dairy Research Institute at Melle in Belgium to find out how variations in one processing parameter, all other conditions constant, affect the properties of whole milk powder. The results have been published by De Vilder *et al.* (1976, 1979) and are summarised below.

In the first experiment (Table VII) the influence of raising the inlet temperature of the drying air was studied. The inlet temperature was raised from 170 °C to 225 °C; the outlet temperature was kept constant at 90 °C by gradually increasing the feed rate. The results show that solubility is affected by raising the air inlet temperature when the outlet temperature is kept constant. The moisture content increases because the air outlet temperature is kept constant. When the air inlet temperature is increased, the outlet temperature will also have to be increased if the moisture content is to remain constant. As a rule, the outlet temperature

TABLE VII

Influence of raising inlet temperature at constant outlet temperature 90 °C.
Homogenised whole milk concentrate, 48% solids

Inlet temp. of the air (°C)	Moisture content (%)	ADMI-solubility index	Free fat content % of total fat	Bulk density (g cm⁻³)
170	2·58	0·05	6·62	0·68
180	2·69	0·05	5·55	0·66
187	2·92	0·05	4·72	0·66
195	2·99	0·10	4·14	0·65
205	3·05	0·15	4·10	0·63
215	3·19	0·30	4·00	0·63
225	3·30	0·30	4·05	0·62

must be increased by 1 °C for each 5 °C increase in inlet temperature. The free fat decreases with increasing inlet temperature – probably because a hard particle surface is formed as a result of the high drying temperature. The hard surface prevents the fat dissolver from penetrating into the particles, so that less fat is extracted when the free fat is determined. The density decreases with increasing inlet temperature in spite of the increased moisture content – probably because of an increase in entrapped air in the powder particles.

In the second experiment (Table VIII) the influence of raising the air outlet temperature was studied. The inlet temperature was kept constant at 195 °C; the outlet temperature was raised from 75 °C to 105 °C by gradually reducing the feed rate. The moisture content decreased with

TABLE VIII

Influence of raising outlet temperature at constant inlet temperature 195 °C.
Homogenised whole milk concentrate, 48% solids

Outlet temp. of the air (°C)	Moisture content (%)	ADMI-solubility index	Free fat content % of total fat	Bulk density (g cm⁻³)
75	4·75	0·05	2·15	0·64
80	4·41	0·05	2·64	0·63
85	3·93	0·05	3·04	0·61
90	3·23	0·50	2·78	0·60
95	2·59	0·50	4·49	0·58
100	2·18	2·90	4·38	0·57
105	1·76	3·30	5·43	0·55

increasing outlet temperature. The solubility is significantly affected by high outlet temperatures. At outlet temperatures of 100 °C and above, the solubility is very poor. The free fat increases with increasing outlet temperature, which may be ascribed to crack formation in the surface of the particles due to overheating. It is a normal feature that the density falls with falling moisture content. The low outlet temperatures (75, 80 and 85 °C) correspond to the outlet temperature range when two- or three-stage drying is used. The results clearly illustrate the beneficial effect of multiple-stage drying on solubility and free fat.

The third experiment (Table IX) was made to illustrate the influence of the degree of atomisation at constant air inlet and outlet temperatures. The rotating speed of the atomiser was varied from 19,600 to 31,300 rpm, corresponding to a circumferential speed of the disc of 106–169 m s^{-1}. The inlet temperature was constant at 195 °C and the outlet temperature was kept constant at 90 °C by gradually increasing the feed rate.

TABLE IX

Influence of rotating speed of the atomiser at constant inlet temperature 195 °C and constant outlet temperature 90 °C. Homogenised whole milk concentrate, 48% solids.

Revolutions of the atomiser (min^{-1})	Peripherential speed of atomiser disc (m s^{-1})	Moisture content (%)	ADMI solubility index	Free fat content % of total fat	Bulk density (g cm^{-3})
19 600	106	3·27	0·90	7·52	0·62
22 250	120	2·82	0·74	5·06	0·59
25 000	135	2·65	0·05	5·57	0·59
28 300	153	2·55	<0·05	6·48	0·58
31 300	169	1·99	<0·05	4·22	0·58

By increasing the degree of atomisation, the concentrated milk is given a large surface area so the water is more readily evaporated. Therefore, a lower moisture content is achieved at the same outlet temperature. The solubility shows a marked improvement with increased speed of the atomiser, even where the powder is overdried, but as the low moisture content is not produced by high outlet temperature (as was the case in test 2), it has no adverse effect on the solubility. The improvement in solubility is due to the short drying time of the milk when it is atomised into smaller particles. One would expect the free fat to rise due to the increased surface of the powder. But by increasing the speed of the atomiser, a certain homogenising effect is obtained in the disc, which may explain the modest drop in free fat. The density is hardly affected but declines slightly due to the lower moisture content.

The last experiment (Table X) shows the influence of heating the concentrate. The air inlet temperature was constant at 195 °C, and the outlet temperature was kept constant at 90 °C by gradually increasing the feed rate. The atomiser speed was constant at 25,000 rpm.

TABLE X

Influence of concentrate heating at constant inlet temperature 195 °C, constant outlet temperature 90 °C, and constant atomiser speed 25,000 rpm. Non-homogenised whole milk concentrate, 49·5% solids.

Concentrate temperature (°C)	Moisture content (%)	ADMI- solubility index	Free fat content % of total fat	Bulk density (g cm^{-3})
50	3·18	1·20	13·8	0·58
60	2·87	0·60	14·5	0·57
70	3·04	0·30	15·9	0·55
80	3·02	0·20	21·1	0·54

Heating the concentrate lowers its viscosity and atomises it into smaller droplets. In the previous experiment, this change was achieved by increasing the atomiser speed but the result is the same, so there is good agreement between the results in Tables IX and X with regard to the change in moisture content, solubility and bulk density. The free fat started at a higher level because the concentrate was not homogenised. The increase in free fat with increase in concentrate temperature is due to the increase in powder surface; this effect has not been offset by increased homogenisation at the atomiser disc. In the previous test, the disc speed was not kept constant.

Summarising the results of the last two experiments, the following conclusions can be recorded. When milk is atomised into smaller droplets, the outlet temperature can be reduced and the solubility is improved. Smaller particles are obtained by increasing the speed of the atomiser, or in case of nozzle atomisation by increasing the pressure. Smaller particles can also be obtained by reducing the viscosity of the feed – either by reducing the total solids or by heating the concentrate. Concentrate heating has the greatest effect at high concentrations. The tendency of the free fat content in whole milk powder to increase due to the larger powder surface can be counteracted by increased atomiser speed (or increased nozzle pressure), or by homogenisation of the concentrate. However, homogenisation does increase the viscosity, so these two contradictory parameters have to be optimised when whole milk is dried.

The experiments clearly illustrate that a slight change in process conditions significantly affects the characteristics of the powder.

Instant Milk Powder

Instant milk powder is powder manufactured in such a way that it has better reconstitution properties than normal powder. It is not unusual for instant powder to be described as a product with improved solubility; this is not correct. The instantising process will never improve the solubility – in fact the process can have an adverse effect on the solubility if the correct conditions are not used.

The purpose of instantising is to improve the rate and completeness of the reconstitution of the powder. This change is especially important for domestic use because the industry normally uses mechanical equipment for reconstitution, hence the 'instant' properties are less important. The concept of an instant powder is that placed on the surface of unheated water, the powder will quickly sink and disperse without stirring. The powder must have a good wettability (short wetting time), good sinkability and good dispersibility. The solubility should be as for normal powder.

Dispersibility, as defined and determined in IDF Standard 87 (1979), is probably the best single criterion for assessing the instant properties because a condition for dispersion of the powder in a defined time is that the particles have been wetted and have sunk into the water. The temperature of the water significantly influences the rate of reconstitution with a maximum reached when the water is at 70–80 °C. The water temperature used for the determination of dispersibility and wettability according to IDF's standard is 25 °C, assumed to match the temperature used in households, although the temperature of tap water is generally lower.

The instantising process is different for skim-milk as against whole milk, and the reconstitution rate of skim-milk powder is governed by the texture of the powder. The instantising process for skim-milk powder consists of agglomeration of the particles into porous aggregates of sizes up to 2–3 mm. This agglomeration increases the amount of interstitial air (i.e. the air between the particles) and reconstitution commences only when the interstitial air has been replaced by water. As a result, the powder particles are wetted and dispersed before the actual dissolution begins.

In practice, agglomeration is accomplished by wetting the powder – usually in a fluid bed with steam or water, sufficiently to cause the surface to be tacky so the particles will adhere. During the wetting, the moisture content may rise to 6–10 per cent, and the agglomeration is followed by re-drying to the normal moisture content using hot air. The re-wet method is a separate process divorced from the drying process. Low to medium heat powder should be used for two reasons. The first is that, as the instantising process includes further heat treatment of the milk solids, it is essential to use low heat powder to avoid a cooked flavour of the reconstituted milk. Secondly, it has been found that the stability of the agglomerates is poor when they are manufactured from a high heat powder; less shattering of the agglomerates takes place during subsequent handling when a low heat powder is used. The bulk density of the agglomerated powder is as low as 0.35–$0.40 \, \text{g cm}^{-3}$.

Straight-through manufacture of instant skim-milk powder discharges powder with a high moisture content from the drying chamber into a fluid bed, where it is agglomerated and dried to its final moisture content. The fine powder fraction, separated from the drying air and the air from the fluid bed, is fed back to the atomisation zone to agglomerate with the moist product from the atomiser. However, the benefit is limited in this context, as discharging the powder with a high moisture content from the chamber with the intention that it should agglomerate in the fluid bed, means that the surface of the particles will be dry and the interior wet. Exactly opposite conditions are required for agglomeration. If attempts are made to take the powder out with a moisture content sufficiently high for the particles to agglomerate, i.e. with a tacky surface, great difficulties are encountered in getting the powder out of the chamber.

The straight-through process does not render a product sufficiently agglomerated to be designated instant. The bulk density of the powder is very similar to the bulk density of normal powder, and this indicates a low level of agglomeration. The powder is classified as 'semi-instant' or 'dust-free' powder due to the absence of fines. For industrial use, the non-dusty property is often more important than any instant properties.

The manufacture of instant whole milk powder is more complicated and was not commercially successful until many years after the introduction of instant skim-milk powder. The basic method is the same as for skim-milk – to convert the powder to porous agglomerates to permit the single particles to be wetted and dispersed before dissolution starts. This is the essence of all processes for instantising milk powder. However, due to the hydrophobic nature of the fat, reconstitution of the agglomerates will not take place in water at temperatures below 45 °C (about 10 °C above the melting point of the fat), unless the powder particles have been coated with a surface-active or wetting agent; the use of lecithin as a surface-active agent is accepted worldwide. A special grade is used, consisting mainly of the natural phospholipid fractions extracted from soya lecithin (available under the trade name 'Metarin'®).

The instantising process for whole milk powder starts with the manufacture of a suitable base powder, i.e. a powder which after lecithination has instant properties in water at 25 °C. The base powder must meet the following specifications above the normal grading requirements.

(1) The free fat content should be as low as possible. This can normally be achieved by homogenisation of the concentrate immediately before atomisation.

(2) The particle density should be as high as possible to improve the sinkability. Concentrate of high total solids should be used and the amount of entrapped air in the particles should be reduced to a minimum. Decreasing the drying air temperature to 170–180 °C will contribute to that.

(3) The powder should consist of porous agglomerates with an absence of fines. The greatest portion should have particle sizes in the range 100–250 μm, with a maximum of 15–20 per cent below 90 μm. The bulk density should be 0·45–0·50 g cm^{-3}.

In order to meet the last requirement, the concentrate should have a high viscosity, i.e. it should be an unheated concentrate with high total solids. Atomisation should take place using the lowest atomiser speed consistent with drying efficiency. These conditions produce large particles which require extended drying time so that the chance of partly dried (and hence sticky) particles colliding and adhering together is increased. To offset the adverse effect of the extended drying time, two- or three-stage drying should be employed, so the temperature in the drying chamber is lower than for single-stage drying, and the free fat content is also lower.

The powder is discharged into a fluid bed for final drying and for removal of the fines. The amount of fine particles blown off is governed by the quantity of air passing through the perforated plate. The fines, separated from the drying air and from the fluid bed air, are fed back to the atomisation zone to agglomerate with the moist product from the atomiser. When the powder has been dried to its final moisture content in the fluid bed, it is ready for lecithination, either in the fluid bed while the powder is still hot, or later in a separate fluid bed, after the powder has been cooled and stored intermediately. The method to use largely depends on how the powder is to be packed.

A 1:1 mixture of lecithin and butter oil heated to 60 °C is sprayed onto the powder in the fluid bed by means of nozzles, and in a quantity to give a lecithin content of 0·2–0·3 per cent in the powder. The temperature of the powder during lecithination should be approximately 50 °C. The best result is obtained when the powder is packed at this temperature. The lecithin continues to coat the particles and any non-lecithinated surface exposed by particle breakdown during packing becomes coated by the continuing spread of the lecithin. The hot powder is highly sensitive to oxidation of the fat and must therefore be packed with inert gas to reduce the oxygen level in the pack to a maximum of 2 per cent. If the lecithination process is linked to the dryer (the straight-through process), the lecithin is sprayed onto the powder immediately after it has reached its final moisture content.

Hot packing must synchronise with the dryer, but it will be very costly to pack metal cans by installing a packing line to match the dryer capacity. Although the powder can be packed hot into tote-bins for later transfer into consumer packs, the tote-bins must be evacuated and filled with inert gas. This intermediate stage is rarely justified because the benefit of the initial hot packing into tote-bins is lost by further handling. This inevitably results in some breakdown of the agglomerates. So the powder is normally lecithinated as it comes out of the dryer for later packing in a cold condition, but it must be held in the fluid bed after lecithination for at least 5 min at 50 °C before it is cooled. Intermediate storage must be in tote-bins or drums, but as the powder is cooled, gas packing is not necessary at this stage provided the powder is stored for only a few days. Pneumatic transport and silo storage will cause particle breakdown and must be avoided.

IDF Standard 87 contains methods for the determination of the dispersibility and wettability of instant skim-milk powder and whole milk powder, but gives recommended values only for dispersibility; no values are given for wettability. When the IDF method is used for determination of the wettability, it is realistic to require a wetting time for skim–milk powder not exceeding 15 s, and for whole milk powder not exceeding 10–15 s in order to designate the powder as instant.

Today, producers of instant whole milk powder, especially producers in New Zealand, use a number of analytical methods describing consumers' impressions of the instant properties when using milk powder for cooking. They are an imitation of how consumers use instant whole milk powder in cold and hot water as well as coffee. The methods comprise

> Determination of slowly dispersible powder particles and deposits after gentle stirring with a spoon when dissolving instant whole milk powder. The analysis is carried out in both cold and hot water.
> Determination of solubility in hot water.
> Determination of solubility in hot coffee.

The accuracy and reproducability of the methods should be improved, but they do give a more comprehensive description of the instant properties of milk powder than the IDF standard for dispersibility and wettability.

Another important parameter is flowability, an expression for how easily the powder may be conveyed. Improved flowability is obtained by agglomeration; decreased flowability is obtained by instantising – spraying with lecithin. This is due to the fact that lecithination gives the powder particles more free fat on the surface, thus resulting in a higher degree of 'sticking together'.

Packing and handling of instant powder is critical. The packing must protect against any pick-up of moisture and against physical breakdown of the agglomerates. In the case of whole milk powder, the packing must also protect against oxidation of the fat. As agglomeration is used to achieve good instant properties, it is essential to preserve the high level of agglomeration until the powder is reconstituted. Even a small amount of agglomerate breakdown results in the formation of fine particles, which decrease both wettability and dispersibility. Agglomerate breakdown in whole milk powder also exposes powder surfaces not coated with the wetting agent, and these will further reduce the wettability.

Baby Food Powder

The manufacture of baby food powder based on cow's milk has shown a steady increase since the beginning of this century. Baby food is nowadays a highly developed product, and the manufacture of infant foods is an important part of the dry milk industry.

The industry has been accused of advertising baby food powder as better than human breast milk. These accusations are unfounded. The manufacturers of baby food powder emphasise that breast milk is the best food for infants. But some women are unable to feed their children, and to these mothers, today's highly developed powdered baby foods are invaluable. It is reported from developing countries that infants have died from artificial milk. These deaths are not caused by the powder, but by contaminated water for reconstitution or poor bottle hygiene.

The incapacity of some mothers to feed their children is not something that belongs only to our time; it has always existed. Records of feeding infants by others go back as far as the seventeenth century. In those days, some children were breast-fed by a woman who was not their mother – the wet nurse. Wet nurses were very carefully selected, as there was a widespread belief that the milk of the nurse would transmit her character to the child. During the nineteenth century, artificial feeding with cow's milk became more accepted, but there was still widespread objection to the use of boiled milk. Whole milk diluted with water was used, and sometimes sugar was added.

At the beginning of the twentieth century, roller-dried milk was produced for infant feeding, and the death rate of artificially fed babies decreased dramatically – probably due not to the formulation of the roller-dried powder, but to its improved bacteriological quality. Roller-dried, full cream and half cream milk with minor additives were used successfully in many countries until about the middle of this century. One of the pioneers was the British company, Cow & Gate. Its products became known in many parts of the world, and company records describe the excitement caused by an urgent order in the 1930s from a famous maharajah for a consignment of baby food powder. The company's export manager visualised the tremendous potential in India, but was somewhat disappointed when it was later learned the powder was required for the maharajah's racing stables.

Since the middle of the twentieth century, emphasis has been placed on advice from paediatricians that baby food powder should approximate to the composition of breast milk, as nearly as is practicable. Medical authorities advise against feeding unmodified cow's milk to infants younger than 4 months; this has led to the development of humanised baby food powder, suitable for feeding from birth. From 4 months old, infants can be fed on cow's milk modified only slightly.

The total solids of breast milk and cow's milk are about the same, but there is a considerable difference in composition. It is therefore necessary to modify cow's milk to obtain a product similar to breast milk. But breast milk is not well defined; it varies widely in composition, particularly during the first 14 days after parturition. In practice, humanised baby foods are produced to match the average analysis of breast milk produced some 14 days after parturition – matured milk. This analysis should not be interpreted too rigidly because natural variations mean that many mothers produce milk 'out of specification'. Table XI shows the composition of average matured breast milk and cow's milk. When infants are fed with normal full cream, cow's milk instead of breast milk, they receive a higher protein intake with a different protein composition, different fatty acid composition, less carbohydrate and a higher mineral intake.

TABLE XI
Composition of breast milk and cow's milk

		Average matured breast milk per 100 ml	*Cow's milk per 100 ml*
Total protein	g	1·5	3·3
Protein composition:			
Whey proteins	%	70	20
Casein	%	30	80
Fat	g	3·3	3·5
Fat composition:			
Unsaturated fats	%	52	43
Saturated fats	%	48	57
Carbohydrate	g	7·0	4·7
Minerals	g	0·21	0·72
Energy	kcal	72	66
Protein calories	%	6	21

The high content of protein and minerals in cow's milk puts a high load on the infant's kidneys. It is essential that humanised baby foods do not have too high a protein content and have a low mineral level, as the kidney capacity of a newly born child is very limited; this becomes particularly critical when loss of water is increased by fever or diarrhoea. The simplest modification of cow's milk is the addition of carbohydrate to dilute the proteins and minerals.

When producing humanised baby food, a casein to whey protein ratio of 40:60 is desirable compared with the cow's milk ratio of casein to whey protein of 80:20. The correct protein ratio can be obtained by the use of whey protein concentrates with a high content of protein (WPC 80) and demineralised whey. The raw materials used must be of a high bacteriological standard. The fatty acid composition of breast milk is also very different from that of cow's milk. The correct fatty acid composition is obtained by the use of mixtures of butter oil and vegetable oils. It is especially important to have a minimum content of unsaturated fatty acids, such as linoleic acid and linolenic acid. The carbohydrate content – lactose in breast milk – is adjusted by means of demineralised whey or directly by lactose. The content of minerals and vitamins is finally adjusted by special mineral and vitamin mixtures.

The minerals calcium, magnesium and phosphorus are desirable in certain quantities. Potassium, sodium and chlorine should be avoided. The low mineral level can be achieved by using demineralised milk as well as demineralised whey. Breast milk is low in iron, but iron deficiency can be avoided by the addition of iron to baby foods. The only vitamins added to those naturally present in cow's milk are vitamin A, vitamin D and vitamin C.

Possible contaminants in cow's milk, such as trace metals, traces of pesticides, herbicides and antibiotics must be kept under strict control. It is not unusual to specially select the milk to be used for the manufacture of infant foods to ensure the absence of contaminants.

To summarise, modern techniques have made it possible to modify cow's milk to produce a baby food which has properties similar to breast milk, and is therefore safe to use from when the child is born. The only shortcoming of the manufactured baby food is that it does not possess the immunological properties of breast milk. It is well known that breast milk has anti-infective properties, and no method exists today to give manufactured baby foods these properties. Although manufactured baby foods give the infant all necessary nutrients, breast feeding should always be preferred when it is possible.

Great care must be taken in the manufacture of infant foods to safeguard the bacteriological quality of the finished product. It is not unusual to require a plate count as low as 50 organisms per gram of powder with pathogens absent. Dry mixing of ingredients into the powder after spray drying should be avoided, as it can lead to bacteriological contamination. Further, by dry mixing it is difficult to obtain a uniform distribution, and the added ingredient, e.g. lactose, may segregate out in the cans when they are subjected to vibrations.

Carbohydrate should be added to the milk before evaporation, or to the concentrate in the form of a 50 per cent solution, which must be pasteurised before it is mixed with the product. The fat soluble vitamins (vitamin A and vitamin D) should be added to the milk before evaporation, and the same applies to the vegetable fat and other possible ingredients. The mix should be homogenised, pasteurised at 100 °C with 1 min holding, and evaporated to 47–48 per cent total solids. The concentrate should have a temperature of 60–70 °C, and should be spray dried using an air inlet temperature of 180 °C.

The heat sensitive vitamins (vitamin C) should be added to the concentrate. It is necessary to overdose with vitamin C as a portion is destroyed during drying. However, it is more costly to dry mix the vitamin into the dried product, and it is difficult to obtain a uniform distribution. Further, by adding vitamin C before drying, the manufacturer can guarantee the end-user that the child receives the correct amount of vitamins, because the portion with the lowest heat resistance, which in the case of dry mixing may be lost when the powder is reconstituted in hot water, has already been lost in the dryer and substituted by overdosing the concentrate.

During recent years, baby food has also become instantised by agglomeration either in the spray tower, with or without fines return, or by re-wetting in fluid beds (Fig. 24). The advantage of instant powder is the feeding bottle needs to be shaken only a little to dissolve all the powder, and the risk of blocking the teat is eliminated.

Buttermilk

Buttermilk can be evaporated and spray dried, but there is a great difference between sweet buttermilk and sour buttermilk. Sweet cream buttermilk behaves like skim-milk, and can be evaporated and dried under the same conditions.

Sour cream buttermilk is difficult to deal with due to the high content of lactic acid. It can only be evaporated to 26–28 per cent solids, as the concentrate becomes very viscous. The flow in the evaporator should be backwards so that the end-concentration is reached in the effect where the temperature is highest; this reduces the viscosity. Spray drying is difficult due to the stickiness and hygroscopicity of the product. A low temperature should be used and, as deposits in the chamber cannot be avoided, frequent cleaning is necessary. It is useful to equip the drying chamber with a rotating air sweep which continuously blows off wall deposits. The air sweep consists of a tube that follows the vertical and conical shape of the chamber. It rotates at slow speed at a short distance

Fig. 24. Fluid bed agglomerator (courtesy APV Anhydro AS, Denmark).

from the interior surface of the chamber. High pressure air is supplied through nozzles attached to the tube and facing the chamber wall.

Whey

One of the problems associated with cheesemaking is the disposal of whey. For each tonne of cheese produced, an outlet has to be found for about 8 tonne of whey. Whey is also the by-product from the manufacture of casein and quark, but cheese whey is by far the most important in quantity. Over many years, the traditional methods for the disposal of whey have been to return it to the farmers for animal feeding or to dump it, either by spreading it on fields or by discharging it to natural water courses. These methods are no longer possible. The concentration of cheese manufacture in large factories has aggravated the whey problem, making it imperative to dispose of the whey at the point of manufacture.

The polluting effect of whey is illustrated by the fact that the biological oxygen demand (BOD) for the degradation of 1 l of whey is equal to the BOD of the effluent water generated during 24 h by one person. As the quantity of whey from a large cheese factory is in the order of 500,000 l per day (corresponding to cheese manufacture of about 60 tonne per day), it will be seen that an effluent plant for biological treatment of the whey corresponds to a plant for the biological treatment of the effluent water from a population of half a million. This clearly illustrates that biological degradation of the whey solids is out of the question; there is only one choice left – to recover the solids.

Utilisation of whey solids has attracted great attention. Robinson and Tamine (1978) describe a number of proposals for employing whey solids in the production of human food, and emphasise that investments should be spent on the recovery of the solids rather than on disposal plants for the whey. Recognition of the values of the undenatured whey proteins is fairly recent, as their recovery was first made possible by the introduction of ultrafiltration.

Today a considerable part of the whey is used for production of whey protein concentrates (WPC) by ultrafiltration with very specific properties. The permeate produced by ultrafiltration can either be concentrated and dried, or used for production of lactose. In lactose production there is a further by-product – mother liquor. It too can be concentrated and dried.

Evaporation and spray drying is also a generally accepted method for the disposal of large quantities of whey, and a whey drying plant is often an integral part of a cheese factory.

In the following, the various constituents of whey solids will be examined, and the conditions will be explained for the conversion of whey into an edible powder of high quality.

Whey is not a product of defined chemical composition. It varies considerably with the product from which it is derived, and with the manufacturing process. Different kinds of cheese yield different types of whey. The whey normally contains 6 per cent solids, i.e. half the total solids of the original milk. The most valuable constituents of the whey solids are the proteins. Whey proteins amount to 20 per cent of the total protein content of the milk. The whey contains almost all the water soluble vitamins and minerals of the original milk. Table XII shows a typical composition of cheese whey solids from different sources.

TABLE XII

Typical composition of cheese whey solids (per cent) from different cheese productions

	Denmark	USA	Switzerland	Netherlands	Poland	USA
Type	Danbo	Cheddar	Emmenthaler	Gouda	Quarg	Cottage cheese
Lactose	74·2	75·1	73·3	75·6	67·5	76·8
Protein	13·2	12·8	13·8	12·9	12·1	10·9
Mineral	10·1	8·8	8·5	7·9	12·5	8·2
Fat	0·7	0·8	0·8	0·9	0·8	1·8
Lactic acid	1·8	2·5	3·6	2·6	7·3	2·3

More than 70 per cent of the whey solids is lactose. The drying process and the properties of the whey powder are chiefly governed by the amount of lactose converted into crystals before drying. The best results are obtained when nearly 100 per cent of the lactose is present in the powder as crystals, rather than in the glassy or amorphous condition.

In skim-milk powder, 51 per cent is lactose, and the protein content is 36 per cent, which is nearly 2–6 times the protein content of whey solids. The high protein content of skim-milk solids acts as a carrier for the lactose, therefore the condition of the lactose is of no importance in relation to either the drying process or the properties of skim-milk powder. This is in spite of the fact that more than half of the powder is lactose.

The lactose in whey is present in two forms – α-lactose (40 per cent) and β-lactose (60 per cent). When whey is concentrated, the concentrate becomes supersaturated with lactose. In the supersaturated solution at temperatures below 93·5 °C, the α-lactose crystallises as α-monohydrate. When the α-lactose crystallises, the β-lactose is transformed into α-lactose by mutarotation in order to keep the balance.

The degree of supersaturation, at which the solids in the concentrate are maintained, increases with falling temperatures, whereas the crystallisation velocity and the mutarotation velocity decline with falling temperatures.

To obtain the best possible crystallisation process, these conditions must be optimised. In a typical crystallisation process, whey is concentrated to 55–62 per cent solids in an evaporator with a flow to prevent crystallisation in the evaporator. After the last effect in the evaporator, the whey is flash-cooled to 30–45 °C, dependent on the solids of the concentrate. Then the concentrate is pumped to a crystallisation tank (Fig. 25) and kept at the flash temperature until the tank is full. The concentrate is cooled to 12–18 °C at a speed of 1·5–3·0 °C h^{-1}. Lactose may remain in the solution in extremely supersaturated solutions, and another purpose of the flash-cooling is to develop nuclei, after which the crystallisation can continue. The reason why the crystallisation is carried out in two steps is that both crystallisation and mutarotation occur rather fast at 30–45 °C. At the same time, fresh whey concentrate is added to maintain the necessary supersaturation with lactose. When the tank is full, the temperature of the concentrate must be lowered to maintain the supersaturation of the solution. At lower temperatures, the crystallisation velocity and the mutarotation velocity decrease. It is important to control

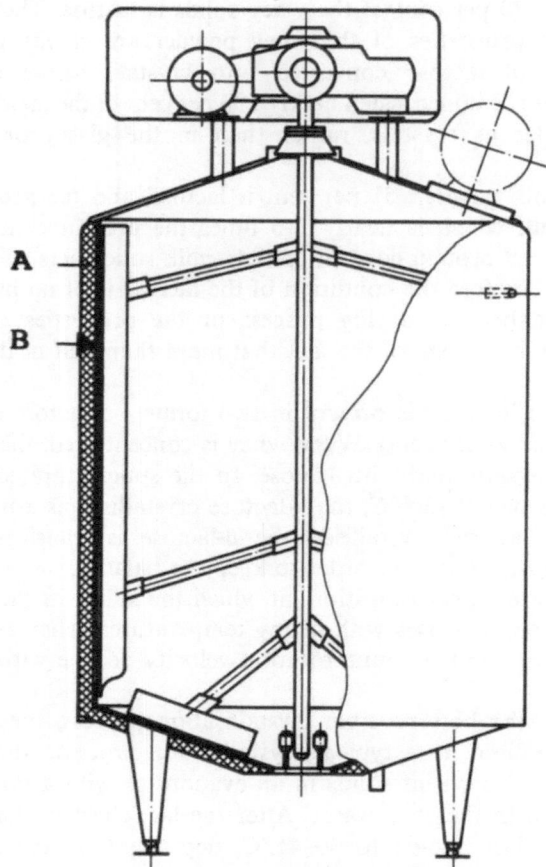

Fig. 25.   Crystallisation tank. (A) insulation (B) water jacket.

the cooling so that the crystallisation does not stop because supersaturation is too low or because crystallisation and mutarotation are
too slow. As further lactose crystallises and the temperature in the
concentrate drops, the viscosity of the concentrate will increase. The
crystallisation tank needs to be equipped with a strong agitator to
keep the crystals floating and ensure they are always surrounded by
supersaturated concentrate. Some concentrate tanks are equipped with
recirculation from bottom to top by means of a high capacity centrifugal pump with reduced efficiency to create still better agitation and
mixing.

The development of nuclei can also be promoted by adding small crystals of lactose to the supersaturated solution. This is known as seeding, and it provides the solution with sufficient small crystal surfaces so that crystallisation can continue. The seeding material is either finely-ground lactose or well-crystallised whey powder.

The proteins lactoglobulin and lactalbumin account for 90 per cent of the proteins in whey. Nutritionally they are superior to most other proteins in human and animal nutrition, due to their favourable amino-acid composition. They should be preserved in the undenatured form, i.e. the heat treatment of the whey must be gentle.

Whey powder of good quality implies the use of whey of good quality; this means there must be no further development of lactic acid once the whey is drawn from the cheese vat. The properties of whey powder depend on the content of lactic acid and thus on the cheese process from which the whey originates. It is therefore necessary to treat the whey in the following manner.

The whey is drawn from the cheese vat and led to a fines saver where the casein particles are removed. It is pasteurised, to stop bacteriological and enzymatic processes, then centrifuged to remove the fat. Thereafter the whey is cooled to below 10 °C if it is to be further processed within 2 h, otherwise it is cooled to 4–5 °C. The whey can then be led to the evaporator and be further processed to powder.

If lactic acid has developed in the whey, it can still be processed to concentrate and powder. But it will cause more fouling in the evaporator and more deposits in the spray tower.

The feed pump to the drier should be low speed, as high speed pumps quickly wear out. The atomiser disc should have interchangeable inserts, as shown in Fig. 26, to avoid wear of the disc. The inserts are made of aluminum oxide, and as they get eroded only at the side where the material leaves the disc, they can be used in different positions and finally replaced by new ones.

Certain drying conditions cause the semi-dried material to stick to the hot metal surface inside the dryer; the powder particles may also stick to

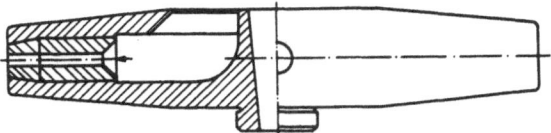

Fig. 26. Atomiser disc with inserts.

each other and form lumps. This phenomenon, known as thermoplasticity or sticking, was first explained by Pallansch (1972), who designed a so-called stickometer to determine the sticking temperature of whey powder under certain conditions. Pallansch found that the sticking temperature depends on the composition of the whey solids, and that it is a function of the content of amorphous lactose, lactic acid and moisture. Figure 27 shows the increase in sticking temperature with increased degree of lactose crystallisation; all other conditions constant. It shows that the sticking temperature of the powder is raised from 60 °C to 78 °C when the level of lactose crystallisation is increased from 45 to 80 per cent. Figure 28 shows the drop in sticking temperature as the content of lactic acid in the powder is increased, with all other conditions constant. The gradual increase in lactic acid up to 16 per cent in the powder linearly decreased the sticking point from 98 °C to 57 °C. A graph similar to Fig. 28 illustrates how sticking points decrease in direct proportion to increases in the moisture content of the powder. At 2·5 per cent moisture, the sticking point was 92 °C, but dropped to 70 °C at 4·5 per cent moisture in the powder.

To achieve trouble-free drying, the powder temperature should be below the sticking point. The lactic acid content can often be reduced by

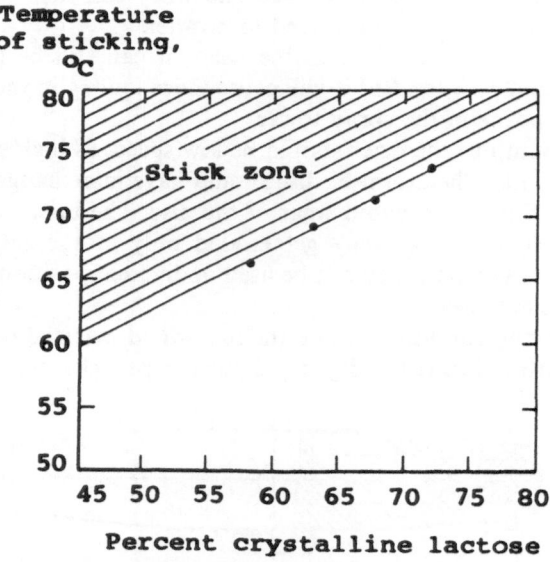

Fig. 27. Sticking temperature in relation to degree of lactose crystallisation.

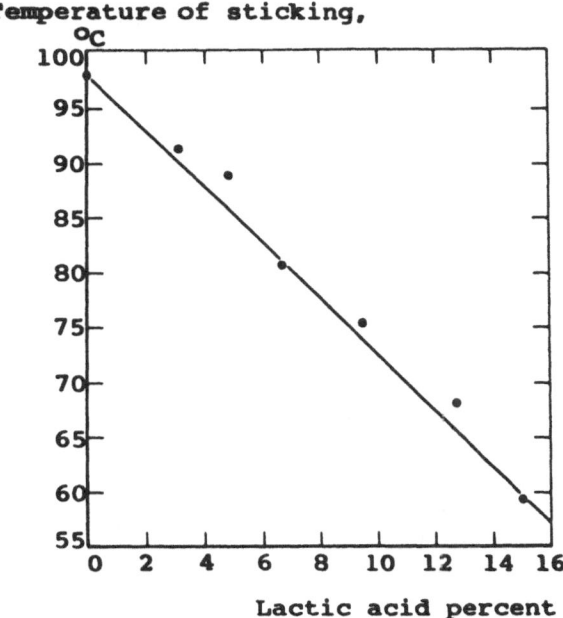

Temperature of sticking, °C

Lactic acid percent

Fig. 28. Effect of lactic acid on the sticking temperature.

changing the process after the cheese vats, i.e. by treating the whey in such a manner that the developed acidity is reduced to a minimum. If that does not raise the sticking point sufficiently, the degree of crystallisation should be increased, if necessary, by increasing the concentration. A decrease in moisture content increases both sticking point and product temperature; an increase in moisture content decreases both sticking point and product temperature. It is therefore difficult to obtain better conditions in the dryer by variation of the moisture content.

Powders that have a high content of amorphous lactose and a high content of lactic acid are particularly difficult to process and should be avoided.

The whey concentrate can be dried by two methods: the double-crystallisation method or the straight-through method. In the double-crystallisation method, the crystallisation takes place both before and after the drying. The more lactic acid the whey contains, the more necessary it is to have double crystallisation to get a non-hygroscopic powder. The powder is dried to 12 per cent moisture with an air inlet temperature of 150–160 °C and an outlet temperature of 50–60 °C. The powder drops from the base of the drying chamber onto a conveyor belt to give it a residence time of approximately 3 min. This intermediate holding of the product with high moisture content allows lactose crystallisation to continue in the damp powder mass. Further, any lactose anhydride is converted to the non-hygroscopic α-lactose hydrate by taking up one molecule of water. In spite of the low outlet temperature, the product temperature may be slightly above the sticking point, as the sticking temperature is low due to the powder's high moisture content. The effect of this is that the chamber wall gets coated with semi-dried powder, which drops off as the moisture content is reduced. Furthermore, the powder particles stick together to form agglomerates, which create a coarse powder structure.

The after-crystallisation process is completed in a few minutes, and the conveyor belt feeds the product into a secondary dryer for removal of the excess moisture. The secondary dryer can be a rotary drum dryer, i.e. a kiln through which hot air is blown, or it can be a fluid bed dryer. The conditions in the secondary dryer must be such that the chemically bound water in the lactose is not removed. In this way, a fully non-hygroscopic powder is obtained, as 95 per cent of the lactose is in the crystalline form. Furthermore, a coarse, free-flowing, non-dusty powder with excellent wetting properties is obtained. These powder properties are required for whey powder to be used in the food industry. The process is no longer used much as it is troublesome and unsanitary.

Today the whey concentrate is dried in a two-stage or three-stage plant with internal and/or external fluid beds. This implies that whey of a good quality is used and that a high degree of crystallisation in the concentrate has been obtained before drying. In these plants, higher drying temperatures can be used, and drying is finished in the chamber and the subsequent fluid bed(s). In particular, use of an integrated fluid bed has yielded powders of improved quality. The powder structure is like skim-milk powder. If a non-dusting, agglomerated product is required, fines from the cyclones can be led back to the atomiser zone, so it is also possible to adjust the bulk density of the powder. Powder produced in this way has properties just as good as or even better than powder produced according to the double-crystallisation method.

CONCLUDING REMARKS

Developments in milk drying since the Second World War have been very significant, and dry milk has achieved an importance similar to traditional dairy products.

Milk drying has developed into a highly sophisticated industry, technically as well as technologically. The manufacture of special powders, i.e. powders with specific properties, may be of increasing importance in the future, and will no doubt be more rewarding than the manufacture of standard skim-milk and whole milk powders.

Energy costs are high and, in future, the energy required to convert milk to powder needs to be reduced as much as possible. An obvious method would be to make powders more concentrated before drying, but means must then be found to counteract the adverse effect from the physical and chemical changes that occur in concentrates of high total solids.

The safety of the equipment needs further improvement, and the same applies to protection of the environment from effluent, noise and dust.

REFERENCES

ADMI, Chicago (1971). *Standards for Grades of Dry Milk including Methods of Analysis*, Bulletin 916.

Challis, B. C. *et al.* (1982). *Amine Nitration and Nitrosation by Gaseous Nitrogen Dioxide*, IARC Scientific Publications 41, pp. 11–20.

De Vilder, J., Martens, R. and Naudts, M. (1976). *Milchwissenschaft*, **31** (7), 396.

De Vilder, J., Martens, R. and Naudts, M. (1979). *Milchwissenschaft*, **34** (2), 78.

Hall, C. W. and Hedrick, T. I. (1966). *Drying Milk and Milk Products*, The AVI Publishing Co., Westport, Connecticut.

Harding, F. and Gregson, R. (1978). In: *Proceedings of the XXth International Dairy Congress*, pp. 356–7.

International Dairy Federation (1979). IDF Standard 87, Square Vergote 41, 1040 Bruxelles.

Leaver, G. H. (1984). *Gas Fired Low NO$_2$ Combustion Systems and Air Heaters for Direct Food Processing*, Urquhart Engineering Co., Greenford, Middlesex, UK.

Muir, D. D., Abbot, J. and Doyle, B. (1981). *Journal of the Society of Dairy Technology*, **34** (1), 15.

O'Sullivan, A. C. (1971). *Journal of the Society of Dairy Technology*, **24** (1), 45.

Pallansch, M. J. (1972). In: *Proceedings of the Whey Products Conference*, Dairy Products Laboratory, Agricultural Research Service, US Department of Agriculture, Washington, D.C. 20250, Eastern Region Research Laboratory Publications No. 3779.

Robinson, R. K. (1982). *Dairy Industries International*, December, pp. 19–23.

Robinson, R. K. and Tamine, A. Y. (1978). *Dairy Industries International*, March.

Rothery, G. B. (1968). *The Australian Journal of Dairy Technology*, **23**, 76.

Wright, N. C. (1933). *Journal of Dairy Research*, **4**, 122.

FURTHER READING

Beaver, P. (1984). *Journal of the Society of Dairy Technology*, **37** (2), 68.

Palmer, K. N. (1973). *Dust Explosions and Fires*, Powder Technology Series, Chapman & Hall, London.

Chapter 5

Protection against Fire and Explosion in Spray Dryers

O. Skov

APV Anhydro AS, Søborg-Copenhagen, Denmark

When spray drying milk products, there is a potential risk of fire and explosion. The greatest risk of fire in the spray drying plant is that an explosion of the powder floating in the air may be initiated, provided the correct conditions for an explosion are present.

This risk is increasing, as modern spray drying plants are loaded ever more heavily, i.e. high powder concentrations. The trend can also be seen from the casualty and accident statistics, where fire in particular, but also explosions are increasing. Insurance companies are increasingly taking an interest in the security of spray drying, and the milk industry is one example.

MILK POWDER AS A POWDER LAYER/DUST CLOUD

Milk powder is an organic material, therefore it is burnable. Milk powder in the spray drying plant is present in two physical states: finely distributed powder in the air in the form of a dust cloud, and deposits on the inner surface of the spray dryer. Both states constitute a danger; the condition of the spray drying plant may allow deposits to burn and the dust cloud to explode.

CONDITIONS NECESSARY FOR A DUST EXPLOSION

In order for a dust explosion to arise, three fundamental conditions must be fulfilled:

ignition source of a certain strength;

oxygen exceeding a certain concentration;
burnable dust at a concentration within the explosion limit.

IGNITION SOURCES

It is difficult to make a full specification of all potential ignition sources, but the most common ones are:

open fire;
hot surfaces and hot gases;
burning materials;
self-ignition;
hot work (welding or cutting operations);
mechanical friction (frictional heat and contact sparks);
electric sparks;
electrostatic discharge.

Self-ignition in deposits of milk powder in milk drying plants must be considered a dangerous and probable ignition source as regards fire and explosion, as deposits containing glowing material may, if the occasion arises, fall down and initiate an explosion in the lower part of the drying chamber (cone) where the dust concentration often considerably exceeds the lower explosion limit. Fatty products are particularly dangerous due to their relatively low self-ignition temperatures.

MINIMUM IGNITION TEMPERATURES

Ignition Temperature for a Dust Cloud

A cloud of milk powder requires a temperature of approximately 400–450°C to be ignited into an explosion – such high temperatures do not occur in a milk drying plant.

Ignition Temperature in a Powder Layer

The ignition temperature (glowing temperature) for deposits of skim-milk powder is approximately 300–350 °C.

Critical Self-ignition Temperature

The critical self-ignition temperature can be defined as the minimum ambient temperature at which self-ignition will occur in a given powder layer of a defined thickness and form.

Self-ignition is caused by oxidation reactions resulting in a temperature increase in the powder layer – exothermic processes. Increase in temperature normally increases the chemical reaction rate. This causes further temperature increases, and so on. When the generation of heat exceeds the thermal loss from the surface of the powder layer, the powder temperature will increase to ignition temperature. Many products, including milk powder, can behave in this manner.

Operation temperature, glowing temperature, layer thickness and form, generation and emission of heat are some of the important parameters for the development of self-ignition in an existing powder layer (deposit).

In general, conditions for self-ignition of deposits exist in milk drying plants, so it is important to prevent or to limit their formation, especially where the temperature is high, e.g. at the chamber ceiling where the hot air enters. Other parts with a high surface temperature are the atomiser and air distributor, so deposits there will also be critical. In deposits of fatty powder, oxidation can take place at approximately 80–90°C, and this corresponds to the outlet temperature. Table I gives the critical

TABLE I

Temperature	Critical layer thickness for self-ignition
100°C	170 mm
200°C	17 mm

thicknesses for self-ignition of skim-milk powder according to Beever (1985). Other ignition sources can be sufficiently strong to initiate an explosion. These are:

frictional heat that typically arises in bearings of rotating atomisers;
electric sparks that may occur by activating switches;
electrostatic discharge, normally low energy and often below 50 mJ – the approximate threshold for ignition of milk powder – but sometimes large enough to cause an explosion.

Although the discharge energy is normally low, there are variations in the actual electrostatic spark energy and in the actual lowest dust ignition energy. These are often so large that an explosion could be initiated. Electrostatic ignition should be taken very seriously.

PREVENTIVE MEASURES

Here is a list of preventive measures:

exclusion of ignition sources;
introduction of inert systems;
reduction of dust concentrations to below the explosion limit.

Figure 1 shows a three-stage spray drying plant with specification of the respective plant components in relation to hazard potential.

Earthing of Plant Components

The plant should be earthed with a copper cable to prevent charging with static electricity, and the individual plant components should be interconnected by means of cable to ensure effective earthing.

Inlet Temperature to the Drying Chamber

The controller feedback circuit of the inlet temperature control should be equipped with an alarm circuit that activates when the temperature exceeds a preset limit – perhaps 30 °C above the normal inlet temperature. At the same time, the heat intake should be shut down.

Fire-fighting in the Drying Chamber

The drying chamber should be equipped with fire-fighting equipment built into the chamber ceiling. This fire-fighting equipment can be activated through a separate alarm circuit with a temperature sensor in the duct between the chamber and the cyclones, and triggered if the outlet temperature exceeds a preset limit – perhaps 50 °C above the normal outlet temperature.

At the same time, the fire-fighting equipment in the external fluid bed (two- or three-stage plants) should be activated. In addition, the

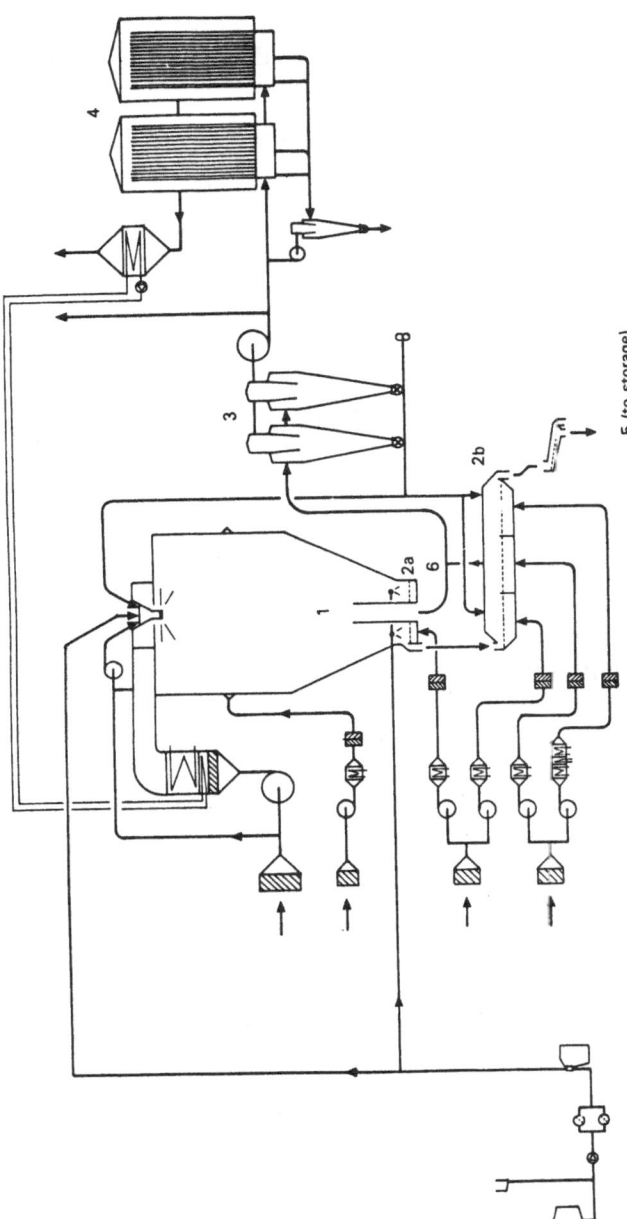

Fig. 1. Three-stage spray drying plant. Primary dust explosion potential in: (1) drying chamber (lower part) (2a, b) fluid beds (above sieve plate) (3) cyclones (4) bag filters (5) powder silos, dependant on degree of emptying (6) ducts carrying powder and air. (Courtesy APV Anhydro AS, Denmark.)

controller circuit of the outlet temperature should be equipped with an alarm that switches off the heat supply to the drying chamber.

Fire-fighting in the External Fluid Bed

As mentioned above, the external fluid bed should be equipped with fire-fighting equipment, either activated by the alarm in the duct between the chamber and the cyclones, or by a separate alarm circuit with a temperature sensor built into the duct between the fluid bed and the cyclones. A typical alarm limit could be up to 50 °C above the normal outlet temperature.

Fire-fighting in the Bag Filter

The filter should be equipped with fire-fighting equipment. The filter can be protected through two alarm circuits. One has a temperature sensor built into the duct before the filter. This automatically activates a flap-bypassing system to prevent the inlet air from entering the filter if the temperature in the duct to the filter exceeds a preset value approximately 30 °C above the normal temperature. The other alarm circuit has a temperature sensor built into the duct after the filter. This activates the fire-fighting system if the temperature after the filter exceeds a preset value approximately 50 °C above the normal temperature.

Centrifugal Atomiser

The atomiser should be equipped with vibration control and bearing temperature control. Vibration control normally consists of a vibration transmitter built into the atomiser and a control device built into the instrument panel. When vibrations exceed a first preset limit, an alarm is given; if a higher limit is exceeded, the atomiser should be stopped.

The control system for the bearing temperature (covering the two bearings) consists of two thermo-elements and two transmitters. The thermo-elements are built into the bearing housings. The limit temperature is normally around 105 °C. Stationary parts of the atomiser should be earthed.

Fans

Any fans that are not placed on the clean air side can be equipped with vibration control. This consists of a vibration transmitter built into the filter fan and a control device built into the instrument panel. When vibrations exceed a first preset limit, a signal lamp or another warning signal is generated; if a higher limit is exceeded, the plant has to be stopped. If the vibration transmitter malfunctions, the upper limit warning will be generated.

Any fans that are not placed on the clean air side, should be made resistant to pressure shock up to the same $P_{red,max}$ as the connected vessels.

PROTECTIVE MEASURES

Dimensioning for P_{max}

A plant can be built sufficiently strong to resist the maximum explosion pressure P_{max}, often 6–10 bar over-pressure. However, it will be unrealistically expensive in large plants because of the high strength demands.

Explosion Suppression

An explosion suppression system primarily consists of explosion detectors, a number of pressure bottles containing a suppression agent and a central control unit. It registers the explosion in its initial phase then immediately injects the suppression agent to reduce the explosion pressure $P_{red,max}$ to an acceptable level. P_a is the activation pressure of the suppression system, i.e. the pressure in the vessel at which the explosion is detected and the suppression system activated.

Explosion suppression is based on the fact that the destructive explosion pressure is not reached instantaneously, and that it will typically take 30–150 ms to obtain the maximum explosion pressure P_{max} in a vessel of a certain size. Figure 2 shows the pressure history in terms of time during non-suppressed and suppressed dust explosions. The characteristic points are:

the ignition occurs;
the explosion is detected, $P_{ex} = P_a$;

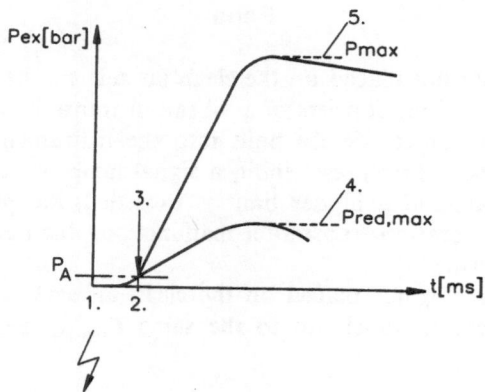

Fig. 2. Pressure history in terms of time at a non-suppressed and a suppressed dust explosion: (1) ignition occurs (2) explosion detected, $P_{ex} = P_a$ (3) suppression agent injected, $P_{ex} \geqslant P_a$ (4) explosion suppressed, $P_{ex} = P_{red,max}$ (5) without suppression, pressure would have been $P_{ex} = P_{max}$.

the suppression agent is injected, $P_{ex} > P_a$;
the explosion has been suppressed, $P_{ex} = P_{red,max}$.

Without suppression, the result would have been ($P_{ex} = P_{max}$), i.e. point 5. Explosion suppression may be appropriate when venting to a safe place is not possible without damaging the surroundings. For example, explosion suppression is particularly important in cases where poisonous compounds are formed during an explosion. The elements of an explosion suppression system are:

pressure detector;
suppression agent stored in steel bottles under pressure;
detonator;
nozzles;
central control unit.

The suppression agent often consists of a powder, e.g. ammonium phosphate, under a propellent pressure from CO_2 or N_2.

Explosion Pressure Relief

An explosion pressure relief device allows an unburned dust–air mixture and the combustion gases to be led out from the component where the explosion takes place. The maximum explosion pressure is thus reduced to a value the component can resist – called the reduced maximum explosion pressure $P_{red,max}$.

P_{stat} is the opening pressure of the pressure relief device, i.e. the pressure in the component that causes the pressure relief device to open. P_{stat} can be chosen by the design engineer.

$P_{red,max}$ has to be chosen in accordance with the strength of the component or the strength it is to be dimensioned for. It is essential to achieve a correctly dimensioned pressure relief area in relation to the actual P_{stat}. Figure 3 shows the pressure history of non-vented and vented dust explosions. Characteristic points are:

the ignition occurs;
the pressure relief device begins to open, $P_{ex} = P_{stat}$;
the explosion is vented, $P_{ex} = P_{red,max}$;
without venting, the explosion pressure would have been $P_{ex} = P_{max}$.

Pressure relief of dust explosions is the oldest and, until now, the usual kind of explosion protection. This is because it is the simplest, sturdiest and least expensive protective measure. But explosion suppression is gaining more and more ground, not least because suppressed explosions do not influence the surroundings.

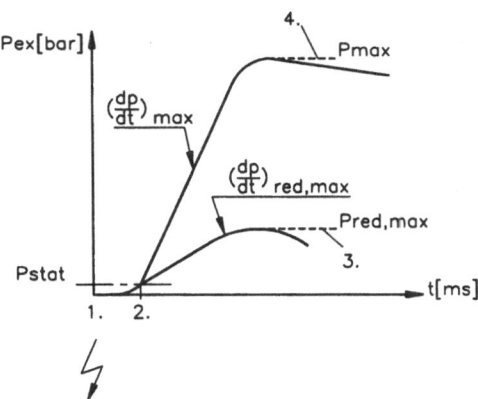

Fig. 3. Pressure history in terms of time at a non-vented dust explosion and a vented dust explosion. (1) ignition occurs (2) venting device begins to open, $P_{ex} = P_{stat}$ (3) explosion vented, $P_{ex} = P_{red,max}$ (4) without venting, pressure would have been $P_{ex} = P_{max}$; $\frac{dp}{dt}$ = velocity of pressure increase.

Primary explosion

A primary explosion is an explosion that takes place in the process plant itself.

Secondary explosion

A secondary explosion is a subsequent ignition of the unburned dust coming out through the pressure relief system. The ignition source is the flames from the primary explosion, and the force of the secondary explosion can be increased considerably if burnable dust is lying around the process plant.

As the secondary explosion may contain very long flames and cause a considerable pressure wave, it can be rather dangerous. This is the reason why most countries have laid down standards and guidelines for indoor plants and demanded the use of vent ducts. Vent ducts are connected to the pressure relief device and lead into the atmosphere at a safe place.

Vent ducts

When using vent ducts, an increase of $P_{red.max}$ may normally be expected, and the specifications of most codes of practice are modified accordingly. The pressure increase is one of the effects explained by the secondary explosion, which normally occurs in the vent ducts.

Pressure relief devices

Rupture discs

A rupture disc is designed to burst at a preset pressure P_{stat} of approximately 0·1 bar over-pressure – the most common value. The total area is released almost instantaneously; a rupture disc cannot be reused after an explosion.

Pop-out panels in a rubber clamping profile

The pop-out panel is situated on the circumference of a rubber clamping profile. It is equipped with a wire or some hinges so it is not ejected far during an explosion. A hinged pop-out panel can often be reused after an explosion.

Explosion doors

Explosion doors are available in different designs; they can be round, square or rectangular. An explosion door consists of a hinged cover with a closing device and a seal between the cover and the vessel to which it is mounted (Fig. 4). The door is normally equipped with a buffer device to catch the cover when it is thrown out during the explosion. Explosion

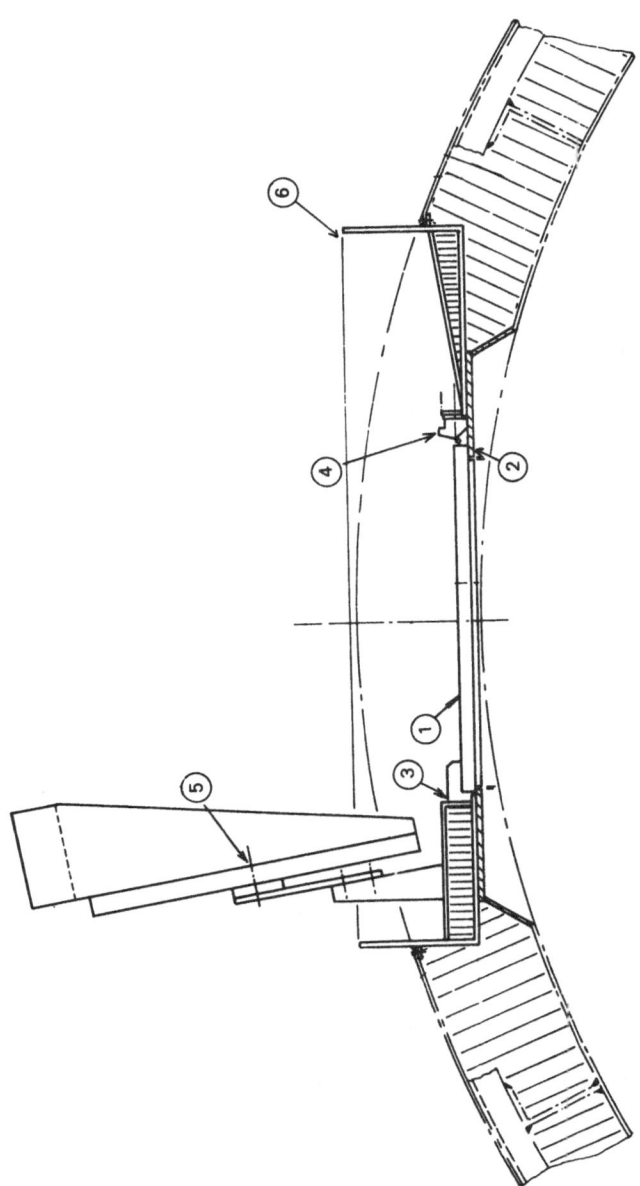

Fig. 4. Explosion door to be placed on the drying chamber wall: (1) cover with internal insulation (2) sealing (3) hinges (4) locking pawl with adjustable opening pressure (0·05–0·1 bar) (5) buffer device with pneumatic damper (6) insulation insert, vent duct can be mounted. (Courtesy APV Anhydro AS, Denmark.)

doors are normally designed to be reused. For less violent explosions and low $P_{red,max}$, lightweight doors are often used, whereas doors for violent explosions and large $P_{red,max}$ are made to heavier specifications.

Efficiency and proper function of explosion doors

The efficiency and proper function of the doors are normally determined by an explosion test (Fig. 5). This test and type approval are required to

Fig. 5. Explosion door being tested in a 25 m³ test vessel at Ciba Geigy, Basel (courtesy APV Anhydro AS, Denmark).

meet VDI 3673 (1979). Door efficiencies vary. Efficiency normally means the size of the effective area of the door compared to its nominal area expressed as a percentage. Efficiencies are normally less than 100 per cent, due to the inertia of the cover, so that only a nominal area of the door is released over a period of time. This time is normally considerably longer than the time for a rupture disc to burst. In addition, the door quality has to be such that no fractions break away during an explosion.

The following parameters are of great importance to the efficiency of the doors:

door shape;
cover weight;
$P_{red,max}$;

$$\left(\frac{dp}{dt}\right)_{red,max} - \text{velocity of pressure increase.}$$

The efficiency of the doors can be precalculated in order to obtain the best possible design before the final explosion test (Siwek and Skov, 1989).

As opposed to rupture discs, explosion doors can also be designed so that relatively long vent ducts (up to 8 m) can be placed after an explosion door without any pressure increase; the unburnt dust is dosed in the vent duct so that the entire cross-section of the duct is not filled up with combustible material.

Explosion reaction force

During pressure relief, a reaction force is generated in the opposite direction to the outflowing combustion products.

Previously, some thorough examinations of its size have been made, i.e. by weigh-cell measurements.

Calculation of pressure relief area

There are various methods for calculation of the necessary pressure relief area; some of them out of date and inaccurate. The best-known methods are:

VDI 3673 (1979)	(Cubic law method);
NFPA 68 (1988)	(Cubic law method);
Vent ratio method	(Palmer, 1973);*
K-factor method	(Simmonds and Cubbage, 1960);*
Rust method	(Rust, 1979).*

*Cited in Schofield (1984).

VDI 3673 (1979). One of the most acknowledged methods is the calculation according to VDI 3673. It is mainly due to the fact that this method is the most scientific one. To be able to determine the pressure relief area, A (m^2), the following parameters have to be known:

active vessel volume;
opening pressure P_{stat} (bar);
reduced maximum explosion pressure $P_{red,max}$ (bar);
dust class St. or K_{st}-value (bar m s^{-1});
$P_{max} = 10$ bar.

The pressure relief area is found by means of the nomograms in VDI 3673. Recent examinations have shown the size of P_{max} is very important when calculating the necessary pressure relief area. Many organic dusts have a $P_{max} \leqslant 9$ bar, which is also often valid for milk powder.

Vent ratio method (1973). The vent ratio method was relatively common in the UK and the USA, but has increasingly been replaced by the cubic law method.

$$\text{Vent ratio} = \frac{\text{pressure relief area}}{\text{vessel volume}} \ (\text{m}^{-1})$$

The method is based on a maximum velocity of pressure increase $(\frac{dp}{dt})_{max}$ measured in Hartmann's apparatus.

Active volume of drying chambers for drying plants within the milk industry. The active volume for drying chambers is to be understood as the size of the explosive dust cloud that may be expected. This implies that the dust concentration is higher than the lower explosion limit. The active volume creates the basis for calculation of the pressure relief area.

It has been a tradition to calculate the size of the active volume to be one third of the total chamber volume, corresponding to the cone volume, i.e. the dust concentration in the cylindrical part of the drying chamber has been considered to be under the lower explosion limit. The dust in flat-bottomed drying chambers has thus been considered to be under, or about on, the lower explosion limit.

Discussions about the size of active volumes for drying chambers are still taking place. In the UK, the whole drying chamber volume has to be calculated as active volume according to Anon (1987) and Schofield (1984). However, according to IDF (1987), Anon (1990) and Zockoll

(1985), active volume for a drying chamber may still be calculated to be approximately one third of the total chamber volume.

For drying chambers with internal fluid beds, the size of the active volume is uncertain because there is a larger quantity of powder in the chamber.

Process Isolation

Vessels interconnected through ducts
In vented vessels interconnected through a duct, an explosion initiated in one vessel will propagate through the duct and generate an explosion in the other vessel. The total explosion violence will be considerably larger than for an individual vessel without any duct connection. Longer ducts also tend to produce more violent explosions. When using process isolation, also called explosion limitation, the following are normally possible:

> physical blocking of the duct using a quick-closing valve;
> explosion suppression in the duct;
> the use of rotary valves as flame arresters.

Quick-closing slide valve
At an explosion, the slide valve is normally activated by a signal from a flame indicator installed close to the expected source of ignition at the end of the duct; with short ducts, a pressure detector installed on the vessel can be used.

'Ventex' protection valve
The valve can be explosion pressure or detector controlled, and is a type of stop valve with a conical valve seat.

Ventex valve (explosion pressure controlled)
At an explosion, the valve is closed automatically by the explosion pressure.

Ventex valve (detector controlled)
At an explosion, the detector will ensure closing of the valve by means of compressed air; the detector can be of the same kind as for sliding valves.

Explosion suppression in duct

In principle, the suppression system is the same type as used for suppression in vessels. However, the system is normally activated by a flame indicator installed close to the source of ignition at the end of the duct, or with short ducts (not long enough for use of a flame indicator) activation can take place by means of a pressure detector installed on the vessel.

Suppression in a duct is especially suitable for ducts with large diameters where quick-closing slide valves cannot be used.

Rotary valves as flame arresters

A rotary valve is acceptable as a flame arrester when fulfilling these conditions.

(1) The valve has to resist a defined reduced explosion pressure $P_{red,max}$.
(2) There have to be at least two cell divisions in mesh along the stator wall all the time.
(3) The maximum gap between stator and rotor has to be 0·2 mm.
(4) The rotary valve still has to operate after a dust explosion.

PERSONNEL AND GENERAL

As regards fire-fighting and how to act in case of fire, information should be available for all operators. Frequent exercises should be held on the prevention and fighting of fire, so that the operators are familiar with the handling and testing of the fire-fighting equipment.

During operation, an operator should always be present in the control room, and it should always be possible to get in contact with the operator should an emergency occur.

Inspections around the plant should be made frequently, and special attention should be paid to powder deposits in areas with hot surfaces, as well as to any tendency to blocking.

Instructions should be available concerning conditions for maintenance and repair, and certain activities should only be carried out with the permission of the factory management. Hot work, e. g. welding, must only be carried out under satisfactory conditions; never while the plant is operating. Work areas should be kept free of dust, so should the internal parts of the component.

Fire prevention and fire-fighting during powder production should be included in training programmes for operators. Operators should obtain a basic knowledge of self-ignition and dust explosions, as well as the relationship between powder properties and these phenomena.

Smoking should not be allowed within the spray drying, powder storage or packing areas. Notices on display should be an adequate measure to create non-smoking zones.

When cleaning the plant, the supplier's instructions have to be strictly observed. This also applies to instructions for general maintenance of the plant.

NATIONAL STANDARDS AND GUIDELINES FOR EXPLOSION PROTECTION

There are several standards and guidelines, and these may differ over details and restrictiveness. Not all countries have their own national standards or guidelines to cover explosion protection.

Recently guidelines (codes of practice), have been laid down for spray drying plants in the dairy industry: IDF (1987), Anon (1987) and Anon (1990). Some literature about explosion protection of spray drying plants producing milk powder has been published: Beever (1985), Beever and Crowhurst (1989). Furthermore, a book about protection of spray drying plants has been published in a more general form (Abbott, 1990).

Other national standards and guidelines are more general; they do not specifically treat spray drying plants. For pressure relief, reference can be made to the German VDI 3673 (1979) and the American NFPA 68 (1988).

In the EC, existing national standards will be gradually replaced by common European standards concurrently under preparation. A European standard for explosion protection is also being prepared, but it is a more general standard. It will probably take some time before detailed European standards for explosion protection have been prepared.

REFERENCES

Abbott, J. A. (Ed.) (1990). *Prevention of Fires and Explosions in Dryers – a User Guide.*

Anon (1987). *Prevention of Fire and Explosion in Spray Drying Plant. A Code of Practice for Designers, Manufacturers, Suppliers and Users,* Association of British Preserved Milk Manufacturers (PMA).

Anon (1990). *Code of Practice for the Prevention, Detection and Control of Fire and Explosion in New Zealand Dairy Industry Spray Drying Plant.*

Beever, P. F. (1985). *Journal of Food Technology*, **20**, 637.

Beever, P. and Crowhurst, D. (1989). *Journal of the Society of Dairy Technology*, **42** (3).

IDF (1987). *IDF Bulletin*, **219**.

NFPA 68 (1988). 'Guide for Venting of Deflagrations', edition approved by the American National Standards Institute.

Schofield, C. (1984). *Guide to Dust Explosion Prevention and Protection, Part 1 – Venting*, Institution of Chemical Engineers.

Siwek, R. and Skov, O. (198). Modellberechnung zur Dimensionierung von Explosionsklappen auf der Basis von praxisnahen Explosionsversuchen. VDI-Berichte 701, VDI-Verlag, Düsseldorf pp. 529–616.

VDI 3673 (1979). 'Druckentlastung von Staubexplosionen'.

Zockoll, C. (1985). In: *Proceedings of Praktischer Explosionsschutz an Sprühtrockner in der Milchindustrie*, Vortrag Stuvex Symposium, Frankfurt, 1985.

FURTHER READING

Bartknecht, W. (1981). *Explosionen, Ablauf und Schutzmassnahmen*, Springer-Verlag.

Bartknecht, W. (1987). *Staubexplosionen. Ablauf und Schutzmassnahmen*, Springer-Verlag.

NFPA 69. 'Explosion Prevention Systems', edition approved by the American National Standards Institute.

Schofield, C. and Abbott, J. A. (1988). *Guide to Dust Explosion Prevention, Part 2 – Ignition Prevention, Containment, Inerting, Suppression and Isolation*, Institution of Chemical Engineers.

Chapter 6

Membrane Processing of Milk

A. S. Grandison
Department of Food Science and Technology, University of Reading, UK

F. A. Glover
*Formerly of National Institute for Research in Dairying,
Shinfield, Reading, UK*

Membrane processes such as microfiltration, ultrafiltration, reverse osmosis and electrodialysis can be used to modify the composition of milk. In the past 25 years membrane technology has advanced sufficiently to permit operation on an industrial scale.

Microfiltration (MF) and ultrafiltration (UF) are fractionation processes. They separate molecular species on the basis of their size, using membranes that retain larger molecules and allow the passage of water and smaller molecules. The exact boundary between MF and UF is not clear-cut. In terms of milk processing, UF membranes allow the passage of water, lactose and soluble salts, but retain protein and fat; whereas the larger pore size of MF membranes allows the passage of some of the protein and lipid fractions. Reverse osmosis (RO) employs much tighter membranes than UF and MF and allows the passage of water only; RO is, therefore, a concentration process. More recently the term 'nanofiltration' has been introduced. This is a separation somewhere between UF and RO, which could, for example, permit the separation of water and monovalent ions and salts from lactose and the larger molecular components of milk. Practical applications of nanofiltration of milk have not yet been reported, and the process will not be considered further in this chapter. The above processes are pressure driven. Electrodialysis (ED) is a membrane process, but the essential driving force is potential difference; molecules pass through the membranes according to their charge as well

273

as their molecular size. ED is therefore employed for deionisation or ion replacement.

For the dairy industry, the potentialities of UF are greater than RO. MF and ED have fewer potential applications. The majority of this chapter will concentrate on UF and RO, with shorter sections on MF and ED.

Lewis (1982) has published a comprehensive article on the concentration of proteins by UF; Glover (1984) has reviewed the use of RO and UF in the dairy industry.

PROPERTIES OF MILK RELEVANT TO MEMBRANE PROCESSING

For an understanding of the application of membrane processing to milk, it is necessary to note some of its physical properties. The course of ultrafiltration is governed by molecular size, dimensions being more relevant than the more commonly known molecular weights. Table I

TABLE I
Molecular sizes of milk components

Component	Molecular weight	Diameter[a] (nm)
Water	18	0·3
Chloride ion	35	0·4
Calcium ion	40	0·4
Lactose	342	0·8
α-lactalbumin	14 500	3
β-lactoglobulin	36 000	4
Blood serum albumin	69 000	5
Casein micelles	10^7–10^9	25–130
Fat globules	–	2 000–10 000

[a]Taking the molecules as approximately spherical.
With acknowledgement to Kessler (1981).

gives some important weights and diameters, and shows the large step in size between lactose and α-lactalbumin. The pore sizes of UF membranes fall in this gap, which demonstrates how UF can be so conveniently applied to milk for fractionation purposes. Osmotic pressure is important in reverse osmosis but does not concern ultrafiltration, since the small molecules which give rise to the osmotic pressure pass through the UF membranes. Milk has an osmotic pressure of about 0·7 MPa due mainly to lactose and salts, and since RO concentrates all the components, osmotic pressure increases directly with the concentration factor. Viscosity is significant in UF. Milk has a viscosity of about 1 mPa s at 50 °C, due largely to the protein. As concentration proceeds the viscosity increases, at first slowly, and then very rapidly in the region of 14–18 per cent protein, so that the viscosity can reach 200 mPa s even at 50 °C at which a UF plant is commonly operated. This causes high pressure drops within a plant, reduces the rate of flow, reduces the rate of diffusion of molecules in the concentrate and causes heating due to the extra input of pumping energy. Viscosity thus imposes a limit on the degree of concentration, namely about six-fold, when the protein concentration is in the region of 18 per cent.

THE PRINCIPLES OF ULTRAFILTRATION

UF is a pure sieving process, through a membrane having pores in the range 5–35 nm. The driving force is an applied pressure, generally about 500 kPa (5 bar), and the principle of operation is known as cross-flow filtration, since the feed flows over the membrane parallel to its surface at a velocity in the region of 5 m s^{-1}. The performance of the membrane is described by two quantities; the retention R of molecular components in the feed, and the rate of filtration known as the permeate flux J. For any particular molecule

$$R = \frac{(C_f - C_p)}{C_f} 100$$

where C_f = concentration of the molecule in the feed
C_p = concentration of the molecule in the permeate.

For very large molecules $R = 100$ per cent, for very small molecules $R = 0$, while intermediate sized molecules have values of R between 0 and 100 per cent, since membranes have a pore size distribution and hence diffuse cut-off levels. Permeate flux is expressed in units of litres per square metre of membrane per hour $(1 \text{ m}^{-2} \text{ h}^{-1})$, and it is mainly by this quantity that any plant is judged. In practice, fluxes from whole milk start at about $60 \text{ l m}^{-2} \text{ h}^{-1}$ and then fall as concentration increases.

Transport through the membrane is by viscous flow, but in reality, the membrane is not the only barrier. As permeate is extracted from the feed, solids are left behind at the membrane surface, causing a local increase in concentration. The concentration of solids therefore increases in the direction looking towards the membrane. It is the phenomenon known as concentration polarisation (Fig. 1), which is all important in both UF and RO, since it takes over control of the processes, setting up a barrier to filtration and reducing flux. At a constant concentration of the feed, an equilibrium is set up between the collection of solids near the membrane and the diffusion of those molecules back into the main stream assisted by the cross flow. All systems of plant operation aim to minimise concentration polarisation, either by turbulence or high rates of shear in the flow channels between membranes. As filtration proceeds, the concentration of solids in the feed increases, and concentration polarisation becomes more serious, until finally solid deposits form on the membrane and in its pores. The membrane has become fouled, and by this time, the flux has fallen very low.

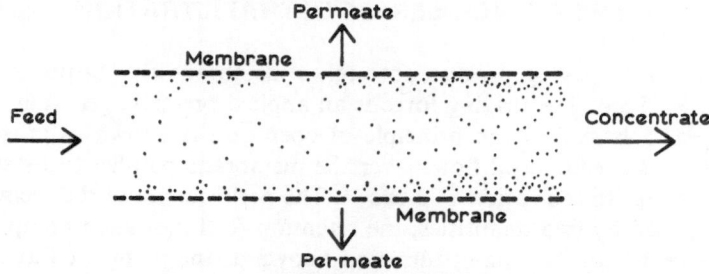

Fig. 1. Concentration polarisation.

The following expression illustrates the factors which determine the rate of filtration.

$$J = \frac{A\Delta p}{32v} \frac{1}{R_m + R_d}$$

where J = permeate flux
        A = area of membrane
        Δp = pressure difference across the membrane
        v = kinematic viscosity of the permeate
        R_m = resistance of the membrane
        R_d = resistance of the layer of solids at the membrane, the gel layer, due to concentration polarisation and fouling.

Since equilibrium is reached through diffusion of the solids back into the feed, the permeate flux can also be described in terms of

$$J = \frac{D}{d} \ln \left(\frac{C_g}{C_f} \right)$$

where D = diffusion coefficient of solids in the feed
C_f = concentration of solids in the feed
C_g = concentration of solids in the gel layer
d = thickness of the gel layer.

This equation is useful in understanding the limit to which concentration can be taken during an ultrafiltration process. When the concentration of solids in the feed reaches the concentration in the gel layer, $C_f = C_g$, $\ln(C_g/C_f) = 0$ and the flux becomes zero.

Yield of Components

The yield of any component is the amount retained in the concentrate as a fraction of that in the initial feed. If

y = yield

f = concentration factor

R = retention coefficient, $\dfrac{C_f - C_p}{C_f}$

In batch operation

$$y = f^{(R-1)}$$

And if

C_0 = initial concentration
C_1 = final concentration
$C_1 = C_0 f y;$

Then

$$C_1 = C_0 f^R$$

This calculation applies only if R remains constant throughout ultrafiltration, which is rarely the case in practice due to concentration polarisation and membrane fouling. An average retention must then be taken, putting

$$R = \frac{\log \left(\dfrac{C_1}{C_0} \right)}{\log f}$$

Milk is a multicomponent system, and this simple analysis does not apply directly to all components. The proteins and fats are completely retained by the membrane and form a considerable proportion of the concentrate, other components are partly retained and others pass freely through the membrane. The reduction in volume during ultrafiltration is from the water phase only. Hence there is a greater loss of water phase components than the concentration factor indicates, resulting in lower concentrations of these components when expressed in the usual analytical terms as fractions of the whole. As ultrafiltration proceeds, the concentration of a component with zero retention will appear to fall. However, if calculated in terms of the water phase only, the concentration will remain constant, consistent with the behaviour of a component having zero retention.

Diafiltration

Diafiltration is a modification of ultrafiltration in which water is added to the feed as ultrafiltration proceeds in order to wash out feed components able to pass through the membrane. With the continued removal of more permeate, more of the small molecules are removed from the concentrate. Diafiltration may thus be regarded as a purification process. Together with ultrafiltration, it is a way of producing milk enriched in protein and fat, and very low in lactose and salts.

The most efficient diafiltration is carried out by adding water at the same rate as permeate is removed. The volume of water to be added to achieve a required level of purity is calculated as follows, taking lactose as the component to be removed.

$$U = \frac{m_w}{1 - R} \ln\left(\frac{l_1}{l_2}\right)$$

where U = volume of water to be added
        R = retention factor for lactose, in practice in the region of 0·1
        m_w = mass of water in the feed before diafiltration
        l_1 = mass of lactose in the feed before diafiltration
        l_2 = mass of lactose in the feed required after diafiltration.

In terms of processing time, the most efficient method for milk is to ultrafilter first to a two-fold concentration, then to start the diafiltration, finishing off with more ultrafiltration if further concentration is required.

MEMBRANES FOR ULTRAFILTRATION

Membrane suppliers can provide a range of membranes with different molecular weight cut-off values, normally in the range 500–300,000 for UF. However, due to factors such as variation in the shape of molecules, or interactions between components of the feed, the value should be considered to be no more than a guideline.

The first successful membranes on a large scale were made of cellulose acetate. Treatment of cellulose with acetic acid, such that an average of 2·5 out of every 3 of the OH groups on the cellulose is replaced with acetyl, CH_3COO groups, produces a material with good membrane forming properties, namely an ability to form films, able to take up water and having a high wet strength. A thin layer of this cellulose acetate is prepared, then one surface is subjected to a controlled time–temperature treatment which produces a thin, tight skin on a much thicker and more porous layer. The tight skin, $0·1–1\,\mu m$ thick having pores 2–20 nm diameter, is the effective ultrafilter, the thicker layer $100\,\mu m$ thick serves as a support.

However, cellulose acetate membranes had severe limitations, particularly for dairy processing. Because cellulose acetate is subject to hydrolysis, its use was confined to a pH range of 3–7, and it would not tolerate temperatures above 35 °C. This placed severe restrictions on cleaning operations so vital in dairying. Polyamide membranes with improved tolerance to temperature and pH were used for some time, but more successful, and now widely used in practice, are the polysulphone membranes. The basic chemistry of this is the diphenylsulphone repeating unit $(C_6H_5)_2\,SO_2$

It forms a membrane with high mechanical strength, capable of withstanding temperatures up to 80 °C and a pH range of 2–12. The structure of a polysulphone membrane is shown in Fig. 2, which is an electron micrograph of the cross section.

The most recent innovation in membrane materials is the development of inorganic membranes – the so-called third generation of membranes. A

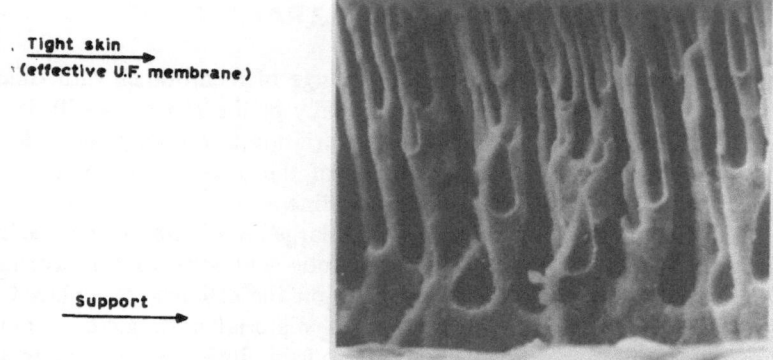

Tight skin
(effective U.F. membrane)

Support

Fig. 2. Ultrafiltration membrane structure. Electron micrograph cross-section of polysulphone membrane magnification × 400. (Micrograph by courtesy of Dr R. F. Madsen, Danish Sugar Corporation.)

range of inorganic materials including compounds of aluminium, zirconium and titanium have been developed for both UF and MF. Some geometries vary radically from conventional membrane designs.

(i) Zirconium oxide membranes supported on a macroporous graphite backing, in the form of narrow tubes (Fig. 3a).
(ii) Multichannel macroporous ceramic supports; the alumina membranes are coated on the inner surface of the channels (Fig. 3b).
(iii) Alumina membranes coated onto a metallic mesh support.

Since inorganic membranes are composed of mineral materials, they are extremely resistant to chemical and abrasive degradation and can

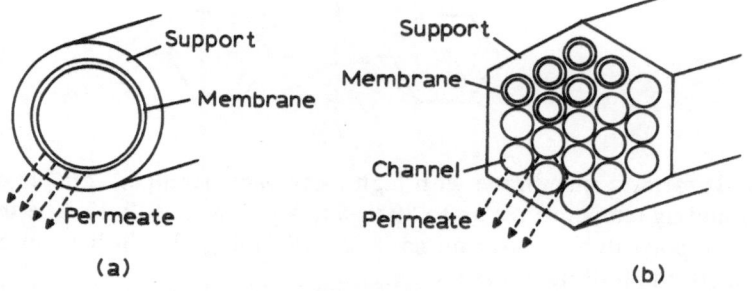

(a)   (b)

Fig. 3. Schematic cross-sections of (a) tube-shaped and (b) multichannel inorganic membranes. (Reproduced with permission from Rios *et al.*, 1989.)

withstand wide ranges of temperature (e.g. up to 400 °C.) and pH (0–14). However, they are expensive compared to traditional organic membranes, and are not widely used for this reason.

Membrane Geometry

Membranes are tubular, flat or spirally wound. The essential difference between these arrangements is the space between adjacent membranes, termed the flow channel, which ranges from 25 mm in the tubes down to about 0·5 mm for the flat and spirally wound types. Membranes are assembled in modules, which are units in the engineering sense used in the building of plants. A module will contain up to 30 m² area of membrane. Wide channel systems rely on high degrees of turbulence to minimise concentration polarisation: the narrow channels achieve this by having shorter lengths, high rates of shear and mixing in the wider sections of the assemblies and the entrances and exits of the channels. The advantages of the tubular systems are that they are easily cleaned and are tolerant of particulate matter, whilst the flat assemblies have the advantage of a low hold-up volume per unit of membrane area, plant occupies a smaller space per unit of membrane area and membrane faults can be detected and replacements made within smaller membrane areas. Spirally wound systems may be classed with the flat assemblies. In energy consumption, the flat and spirally wound systems have the advantage over the tubular arrangements. Another version of the tubular system has membrane tubes less than 1 mm in diameter, known as hollow fibres rather than tubes. It is less common in large-scale plants than the other types.

Membrane Modules

Tubular

The example of the tubular system is the design of Paterson Candy (PCI) (Fig. 4). Membranes are cast onto the internal surface of a synthetic fibre tube, 12·5 mm internal diameter in lengths up to 3·6 m. The fibre tubes are inserted into supporting, perforated stainless steel tubes, and assembled into bundles of 18 tubes fitted into end-caps which direct the flow into two groups, each of 9 tubes. Within each group, the tubes are connected in series, the two groups are in parallel. One module (3·6 m long) has a membrane area of 2·6 m². The feed flows along the inside of the tubes and permeate passes radially through

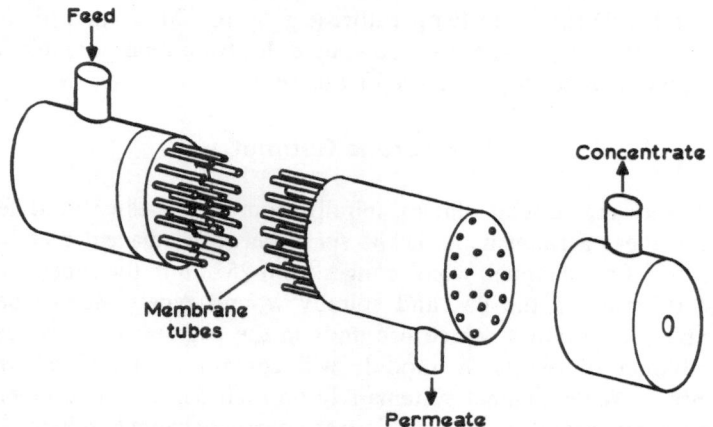

Fig. 4. PCI ultrafiltration module, tubular membranes. Membranes 12·5 mm diameter, 3·6 m long; each module has 18 tubes, total area 2·5 m². Module outside diameter 10 cm. (Illustration from PCI technical literature.)

the fibre backing, through the perforations in the stainless steel tubes, and is collected in stainless steel surrounding shrouds. In the event of a membrane failure, it is not possible to detect an individual faulty tube due to their close assembly inside the shroud. All 18 tubes, 2·5 m² area of membrane must be replaced.

Flat sheet

The modules made by the Danish Sugar Corporation, De Danske Sukkerfabrikker (DDS) are examples of the flat sheet arrangement (Fig. 5). The membranes are in oval sheets, 42 cm × 31 cm, area 0·1 m², clamped in pairs between plates which have a pattern of curved ribs, so that when two plates with membranes in between are pressed together the ribs touch, forming the flow channels in between. The depth of the channels is 0·7 mm, and the flow along the channels is in parallel. Plates are 0·5 cm thick, made of polysulphone and have a cavity inside to receive the permeate through fine slits cut in the surface. Permeate is led out through fine steel tubes (1 mm internal diameter) from the edge of the plates. Many membranes and plates are clamped together in a horizontal frame to form a module containing 27 m² area of membrane.

Recently DDS has improved its modules considerably (Kristensen *et al.*, 1981). By changes to the inlet and outlet ports and pipework, a 50 per cent reduction in energy consumption has been achieved. Even more

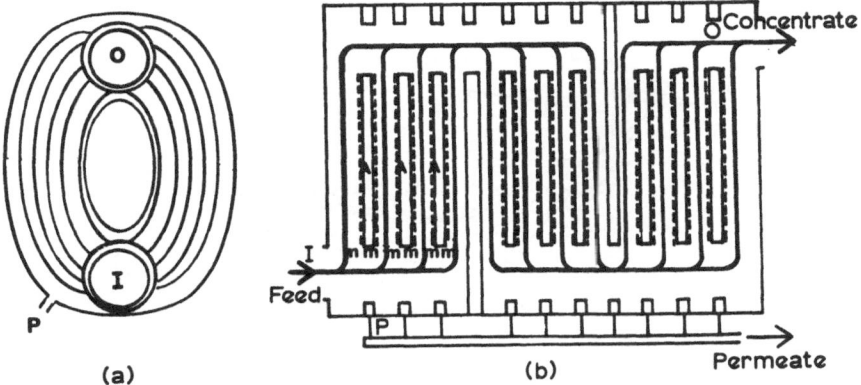

Fig. 5. DDS ultrafiltration module, flat sheet membranes. (a) membrane support plate for module 37; I: inlet port; O: outlet port; P: permeate outlet, dimensions 38 cm × 31 cm, channel height 1·3 mm at centre, 1·8 mm at periphery. (b) membrane assembly; A: support plates; m: membranes. (Illustration from DDS technical literature.)

important has been a modification of the flow channels between adjacent membranes. In order to get a more uniform distribution of flow over the whole membrane area, the shortest centre channels have been removed, thus reducing the ratio of the longest to shortest channels, and channel heights have been increased to 1·3 mm near the centre and to 1·8 mm at the periphery. Rates of flow in the outer channels are now higher, concentration polarisation is less and permeate flux is improved. Whereas the limiting total solids in the concentrates was 42 per cent, the new modules enable 51 per cent to be reached, and even 58 per cent if the initial fat content of the milk is first increased from the standard 3·8 per cent up to 6·1 per cent. Modules of this new design have been produced in three sizes, 4·5 m², 9 m² and 13·5 m² membrane area.

Spirally wound
A layer of plastic mesh (1 mm thick), a flat membrane and a layer of absorbent material to collect and transport the permeate are rolled into a spiral to form a cylindrical module (Fig. 6). At the ends, the spaces between the membrane and absorbent material are sealed, leaving the ends of the mesh open. A perforated, stainless steel tube is inserted along the centre of the roll. The flow of feed is parallel to the axis of the module between membranes, the mesh acting as a spacer and turbulence promoter. Permeate flows radially through the membrane

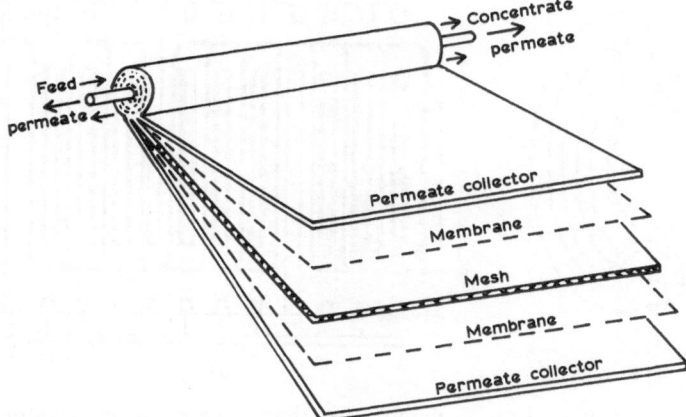

Fig. 6. Abcor ultrafiltration module, spirally wound membrane. Length 84 cm, diameter 11 cm, membrane spacing 0·7 mm, area 5 m². Three such sections are housed in series in a stainless steel tube to make one module, 3 m long, membrane area 15 m². Schematic diagram to show the formation of the spiral. (Illustration from Abcor technical literature.)

and into the absorbent material, then tracks along it towards the centre of the roll, finds its way through the perforations in the central tube and out at the ends. The feed, now concentrated, leaves the module at the further end.

Hollow fibre

A hollow fibre is a narrow tube, internal diameter 0·5 or 1 mm and length 1 m, made of non-cellulosic polymer. The structure is dense on the inner surface, 0·1–1·5 μm thick, and much looser towards the outer surface in a layer 50–250 μm thick. This tube is itself the membrane; the feed passes along the inside of the tube and permeate moves radially outwards. Fibres are assembled in bundles of 1000 or more, and sealed in a clear plastic cartridge to form a module containing 2·5 m² area of membrane.

ULTRAFILTRATION PLANT

There are several factors which determine the design.

   (i) Membrane performance – permeate flux in relation to the concentration required.

(ii) Flow velocity – compromise between high flow velocity to control polarisation and energy afforded for pumping.

(iii) Pressure drop along the plant – choice of flow velocity, arrangement of modules and pumps to ensure an adequate working pressure is available over all the membrane.

(iv) The volume of feed to be processed.

(v) The retention time of the liquid in the plant in relation to bacterial growth.

(vi) Capital and maintenance costs of plant and membranes.

The basic design of an ultrafiltration plant is shown in Fig. 7. The essential components of a plant are:

   (i) Feed tank;

  (ii) Pumps – feed and circulation, usually centrifugal;

 (iii) Flow control valve;

 (iv) Flow meter;

  (v) Thermometer;

 (vi) Pressure gauge – inlet;

(vii) UF module assembly;

(viii) Pressure gauge – outlet;

 (ix) Pressure retaining valve.

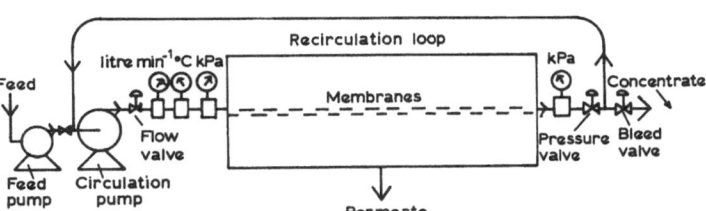

Fig. 7. Ultrafiltration plant design, continuous internal recycle or feed and bleed.

A heat exchanger is usually included, either to heat the milk to the desired operating temperature or to prevent overheating due to the pumping. Fully automated plants are fitted with controls for pressure, flow and temperature and safety devices to stop the pump should the feed cease or the system become blocked.

To achieve a useful degree of concentration, the feed must pass many times, probably 100 or more times over the membrane. Passing once only would require far too big an area of membrane. Plant may then be operated in two ways.

(i) Continuously – recirculating round an inner loop.
(ii) As a batch process – recirculating back to the feed tank.

Continuous Operation

The design in Fig. 7 is for this mode, known as internal recycle, since the feed is pumped many times round an inner loop which consists essentially of the circulation pump and the membrane assembly. The plant is operated so that, at the outlet end of the membrane, the required concentration is reached and a little of the concentrate is bled off; the rest is diverted back to the beginning of the loop at the entrance to the circulation pump, where more fresh feed is added from the feed pump to make up the volume of concentrate removed. An alternative name for this method is, therefore, the feed and bleed mode. It is the most appropriate mode for large commercial plants where the usual pattern is to run continuously for 20 h and clean for 4 h.

The advantages of the feed and bleed operation are short retention times in the plant, measured in minutes rather than the hours of the batch process, and economy of pumping energy since the whole pressure is not lost on each recycle as the internal loop is closed. Its disadvantage is that the concentration in the loop at the membrane is permanently high, close to the final required level, hence the permeate flux is low.

Commercial size plants may contain several hundreds of square metres of membrane area, and one pump cannot provide sufficient pressure for all the necessary modules. Plants are, therefore, made up of repeating

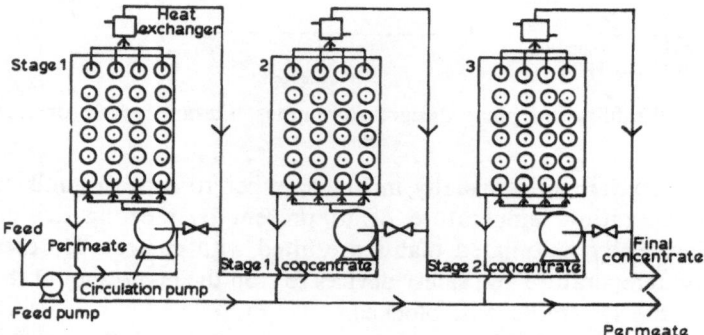

Fig. 8. PCI ultrafiltration plant – commercial scale. Each stage contains 24 modules, 4 banks in parallel each containing 6 modules in series. One module $2.5\,\text{m}^2$ area of membrane, one stage $62\,\text{m}^2$, whole plant $186\,\text{m}^2$. (Illustration from PCI technical literature.)

units, each with its own pump and array of modules which are then all provided with a working pressure. An example of such a plant fitted with tubular membranes is shown schematically in Fig. 8. In large plants, the size of successive units decreases to maintain the flow of the declining volume of concentrate.

Batch Operation

Instead of recycling the feed back to the entrance of the recirculation pump, it is returned to the feed tank, so that it circulates round the whole plant. This mode of operation is suitable for processing small volumes which can be contained wholly within the plant, as in laboratory and pilot plant trials. No concentrate is withdrawn until the end of the process, it is all left in the plant until the required concentration is reached. The method has the advantage of a simple plant with only one pump and the removal of a large volume of permeate while the concentration of the feed over the membrane is low and hence flux is high. Against the method are the long retention times of the feed in the plant and, of course, it is possible to process only the volume which the plant will hold. A modification in operation will extend this capacity by topping up the tank with fresh feed as ultrafiltration is going on, but the limit here is reached when the whole plant is filled with concentrate.

Engineering Equipment

Beyond the specialised component – the membrane assembly, the engineering equipment comprises pumps, pipes, valves and controls, all standard items familiar to the dairy industry. The main component is the pump which must be capable of running continuously. Delivery rates depend on the membrane geometry and size of plant, tubular arrangements requiring the higher flow rates because of their wider flow channels. For example, the tubular membrane (diameter 1·2 cm) requires a flow of about $20 \, \mathrm{l \, min^{-1}}$ to give a velocity over the membrane of $2·7 \, \mathrm{m \, s^{-1}}$, so that through one module containing 18 tubes arranged in parallel, the pump must deliver $360 \, \mathrm{l \, min^{-1}}$ and through one stage of the commercial plant in Fig. 8, the flow must be $1440 \, \mathrm{l \, min^{-1}}$. The delivery pressure has to be in the region of 0·5 MPa. Centrifugal pumps perform this duty very well, they are easy to control and are not expensive. However, at the very high concentrations of milk now being aimed at for cheesemaking, centrifugal pumps are less able to contend with the very high viscosities of the concentrates, hence it is proposed that positive pumps should be used in the final stages of UF plants.

Valves can be of the diaphragm type, one at the outlet of the centrifugal pump to control the flow, another at the outlet of the membrane assembly to maintain the pressure or bleed off the concentrate. Pressure gauges must be of the diaphragm or capsule type, so that milk cannot enter the Bourdon tube. Pressure safety devices are not required in ultrafiltration since pressures are low, and centrifugal pumps are not damaged if inadvertently operated against a closed valve. For the measurement of flow rate, rotameters are used though they are difficult to read through concentrated milk; turbine flow meters are much better, but more expensive. Temperature control is necessary for both product and plant. UF is commonly carried out at 50°C, a compromise between 37°C where maximum bacterial growth occurs, and 60°C where the whey proteins start to denature. A high temperature favours the rate of the process mainly on the grounds of lower viscosities of the feed and permeate. Operation at low temperature causes serious concentration polarisation.

An impression of UF in practice is given in Fig. 9, a photograph of a large commercial plant using flat membranes for the manufacture of whey protein concentrate. The International Dairy Federation (1979) has conducted a survey entitled 'Equipment Available for Membrane Processing'. This is the most comprehensive collection of data on membranes

Fig. 9. DDS ultrafiltration plant – commercial scale, flat sheet membranes. Ultrafiltration of 1 million litres of whey per day for the manufacture of whey powder. (Photograph by courtesy of the Danish Sugar Corporation, DDS-RO Division, 4900 Nakskov, Denmark.)

for UF, RO and electrodialysis, modules, operating parameters, cleaning and disinfection.

ULTRAFILTRATION PLANT PERFORMANCE ON MILK

Performance is usually represented by a permeate flux curve, i.e. a plot of flux against concentration under stated conditions. Few details on the performance of large-scale industrial plants in operation with firms are available, but an appreciation of possible processing rates may be obtained from the many research papers published. They are accounts of experiences mainly on the pilot-plant scale. Figure 10 is a flux curve obtained with zirconium oxide membranes by Goudedranche *et al.* (1981). These are the best fluxes yet seen in the literature, favoured by the new membranes, a small plant and a batch operation. From polysulphone membranes, fluxes commonly start at about $40–50 \, l \, m^{-2} \, h^{-1}$; fluxes in commercial plants operated in the continuous mode are lower than this. The improved performance of the zirconium membranes is a result of their higher mechanical strength over that of the former polymer types. The method of operation is to start with a low pressure in the initial stages of concentration, then as flux decreases with concentration, the applied pressure is increased to keep up the flux. However, not all inorganic membranes produce higher fluxes. For example, flux rates not exceeding $15 \, l \, m^{-2} \, h^{-1}$ have been reported for UF of milk using alumina membranes (Bennasar *et al.*, 1982).

The Effect of Pressure, Flow Velocity and Temperature

As the operating pressure is increased from its lowest value, the flux increases until a region is reached where further pressure increases are of no advantage. As more pressure attempts to increase the flux, the gel layer on the membrane increases, nullifying the effect of the additional pressure. The point at which the flux becomes independent of pressure is related to the velocity of the feed over the membrane, the higher velocities making it profitable to use higher pressures and obtain still higher fluxes.

Increasing flow velocity is more beneficial than increasing pressure, as the action is to assist in dispersing the polarised layer. Continued gain in flux is limited by the energy that can be afforded for pumping, and in the case of whole milk, there is a danger of damage to the fat globules by excessive pumping.

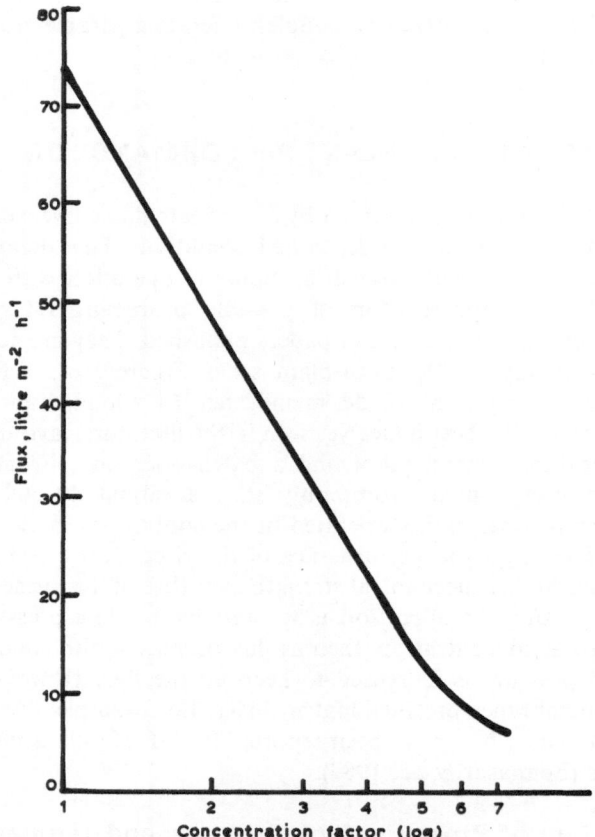

Fig. 10. Utrafiltration of whole milk – flux against concentration factor. Membrane ZrO_2, pilot plant, $1.6\,m^2$ membrane area, batch operation, at 52 °C. (With acknowledgement to Goudedranche *et al.* (1981) and reproduced by courtesy of Technique Laitière.)

Flux increases more than linearly with temperature. Increase in temperature has the dual effect of lowering the viscosity, which assists flow rate, and of increasing diffusion, which assists dispersion of the polarised layer.

Composition of the Product

Membranes retain all the fat and practically all the protein from milk. Retention coefficients of the non-protein nitrogen compounds are generally 20–40 per cent, and higher for the high concentration factors. The losses through the membrane are mainly urea and amino-acids. Retention of lactose may be up to 10 per cent; since some minerals are partly bound to protein, namely calcium, magnesium, phosphate and citrate, they will be retained with the protein, but others are free and will pass through the membrane. Likewise the fat soluble and protein bound vitamins are completely retained.

An example of the compositions of concentrates and permeates in the ultrafiltration of whole milk is given in Table II.

TABLE II

Ultrafiltration of whole milk – example of compositions of the products – concentrate and permeate (Concentrations in per cent w/v)

Volume concentration factor	Total solids	Fat	Protein	Non-protein nitrogen compounds	Lactose	Ash
Concentrate						
× 1	12·9	3·9	3·1	0·18	4·7	0·77
× 3	28·6	12·6	9·8	0·18	4·1	1·3
× 5	43·3	21·8	16·1	0·18	3·2	1·9
Permeate						
× 1	5·7	0	0	0·18	4·8	0·53
× 3	6·1	0	0·06	0·19	5·1	0·53
× 5	6·7	0	0·49	0·19	5·2	0·54

System: tubular membranes, small pilot-plant, batch operation, 50 °C.
Note. See p. 280 on the concentration of components with low retentions.

Ultrafiltration of Skim-milk

Skim-milk behaves in exactly the same way as whole milk in ultrafiltration. The permeate flux from skim is slightly higher than from whole milk by about 20 per cent, indicating that the fat in whole milk, though it forms a large proportion of the total solids and is present in the largest particle sizes, is not a great hindrance to filtration. More important is the fact that skim-milk and whole milk have the same protein content and, as the protein is in smaller particle sizes than the fat, it is more liable to denser packing at the membrane surface. Of all the components in milk, protein exerts the greatest control over the rate of ultrafiltration.

Growth of Bacteria during Ultrafiltration

Microbial growth is not a serious problem if the process is carried out at a high temperature (50 °C) and on the feed and bleed principle, such that retention times are below the lag phase of the growth. Some plants claim a retention time as low as 10 min, well below lag phases of around 2 h at 50 °C (Kiviniemi, 1974). Design features, such as low hold-up volume, low pressure drops and optimum flow velocity, all help to keep growth in check, and provided the starting milk is of good hygienic quality, bacterial numbers do not increase beyond that due to the concentration of the milk.

Deposits on Membranes

Deposits from milk have not been studied, but an examination of membranes after processing whey illustrates the nature of deposits from milk. The whey proteins are smaller than the casein micelles, so they are the main fouling agents. The electron microscope shows deposits up to 1 μm thick, and particulate matter penetrating the membrane to a depth of 2 μm. On the surface are seen microorganisms, granules, thin strands, coarse fibres and sheets of various thicknesses. They are clearly visible within 10 min of the start of filtration, increasing in thickness thereafter. The worst offender is β-lactoglobulin, since it is able to form sheets, as does blood serum albumin; γ-globulin and α-lactalbumin exist as granules. It is postulated that salts could form bridges between the proteins and the membrane, so building up a fouling layer (Lee *et al.*, 1975).

Cleaning Membranes

Polysulphone and other non-cellulosic membranes are cleaned with highly alkaline detergents and chlorine to break down protein deposits, and acids to remove minerals. Chlorine released from sodium hypochlorite at pH 10·5 rapidly attacks protein. As the chlorine is consumed, it must be made up to the recommended level of 250 ppm until no further decline occurs. Proprietary cleaning solutions containing organic and inorganic sequestrants are available specifically for ultrafiltration plants. Cleaning is assisted by a high flow rate, low pressure and high temperature (50 °C), and it is important to use pure water as recommended by the manufacturer of the cleaning agent. A membrane is clean when the permeate flux from pure water is restored to the level at which it stood before the UF process.

USES OF UF MILK CONCENTRATES

The first use of ultrafiltration in the dairy industry was to fractionate whey, concentrating the protein for food processing, and reducing the pollution load of the carbohydrate portion by making use of the permeate; up to 10 million litres of whey are now filtered daily. Of more direct use to industry is the ultrafiltration of milk for cheesemaking, which promises to become the widest application. Lesser applications are milk standardisation and making fermented milk products. Ultrafiltration of milk on the farm immediately after milking is being investigated.

Cheesemaking

The use of UF technology in cheesemaking has been extensively reviewed by Lawrence (1989). The principle of cheesemaking using UF retentate rather than milk has a number of potential advantages. The major aim has been to increase cheese yields (by incorporating the whey proteins), but other advantages include reduced requirement for starter and rennet, reduced milk volume and, therefore, reduced requirement for process plant. UF retentate can be standardised to give a more uniform product quality, and can be used as part of a continuous cheesemaking process.

During conventional cheesemaking approximately 20 per cent of the protein in milk is lost in the whey. If milk is ultrafiltered first to a concentration such that during cheesemaking no further liquid drains, the extraction of permeate during the ultrafiltration may be regarded as similar to the drainage of whey in the traditional process. The difference is that permeate does not contain protein; this has been retained in the cheese and has increased the yield by as much as 15–30 per cent of the traditional yield according to the type of cheese. Since the milk concentrate, termed the 'precheese', must be similar in composition to the final cheese, the process is more easily applied to those types having the lower solids content, for example cottage cheese (21 per cent total solids), Ricotta (28 per cent total solids) and Feta (40 per cent total solids). Concentration of milk up to five-fold is sufficient for these varieties, and commercial processes have been set up successfully. France is the leader in this field because of the popularity of soft cheeses amenable to manufacture by this process, known as the MMV process – Maubois, Mocquot and Vassal – in whose names the original patent was taken out in 1969. It was reported in 1980 that more than 60,000 tonnes of cheese per year was being made by this method (Maubois, 1980).

Camembert cheese can be made from ultrafiltered milk concentrates, with the advantage of a 20 per cent increase in yield and improved uniformity in the weight of cheese taken from the moulds. The Alfa-Laval company has combined ultrafiltration with its 'Camatic' process to make Camembert cheese in a continuous process (Hansen, 1981). The largest production using ultrafiltration is of Feta cheese, mainly in Denmark for export to the eastern Mediterranean countries (Hansen, 1977). For this type, an extra yield of cheese of 30–35 per cent is obtained, only 20 per cent of the normal quantity of rennet is required, losses of fat in the whey are avoided, continuous production is possible and the quality of the cheese is good.

Many other soft cheeses have been made on a trial basis from ultrafiltered milk concentrates, namely cottage cheese, cream cheese, Danish Blue, Edam, Herve, Mozzarella, Quarg, Queso Blanco, Queso Fresco, Ricotta and Teleme (see Glover, 1984).

However, to make Cheddar cheese, which has a solids content of 62 per cent, would require ultrafiltration to a concentration factor of 8 or 9, and though this has been achieved experimentally, it is not a practical reality because of the very low permeate fluxes at this level. In addition, the incorporation of whey protein has tended to give rise to poor textural quality in Cheddar. The greatest advance towards making Cheddar cheese by UF has been made by the CSIRO in Australia (Sutherland and Jameson, 1981). This has led to the technical reality of continuous automated production of Cheddar by the APV Sirocurd process (Garret, 1987). Steps in their process include ultrafiltration, diafiltration and special care in handling the curd to avoid losses of fat and casein. Further development is needed to achieve the required quality of the final cheese. Related to this process is the production of a cheese base and a processed cheese, in which milk concentrates are taken to a final 60 per cent total solids by evaporation. By complete drying, it is possible to make a cheese powder for subsequent reconstitution and making cheese in an importing, under-developed country.

It should be stressed that UF cheese may differ from traditional varieties. This is not only due to the presence of whey protein, since the high buffering capacity of the retentate results in cheese with higher pH and mineral content than with traditional methods. It is, therefore, necessary to adjust the processing techniques appropriately. For example, by acidifying the milk or adding sodium chloride to the milk before UF, it is possible to release bound calcium from the micelles. The use of UF processing to a concentration factor of only 2 can be used with traditional cheesemaking methods to give increased throughput in the plant, although no yield benefit is obtained.

Fermented Milk Products

The use of milk concentrates generally improves the yield and body of fermented products. There is a report on the production of yoghurt and ymer by Morgensen (1980).

Trials on Other Uses of UF Milk Concentrates

Adjustment of protein levels in milk up to 6·5 per cent are reported to have no effect on the organoleptic properties (Roenkilde Poulsen, 1978). The use of UF concentrates in making sweetened condensed milk has been found to avoid sandiness in the texture (Serpa Alverez *et al.*, 1979). If milk is first concentrated by UF, it is possible to prepare low-fat, high-protein whipping cream, which is attractive to the producer since the yield of product per unit volume of milk is increased (Scurlock and Glover, 1983). Such a product could be regarded as nutritionally preferable when the consumption of milkfat is questioned. Trials on the ultrafiltration of the milk of other species, namely the buffalo, goat and sheep, have been carried out, again for the purpose of making cheese.

USE OF THE PERMEATE

The permeate from the ultrafiltration of milk is a solution, mainly of lactose, in the same concentration as in the water phase of the original milk, which is about 4·8 per cent. Such a solution has a high biological oxygen demand (BOD) so it cannot be discarded as waste; far better that it should be regarded as a valuable source of carbohydrate. Three uses are possible, after different treatments, for human food, for animal food and as an industrial fuel (Coton, 1980). For human food, the lactose can be hydrolysed enzymically or by ion exchange into a glucose/galactose sweet syrup; this is finding uses in the brewing and confectionery industry. By fermentation the lactose may be converted to alcohol, or lactic acid, or antibiotics also for human use. Other fermentations will produce a biomass or ammonium lactate for animal food, or methane for fuel. Evaporation and crystallisation produces a food for human or animal use. Of these the most economically viable are the production of the glucose/galactose syrup and the alcohol, and indeed both have reached commercial practice.

THE PRINCIPLES OF REVERSE OSMOSIS

Consider the arrangement in Fig. 11 where a membrane is placed between a solution and a solvent. The membrane is permeable only to the solvent, no other molecules in solution or suspension can pass. In an attempt to equalise the concentrations on both sides of the membrane, the solvent, usually water, will pass through the membrane from left to right. The driving force is the gradient of the chemical potential across the membrane. This is a thermodynamic quantity, depending on concentration, pressure and temperature, whose gradient determines the movement of matter. The concentration of water is higher on the left than on the right. Water moves down the gradient from left to right causing the liquid level at A to rise, thus setting up a pressure. Solute molecules cannot move since they are restrained by the membrane. Equilibrium is reached when the chemical potentials, made up of concentration and pressure on both sides of the membrane, become equal. Temperature is not involved since it is the same on each side. The pressure set up by the rise of water at A is the osmotic pressure of the solution in the right-hand compartment. It is directly proportional to the concentration, and inversely proportional to molecular weight.

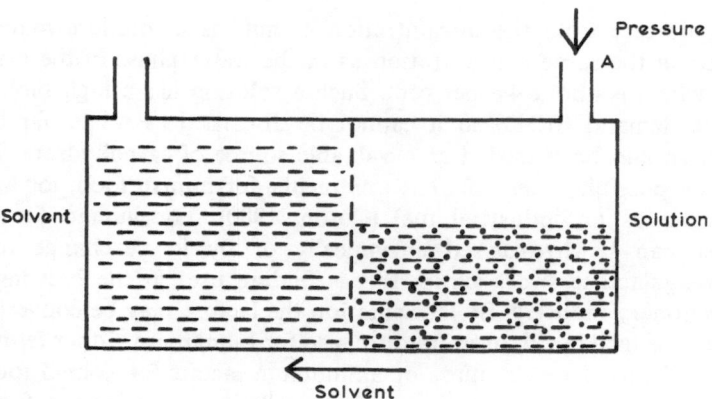

Fig. 11.   The principle of reverse osmosis.

If a pressure in excess of the osmotic pressure is applied at A, the above process may be reversed, and water may be driven from right to left through the membrane achieving a concentration of the solution. This is reverse osmosis. It can be used for concentration or for purification, depending on whether the concentrate is the required product, as in milk processing, or the permeate, as in water purification.

The favoured theory of the passage of water through the membrane considers that the solvent dissolves in the membrane, then moves through by diffusion. The solute is far less soluble in the membrane and is therefore held back. This theory does not require the membrane to have pores; in practice, some small molecules do get through due to defects in the membrane.

The rate of transport of any component through the membrane is a function of the gradient of its chemical potential. At a constant temperature, chemical potential is a function of concentration and pressure.

$$\text{Flux} \propto f(\text{concentration gradient}) + f(\text{pressure gradient})$$

Water is the solvent and it passes easily through the membrane, the difference in the concentrations of water across the membrane is small, hence, in the above expression, the concentration term may be neglected, and the pressure term becomes predominant. The driving pressure is the applied pressure minus the opposing osmotic pressure. Then

$$J_w \propto f(\Delta P - \Delta \Pi), \text{ or } J_w = A(\Delta P - \Delta \Pi)$$

where J_w = water flux
ΔP = pressure applied for reverse osmosis
$\Delta \Pi$ = osmotic pressure of the solution
A = the solvent flow constant.

A contains the membrane area, thickness and void volume fraction, the diffusion constant and the solubility of the solvent in the membrane, the molar volume of the solvent in the solution, and the densities of the solvent and solution.

For the solute, which is largely retained by the membrane, the difference in concentration across the membrane is relatively large. Hence for the solute, the concentration term is predominant.

$$J_s \propto f(\Delta C_s)$$

Or $$J_s = B\Delta C_s$$

where J_s = solute flux
           ΔC_s = difference in solute concentration across the membrane
           B = the solute flow constant, similar in nature to A.

From this analysis it should be noted that

(a) Water is driven through the membrane by pressure, solute is driven through by a concentration gradient, i.e. the transport of solvent and solute are not connected.
(b) An increase of pressure increases the water flux, but has no effect on the passage of solute. Hence the concentration of solute in the permeate decreases with an increase in pressure.
(c) An increase in the feed concentration reduces the permeate flux since the osmotic pressure of the feed is increased, and it increases the passage of solute because of a higher concentration gradient.
(d) An increase in temperature increases the rate of diffusion of both water and solute, both at a rate of about 3·5 per cent $°C^{-1}$.
(e) Ageing of the membrane causes the constants A and B to reduce due to creep and tightening under the applied pressure, hence both permeate flux and passage of solute fall with age. The fall is logarithmic with time, most noticeable in the early stages of membrane life.

This theoretical treatment was developed to fit dilute monocomponent systems, and is offered here to provide an understanding of the fundamental aspects of reverse osmosis. Milk has many components covering a great range of molecular sizes, and some in very high concentrations. The theory could not be expected to fit the practical behaviour of milk.

As with ultrafiltration, the main hindrance to the process is concentration polarisation, which must again be minimised either by high degrees of turbulence or high rates of shear.

MEMBRANES FOR REVERSE OSMOSIS

Cellulose acetate formed the first successful RO membrane as it did for UF. Although nylon membranes appeared for water purification, it was almost twenty years before a successor to cellulose acetate became available for milk processing. PCI has its non-cellulosic membrane ZF99, superior to cellulose acetate in permeability, retention, pH and temperature tolerance. It can be operated safely over a pH range of 3–11, and up to a temperature of 80 °C. DDS has its new RO membrane, designated HR, and described as a thin film, composite membrane consisting of a UF membrane coated with a very thin, polymer layer. The polymer is the effective RO membrane, laid on a polysulphone layer, in turn supported on a polypropylene backing. It has properties similar to ZF99. These new membranes are susceptible to damage by chlorine and must be cleaned only with approved agents. Inorganic membranes have not yet been developed for RO.

Membrane Geometry and Modules

Membrane geometry for RO is the same as that for UF, namely tubular, flat and spirally wound, and the same comments on the different types apply. There is no hollow fibre design suitable for milk as there is for UF, and the Du Pont RO system of bundles of many thousands of very fine, nylon fibres only 90 μm diameter could not take the high solids content of milk.

Modules for RO are also very like UF modules. For the PCI tubular membrane type (Fig. 4) all membrane tubes are connected in series. The flat membrane arrangement of DDS (Fig. 5) has circular plates for RO (30 cm or 40 cm diameter) built into vertical stacks containing 19 m^2 or 28 m^2 membrane area.

REVERSE OSMOSIS PLANT

RO plants look exactly like UF plants, comprising pumps, membrane modules, controls and gauges. The same modes of operation may be used, namely continuous internal recycle and batch operation, and also since concentration factors for RO are lower, single-pass operation has sometimes been possible. Single-pass plants are designed as a series of stages in tapered formation, successive stages having few modules to ensure that high rates of flow are maintained and not reduced by the extraction of permeate in the earlier stages.

Recently, centrifugal pumps with high pressure delivery have become available providing higher flow velocities, and hence more efficient use of membranes. Plant configuration is shown in Fig. 12 for the multistage recycle system in which modules are arranged in a number of stages, each with a short path length and booster pump at the entrance of the stage. Throughout the plant, higher pressures and higher velocities than in

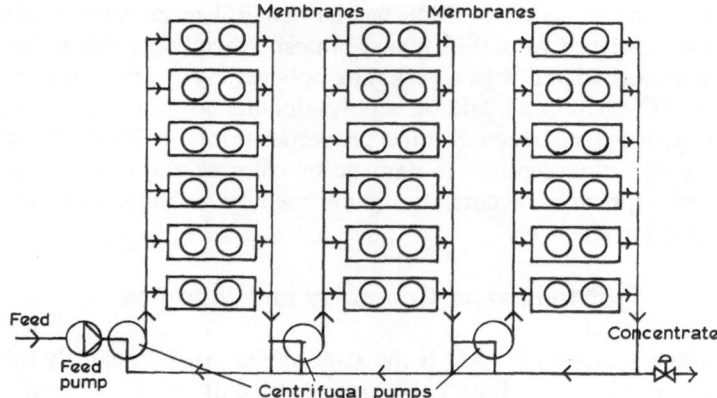

Fig. 12. PCI reverse osmosis plant – multistage recycle system. Three stages each with 6 parallel rows of 2 modules, total area 92 m². (With acknowledgement to Pepper and Orchard (1982) and reproduced by courtesy of the Society of Dairy Technology.)

single-stage recycle are then maintained. In the plant of Fig. 12 working on whey the pressure drop across each stage is only 0·9 MPa from 4·5 MPa to 3·6 MPa and the flow velocity nowhere falls below 2·2 m s^{-1} which is well above the critical velocity of 1·6 m s^{-1} for such tubular membranes at which experience has shown that membrane fouling in whey starts.

Engineering Equipment

There are differences in the engineering equipment for RO and UF due to the much higher pressure demanded by RO. For the tubular system working on milk, it is required to deliver about 20 l min^{-1} into a single module at a pressure of about 5 MPa. Positive displacement pumps (constructed in stainless steel) of the three-piston type, as in the normal dairy homogeniser, serve very well. They are fitted as feed pumps to raise the pressure for entry to the module assembly, then booster pumps are used between stages as described above. Any pressure pulsation can be smoothed out by including a nitrogen-filled accumulator in the pipeline.

The Archimedian screw type of pump is capable of raising the pressure for RO, but cannot maintain high rates of flow due to slipping of fluid backwards past the screw; it can, therefore, be used only on small capacity RO plants. The Pitot pump has a number of attractions for RO and has been used. It has few moving parts, its action is rotary, it can operate against a closed valve and the flow is pulse-free. Its disadvantage is that its efficiency is low.

Since RO feed pumps are the positive type, they will not necessarily require flow controls or flow meters; they will, however, need protection against overload. The normal, spring-loaded, relief valve is not acceptable for dairy processing on hygienic grounds due to the dead space in the valve, and stainless steel, bursting discs are much more suitable. In all other respects, the auxiliary equipment for RO is the same as that for UF.

A commercial RO plant using tubular membranes is illustrated in Fig. 13.

Fig. 13. PCI reverse osmosis plant – commercial scale, tubular membranes. Area 327 m^2, concentrating cheese whey, 10,000 litre h^{-1} to 24 per cent T.S. ($\times$ 4). (Photograph by courtesy of Paterson Candy International, Laverstoke Mill, Whitchurch, Hants, RG28 7NR, England.)

REVERSE OSMOSIS PLANT PERFORMANCE ON MILK

In comparison with ultrafiltration, additional factors come into play to govern the rate of reverse osmosis. The most important of these is osmotic pressure, and its increase both in the main stream of the feed and, more particularly, in the polarised layer at the membrane surface as concentration proceeds, both considerably reduce the effectiveness of the applied pressure. Secondly, because RO membranes retain all solids, calcium phosphate, which is in saturated solution in normal milk, can be precipitated on the membrane if hot milk is concentrated. This precipitation of calcium phosphate may be reduced by pretreating the milk before RO, such as reducing the pH to increase the solubility of the calcium phosphate, preheating to transfer some of the calcium to the casein micelles, or removing some of the calcium by ion exchange. The most practical of these methods is the preheating. Nevertheless, as in UF, the main fouling agent is the protein, mainly because of the lower diffusion coefficients of the large molecules and micelles. The diffusion coefficient of casein is one third of that of the whey proteins and one twentieth of that of lactose, even at normal concentrations in milk.

The reverse osmosis of milk is not yet practised on a large scale. There is some activity in the yoghurt and ice cream business, and some trials on hard cheesemaking, but nothing for the preparation of concentrates for the liquid market. The only information on the behaviour of milk during RO comes from research reports, which are few indeed compared with the vast number on UF in the scientific literature and advertising material. An impression of the rate of RO for whole milk is given in Fig. 14; the results were obtained in experiments using a small pilot-plant. These data show that a concentration factor of 2 is easily attainable and 3 is possible though scarcely practicable, corresponding to total solids contents of approximately 25 per cent and 37 per cent respectively. For skim-milk concentrated under the same conditions, the fluxes are only about $3 \, \text{l} \, \text{m}^{-2} \, \text{h}^{-1}$ higher than those from whole milk. More realistically, skim-milk has been concentrated two-fold in a larger plant as a continuous operation by the Paterson Candy, multistage recycle plant fitted with $16 \, \text{m}^2$ area of cellulosic membrane (Pepper and Orchard, 1982). The operating conditions were: pressure $3 \cdot 8 \, \text{MPa}$, flow velocity $2 \cdot 5 \, \text{m} \, \text{s}^{-1}$, temperature $30°C$ and fluxes at the first and second stages ($13 \cdot 5$ and 18 per cent total solids) were maintained at $20 \, \text{l} \, \text{m}^{-2} \, \text{h}^{-1}$ and $10 \, \text{l} \, \text{m}^{-2} \, \text{h}^{-1}$ respectively over a period of 6 h. A larger plant of the same type having a membrane area of $250 \, \text{m}^2$ is concentrating $11{,}250 \, \text{l} \, \text{h}^{-1}$ of skim-milk by a factor of $1 \cdot 5$.

Effects of Pressure, Flow Velocity and Temperature

The effects of changes in the operating parameters follow the same trends as in UF, with the difference that as pressure is increased the flux passes through a maximum, so that beyond this point, it is definitely dis-

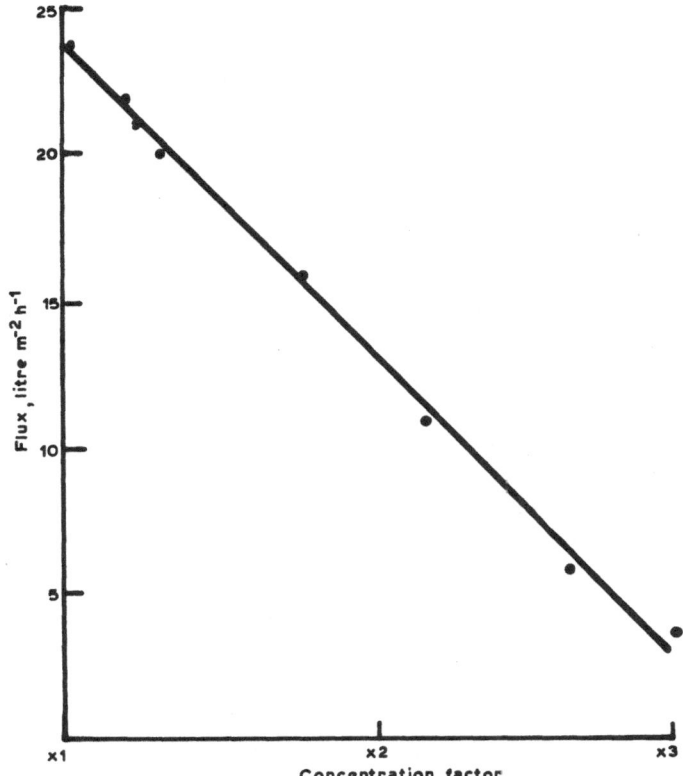

Fig. 14. Reverse osmosis of milk – flux against concentration for a membrane of cellulose acetate type, pressure 4·8 MPa, flow velocity 2·4 m s^{-1}, 30 °C, small pilot-plant, batch operation. (With acknowledgement to Abbot *et al.* (1979) and reproduced by courtesy of the *Journal of Dairy Research*.)

advantageous to apply more pressure as it only decreases the flux. The level of the flux at the maximum rises as the flow velocity is increased.

Composition of the Product

For practical purposes, all the components of milk are retained in the concentrate, and only a small proportion of the smallest ions escapes into the permeate. Retention of the whole mineral content of milk exceeds 99 per cent; individual ions with retentions below this are Na$^+$ 95 per cent, K$^+$ 98 per cent and Cl$^-$ 94 per cent. The permeate contains 1000 ppm total solids made up mainly of lactose 250 ppm, Na^+ 110 ppm, Cl^- 200 ppm (Skudder, 1978); it can only be presumed that defects in the membrane caused the loss of lactose into the permeate. In the experience of Pepper and Orchard (1982), using a skim-milk feed initially with a BOD of 65,000 mg l^{-1} concentrated continuously to 18 per cent total solids in the multistage recycle RO plant, the permeate BOD was only 100 mg l^{-1}; there was also very little bacterial growth in the concentrate.

Deposits on the Membrane

Solids collect at the membrane in two forms: a compacted layer in contact with the membrane, and a gelatinous layer between this and the main stream of the feed. The layer of solids may be up to a few micrometres thick, and may consist mainly of proteins. Electron microscopy of the deposits shows that they are granular in structure, more compact nearer the membrane surface. The granules are mainly casein, some of them linked by bridges forming a network. A few fat globules are trapped between the granules, but it is evident that the compacted casein forms the greatest obstruction to the process. In the deposited layer, the concentration of the casein can be up to 10 times that in the original milk (Skudder, 1978).

Cleaning Reverse Osmosis Membranes

For their cellulose acetate membranes, PCI recommends cleaning by circulating a 1 per cent enzyme detergent solution for 90 min at 35 °C, and then sterilising with 50–100 mg l^{-1} free chlorine for 30 min. Recommended cleaning materials for their new ZF99 membranes are:

0·1–0·3 per cent w/w nitric or phosphoric acid at 45 °C;
0·1–0·5 per cent w/w sodium hydroxide at 50 °C; and
0·1–0·5 per cent proprietary alkaline detergent.

Sterilisation is with hydrogen peroxide or formalin. The introduction of ZF99 membranes has halved the cost and cleaning time of membranes.

USES OF RO MILK CONCENTRATES

The main application of reverse osmosis in dairying is the concentration of whey to facilitate handling, transport and storage. On milk its use is not nearly so widespread, although there is some activity for making ice cream and yoghurt, and trials are in progress for Cheddar cheesemaking, recovery of milk solids from rinse water, and the consideration of RO as an initial step in drying from the point of view of energy saving.

Ice cream
Skim-milk powder is used in making ice cream. Since skim-milk RO concentrate is cheaper to make than the powder, the possibility arises of substituting concentrate for powder. DDS has prepared skim-milk RO concentrate to 15·4 per cent total solids for this purpose, and reports that the quality of the ice cream is as good in all respects as the product made by traditional methods.

Yoghurt
Skim-milk concentrates are used for yoghurt as they are for ice cream – as a substitute for powders. Some yoghurt has been made by this method for several years now, and involves first concentrating milk by RO by a factor of 1·4, yielding 15 per cent total solids.

Cheese
There has been some exploratory work on making cheese from RO milk concentrates. Using 15 per cent total solids concentrates, increases in the yield of Cheddar cheese of 1–3 per cent are reported, and this improvement is attributed to the higher retention of casein fines, fat and whey solids. Cottage cheese and pâte fraîche have also been made from RO concentrates. The quality of all the cheese has been pronounced good.

Recovery of milk solids from dairy rinse water
Experiments in Sweden have shown that in a dried milk factory, rinses of pipelines obtained in the first few seconds contained as much as 8 per cent solids. This level could be concentrated to 18 per cent total solids by RO, and the material dried and sold for animal feed, so reducing pollution and effecting a considerable saving.

Milk powder
Milk powder has been made experimentally in a process in which RO was used to concentrate milk before evaporation and spray drying. The whole milk powders had high free fat levels due to damage suffered by the milk in the RO pumps and valves, but the skim-milk powder was just as good as that made conventionally.

Energy saving

In the initial stages of the concentration of milk, RO consumes less energy than evaporation since RO demands pumping energy only; evaporation involves a phase change. Over the range of concentration for which RO can be applied, for example in the case of concentrating whey from 6 per cent to 28 per cent total solids, the energy required by RO is approximately one quarter of that of the latest mechanical, vapour recompression evaporator. Unfortunately for the drying of milk, RO cannot concentrate to the level of 50 per cent total solids where spray drying starts, hence RO is not a total replacement for evaporation. Concentration of milk by RO in order to save transport costs has been studied in many countries (Gekas *et al.*, 1985). The process is only economically viable if very long distances are involved.

MICROFILTRATION

The distinction between MF and UF is somewhat arbitrary. UF may be considered to involve the processing of dissolved macromolecules, whereas MF considers dispersed particles such as colloidal casein micelles or fat globules. There is no distinction on purely theoretical grounds. MF has been carried out for many years in the dead-end mode, for example for filtration sterilisation. These applications have used depth filters in which particles become trapped in some way within the filter structure. Recent developments in MF membranes with a narrow pore size distribution have led to the possibility of MF in the same cross-flow mode as UF and RO.

MF membranes are available in the pore size range 0·01–10 μm. The materials of construction of the membranes, plant requirements and method of operation are essentially the same as with UF. It is notable that developments in inorganic membranes have often been aimed towards cross-flow MF, and many new applications have appeared in the literature (reviewed by Rios *et al.*, 1989). Some designs of inorganic membrane are available only in the MF range.

The development of cross-flow MF has led to the possibility of applications in the dairy industry. The use of MF to remove undesirable components (such as insoluble protein, lipids and microorganisms) during the processing of whey has proved very successful (Hanemaaijer, 1985). Also the process has been used successfully to recover fat from buttermilk (Rios *et al.*, 1989). Because of the overlap in size of bacteria and milkfat globules, and to a lesser extent with the smaller casein micelles in milk, the application of MF to milk is limited. Two promising applications have the advantage that no heating is necessary.

Fat Separation

Piot *et al.* (1987) demonstrated the use of alumina membranes (pore size 1·8 μm) for fat separation from raw whole milk; 98 per cent removal of fat from the permeate was achieved with a hundred-fold reduction in bacterial count. Obviously this leads to a concentration of bacteria in the fat-rich retentate.

Skimmed Milk of High Microbial Quality

The use of MF for improving the bacterial quality of skimmed milk (Bactocatch process) has been developed by the Alfa-Laval company (Malmberg and Holm, 1988). In this case, the size overlap between fat globules and bacteria is not important. The use of ceramic MF membranes (with pore size 0·5 μm) can produce a permeate with 99·6 per cent removal of bacteria with good permeability to protein, and without heating. The bacteria-enriched retentate may then be heat treated and either returned to the permeate or treated separately. The product could either be pasteurised to produce a market milk with extended shelf-life over conventional pasteurised milk, or used for cheesemaking. The disadvantage of the process is that complete removal of bacteria is not possible while allowing reasonable permeate flux and permeability to protein.

ELECTRODIALYSIS

The principles of ED and its applications in the food industry are reviewed by Lopez Leiva (1988a,b). ED can be used for separation of electrolytes from non-electrolytes, concentration or depletion of ions in solution, and the exchange of ions between solutions. Separation occurs due to electromigration of ions and charged molecules through membranes in an electric field. This depends on their charge combined with their relative permeability through the membranes. Separations are based on the use of ion selective membranes, effectively sheets of ion exchange resins. Composed of polymer chains (usually styrene-divinylbenzene), cross-linked and intertwined into a network, they bear either fixed positive or fixed negative charges:

A, anion permeable membranes: contain quaternary ammonium groups and repel cations;
C, cation permeable membranes: contain sulphonic groups and repel anions.

Counterions (i.e. with opposite charge to the fixed charges on the membranes) are freely exchanged by the fixed charges on the membranes and thus carry the electric current through the membranes. Charged macromolecules, such as proteins, will attempt to migrate in the electric field, but will not be able to pass through either anion or cation membranes due to their size. The basic operation of ED for demineralisation is illustrated in Fig. 15. A and C membranes are arranged alternately with plastic spacers to form thin solution compartments. Commercial membranes may be $1–2 \, m^2$ and 100–200 membranes may be assembled to form a stack. Commercial ED plants may contain several membrane stacks. At the ends of the stack are the anode and cathode, which set up a potential difference across each pair of membranes of 1–2 V and operate at a current density of $10 \, mAcm^{-2}$. The electrical potential causes anions to move towards the anode, and cations to move towards the cathode. However, the ion selective membranes act as barriers to either anions or cations. Hence, in the illustration, it can be seen that cations and anions are removed from compartments 2, 4 and 6, and concentrated in compartments 1, 3 and 5. When a food stream such as milk is to be

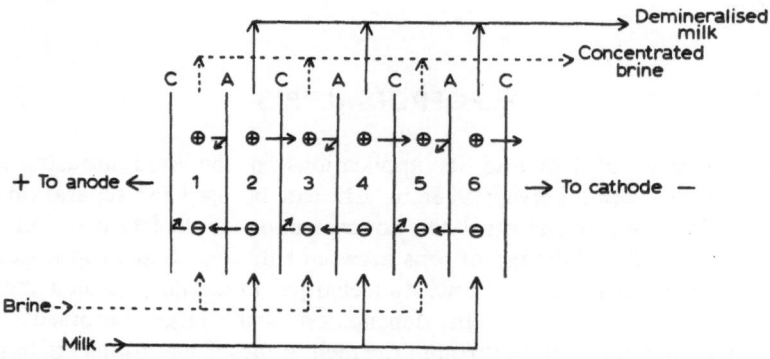

Fig. 15. Schematic diagram of electrodialysis for demineralisation of milk. A: anion permeable membranes; C: cation permeable membranes.

demineralised, it is fed into compartments 2, 4 and 6, and a dilute salt solution is fed into 1, 3 and 5 to maintain electrical conductivity across the stack, and to flush out the mineral ions from the feed. With a feed such as milk or whey, the membranes allow the passage of small ions; the protein, lactose, fat and colloidal minerals remain in the feed. In practice, the feed has to be circulated many times to effect a useful degree of demineralisation; for example when processing whey, a single pass through a stack may remove only 10 per cent of the minerals, so to reach 90 per cent demineralisation, some 30 passes are required. The electrodes are flushed with a separate electrode solution to remove gases produced by the discharge of ions, as well as other ionised species that may cause chemical damage.

The main problem in electrodialysis, as in UF, is concentration polarisation. Where the transport number of ions in solution is less than that in the membrane, at any one membrane there will be a depletion of ions on one side of the membrane and an increase in concentration on the other. This unequal distribution of ions must be minimised by employing high degrees of turbulence between the membranes, and since high current densities increase this inequality, densities must be limited. When processing milk or whey, calcium phosphate collects on the cathode side of a cation membrane, but more serious is the deposition of large organic molecules on the face of anion membranes; denatured protein components with large negative charges are particularly obstructive in this respect.

Alternative configurations of membranes may be used.

(i) Ion replacement can be achieved using only cation exchange or only anion exchange membranes. These replace a particular ion in the feed with another ion of the same sign.

(ii) Complete replacement of ions (ion substitution) can be achieved in which one species of ion is removed from the feed into an acceptor stream, and replaced by a different species from a third donor stream. Instead of the conventional A–C–A arrangement, the membrane sequence A–C–C–A is used, and the feed is passed into the C–C section.

(iii) A possible solution to the problem of fouling of the anion exchange membranes is to replace them with neutral membranes in the 'transport depletion' mode. This permits demineralisation to be achieved with longer processing times and easier cleaning, although the effectiveness of demineralisation is reduced as only one set of membranes is selective.

By far the major application of ED in the dairy industry has been for demineralisation of whey – approximately 3 million tonnes are treated annually. However, ED of milk may have some potential benefits. Demineralisation of milk may be necessary for infant foods, and special adult diets. Removal of calcium from milk by ED can be used to improve the stability of casein during freezing, and hence to improve the quality of frozen milk (Lonergan, 1982). In India, calcium removal from buffalo milk has been used to simulate human milk for infant nutrition (Kuchroo and Ganguli, 1980). Bodor *et al.* (1987) have used ED to demineralise and de-acidify fermented milk products, resulting in improved flavour and texture. It is possible that the process could be used in the manufacture of dietetic milks, or in the detoxification or radioactive decontamination of milks, although large-scale processing of this type has not been described in the literature.

POSSIBLE FUTURE DEVELOPMENTS

The most promising applications of membrane processing to the treatment of milk are in the use of UF retentates for production of cheeses and other products. The use of UF is more widespread in other countries, such as France, where soft cheeses are popular, and in Denmark where the Danish Sugar Corporation has been so progressive in the technology. In the UK, where hard cheeses predominate, UF awaits the development of a method for making satisfactory Cheddar type varieties from UF retentates. In the past, it has been predicted that all cheese would eventually be made from UF retentates. This now looks to be extremely unlikely.

The most revolutionary recent development in membrane technology has been the development of inorganic membranes. How widespread their application to milk processing will become remains to be seen. In general, developments in membrane technology have been less rapid, and applications to the dairy industry less widespread than would have been envisaged ten years ago.

REFERENCES

Abbot, J., Glover, F. A., Muir, D. D. and Skudder, P. J. (1979). *Journal of Dairy Research*, **46**, 663.
Bennasar, M., Rouleau, D., Mayer, R. and Tarodo de la Fuente, B. (1982). *Journal of the Society of Dairy Technology*, **35**, 43.

Bodor, J., Schoenmakers, A. W. and Verhue, W. M. M. (1987). European Patent 0 233 566 A 2.
Coton, S. G. (1980). *Journal of the Society of Dairy Technology*, **33** (3), 89.
Garrett, N. L. (1987). *Journal of the Society of Dairy Technology*, **40**, 68.
Gekas, V., Hallstrom, B. and Tragardh, G. (1985). *Desalination*, **53**, 95.
Glover, F. A. (1984). Technical Bulletin No. 5, National Institute for Research in Dairying, Reading, England.
Goudedranche, H., Maubois, J. L., Ducruet, P. and Mahaut, M. (1981). *Technique Laitière*, **950**, 7.
Hanemaaijer, J. M. (1985). *Desalination*, **53**, 143.
Hansen, R. (1977). *Nordeuropaeisk Mejeri Tidsskrift*, **43** (9), 304.
Hansen, R. (1981). *Nordeuropaeisk Mejeri Tidsskrift*, **47** (5), 147.
Kessler, H. G. (1981). *Food Engineering and Dairy Technology*, Verlag A. Kessler, Freising, p. 86.
Kiviniemi, L. (1974). *Kemia-Kemi*, **1** (12), 791.
Kristensen, S., Nielsen, W. K. and Madsen, R. F. (1981). *Northern European Dairy Journal*, **47**, 268.
Kuchroo, C. N. and Ganguli, N. C. (1980). *Journal of Food Science*, **45**, 1333, 1346.
Lawrence, R. C. (1989). Bulletin of the International Dairy Federation, Document No. 240.
Lee, D. N., Miranda, M. G. and Merson, R. L. (1975). *Journal of Food Technology*, **10**, 139.
Lewis, M. J. (1982). In: *Developments in Food Proteins – 1* (Ed. B. F. J. Hudson), Applied Science Publishers, London, pp. 91–130.
Lonergan, D. A., Fennema, P. and Amundson, C. H. (1982). *Journal of Food Science*, **47**, 1429.
Lopez Leiva, M. H. (1988a) *Lebensmittel-Wissenschaft und Technologie*, **21**, 119.
Lopez Leiva, M. H. (1988b). *Lebensmittel-Wissenschaft und Technologie*, **21**, 177.
Malmberg, R. and Holm, S. (1988). *North European Food and Dairy Journal*, **1**, 2.
Maubois, J. L. (1980). In: *Ultrafiltration membranes and applications. Polymer Science and Technology*, Volume 13 (Ed. A. R. Cooper), Plenum Press, New York, pp. 305–18.
Morgensen, G. (1980). *Desalination*, **35** (1, 2, 3), 213.
Pepper, D. and Orchard, A. C. J. (1982). *Journal of the Society of Dairy Technology*, **35** (2), 49.
Piot, M., Vachot, J. C., Veaux, M., Maubois, J. L. and Brinkman, G. E. (1987). *Technique Laitiere et Marketing*, **1016**, 42.
Rios, G. M., Tarodo de la Fuente, B., Bennasar, M. and Guizard, C. (1989). In: *Developments in Food Preservation – 5* (Ed. S. Thorne), Elsevier Applied Science, London, pp. 131–75.
Roenkilde Poulsen, P. (1978). *Journal of Dairy Science*, **61** (6), 807.
Scurlock, P. G. and Glover, F. A. (1983). Annual Report, National Institute for Research in Dairying, Shinfield, Reading, England, p. 130.
Serpa Alverez, D., Bennasar, M. and De la Fuente, B. (1979). *Le Lait*, **59** (587), 376.
Skudder, P. J. (1978). PhD thesis, University of Reading.
Sutherland, B. J. and Jameson, G. W. (1981). *Australian Journal of Dairy Technology*, **36**, 136.

Chapter 7

Utilisation of Milk Components: Whey

J. G. Zadow*

*Division of Food Research, CSIRO, Dairy Research Laboratory,
Highett, Victoria, Australia*

Whey may be defined, broadly, as the serum or watery part of milk remaining after separation of the curd that results from the coagulation of milk by acid or proteolytic enzymes. Its composition will vary substantially, depending on the variety of cheese produced or the method of casein manufacture employed. On average, whey contains about 65 g kg^{-1} of solids, comprising about 50 g lactose, 6 g protein, 6 g ash, 2 g non-protein nitrogen and 0·5 g fat. The protein fraction contains about 50 per cent β-lactoglobulin, 25 per cent α-lactalbumin and 25 per cent other protein fractions including immunoglobulins. There are wide variations in composition depending on milk supply, and the process involved in production of the whey. Wheys can be conveniently classed into groups:

Sweet wheys: titratable acidity 0·10–0·20 per cent pH typically 5·8–6·6.
Medium acid wheys: titratable acidity 0·20–0·40 per cent, pH typically 5·0–5·8.
Acid wheys: titratable acidity greater than 0·40 per cent, pH > 5·0.

In general, wheys produced from rennet-coagulated cheeses develop low levels of acidity, whereas the production of fresh acid cheeses, such as Ricotta or Cottage cheese, yields medium acid or acid wheys. Whey from caseins produced by acid addition is classed as high acid whey, whereas whey from rennet casein is sweet whey. Typical pH values and titratable acidities for a range of wheys are shown in Table I.

---

* Now of: Z. G. Zadow & Associates, Mordialloc, Victoria, Australia.

313

TABLE I
pH values and acidity of various wheys

	pH	Titratable acidity
Rennet types		
Edam	6·5–6·6	0·09
Gouda	6·5–6·6	0·10–0·12
Tilsit	6·5	—
Gruyere	6·5	0·07
Mozzarella	6·3	0·11
Emmental	6·2–6·5	—
Camembert	5·8–6·0	0·18–0·25
Cheddar	5·7–6·3	0·15–0·19
Danbo	5·3	—
Fresh acid types		
Ricotta (whole milk based)	5·3	0·14
Ricotta (whey based)	5·3	—
Cottage	4·5–4·6	0·50–0·55
Cream-Neufchatel	4·5	0·60
Quarg	4·5	0·70
Caseins		
Rennet	6·5	0·10
Lactic acid	4·5	0·64
Hydrochloric or sulphuric	4·0–4·4	—

Data from Kosikowski (1977)

Composition and Seasonal Variations

The mean composition of dried, sweet and acid whey samples manufactured in the USA is shown in Table II. Standard errors for many of the components are also shown, indicating possible variations. More specific information regarding the composition of rennet whey and whey after production of Emmentaler, Gruyere and Tilsit cheese has been reported by Blanc (1969). Other useful studies covering the composition of types of whey have been reported by Demmler (1968), Hargrove *et al.* (1976), Moulin and Galzy (1981), and Webb (1972). Of particular value is the detailed study of Sloth Hansen and Kjaergaard-Jensen (1977) covering the composition of cheese wheys and a range of caseins prepared by use of different precipitants.

Following the recent nuclear accident at Chernobyl in 1986, a number of studies have investigated the level of radionucleotides in whey. In general, any radionucleotides will behave like their chemically related,

TABLE II
Composition of dried sweet and acid wheys

Component	Content (per kg powder)			
	Sweet whey	Samples	Acid whey	Samples
Proximate				
Water (g)	31·9(0·887)	151	35·1(1·74)	52
Food energy (kJ)	14 760		14 190	
Protein (Nx6·38) (g)	129·3(0·874)	165	117·3(1·819)	57
Total lipid (fat) (g)	10·7(0·334)	148	5·4(0·31)	49
Carbohydrate (total) (g)	744·6		734·5	
Ash (g)	83·5(0·607)	147	107·7(2·226)	48
Minerals (mg)				
Calcium	7 960(314)	66	20 540(1 055)	
Iron	8·8(0·33)	41	12·4(1·78)	19
Magnesium	1 760(51·8)	41	1 990(96·4)	19
Phosphorus	932(39·9)	65	13 480(825·1)	20
Potassium	20 800(500·4)	60	22 880(976·4)	20
Sodium	10 790(671·3)	62	9 680(853)	18
Zinc	19·7(1·56)	43	63·1(5·49)	19
Vitamins				
Ascorbic acid (mg)	14·9(3·35)	41	9·09(3·3)	13
Thiamin (mg)	5·19(0·141)	44	6·22(0·720)	13
Riboflavin (mg)	22·08(0·506)	44	20·60(1·301)	13
Niacin (mg)	12·58(0·506)	43	11·60(1·952)	10
Pantothenic acid (mg)	56·20		56·32	
Vitamin B6 (mg)	5·84(0·159)	41	6·20(0·514)	10
Folacin (mcg)	120(0·94)	40	330(46)	10
Vitamin B12 (mcg)	23·7(0·971)	48	25(1·747)	10
Vitamin A (IU)	440		580	
Lipids				
Saturated,				
total (g)	6·8		3·4	
4:0	0·6(0·6)	4	0·2	
6:0	0·1(0·007)	4	0·1	
8:0	0·1(0·005)	4	Trace	
10:0	0·2(0·01)	4	0·1(0·02)	4
12:0	0·1(0·008)	4	0·1(0·02)	4
14:0	1·0(0·054)	4	0·5(0·05)	4
16:0	3·3(0·16)	4	1·5(0·13)	4
18:0	1·0(0·058)	4	0·6(0·04)	4
Mono-unsaturated,				
total (g)	3·0		1·5	
16:1	0·3(0·02)	4	0·3(0·07)	4
18:1	2·5(0·098)	4	1·1(0·09)	4

TABLE II *Cont.*

Component	Content (*per kg powder*)			
	Sweet whey	Samples	Acid whey	Samples
Polyunsaturated,				
total (g)	0·3		0·2	
18:2	0·2(0·03)	4	0·2(0·03)	4
18:3	0·1(0·02)	4	Trace	
Cholesterol (mg)	60			
Amino acids (g)				
Tryptophan	2·05		2·41	
Threonine	8·17		5·9	
Isoleucine	7·19		5·81	
Leucine	11·86		11·16	
Lysine	10·30		10·08	
Methionine	2·41		2·21	
Cystine	2·53		2·11	
Phenylalanine	4·07		3·86	
Tyrosine	3·63		3·00	
Valine	6·97		5·97	
Arginine	3·75		3·27	
Histidine	2·37		2·30	
Alanine	5·98		5·06	
Aspartic acid	12·69		11·49	
Glutamic acid	22·48		20·96	
Glycine	2·80		2·11	
Proline	7·86		6·99	
Serine	6·22		5·41	

Brackets values are standard errors of means. Data from Posati and Orr (1976).

non-radioactive elements; so caesium partitions mainly to the whey in a manner similar to potassium, and strontium partitions in a manner similar to calcium. Recent studies on this topic include those of Ron (1986), Wilson *et al.* (1988) and Pirhonen *et al.* (1987).

Seasonal variation in the composition of whey is of particular importance in countries such as New Zealand, Australia, and Ireland, where milk production varies widely throughout the year. Matthews (1978) has examined changes in the composition of New Zealand rennet, sulphuric and lactic casein wheys over a 12 month period. Seasonal changes in the composition of the wheys all followed similar patterns; the main change was an increase in protein content and a decrease in lactose concentration towards the end of the season. The true protein content of rennet whey varied from 110 to $140\,g\,kg^{-1}$ of solids, and the lactose content from 750 to $810\,g\,kg^{-1}$ solids throughout the season. The major differences between the wheys were the lower calcium and phosphate levels and higher amounts of protein and lipids in rennet whey; lower lactose and lipid contents, higher non-protein nitrogen contents, and the presence of lactate in lactic whey. Roeper (1971) has also examined seasonal changes in whey composition in New Zealand. It was reported that from August to February, when more than 80 per cent of the whey is produced, there was little change in the composition of Cheddar whey, apart from a gradual decrease in lactose. Ash content, true protein and non-protein nitrogen began to increase from February onwards, peaking in May or June. Drought conditions also had a significant effect on whey composition, with protein and lactose levels changing to typical winter values. Similar broad trends were observed in Australian studies (Dunkerley, unpublished data). However, the seasonal changes in whey composition reported in Australia and New Zealand are not likely to pose major difficulties for most whey processors. Some details of the seasonal variation for a number of components in wheys from New Zealand and Australia are shown in Table III. Detailed studies on the seasonal changes in whey produced in Denmark have been reported by Sloth Hansen and Kjaergaard-Jensen (1977).

Seasonal changes in the β-lactoglobulin, α-lactalbumin, glycomacropeptide and casein content of Cheddar cheese whey protein concentrate (WPC) produced in Australia have been reported by Regester and Smitters (1991).

PRODUCTION

Statistics

The full range of unit processes, such as concentration, drying, fermentation, demineralization, membrane processing and lactose hydrolysis, may be applied to all wheys. However, the commercial viability of many processes depends on specific compositional factors. For example, the use of wheys with high mineral contents for the manufacture of demineralised products may prove uneconomical.

Within the past 20 years, there has been a general realisation of the economic potential of whey. In the past, it has been treated as a waste product, to be disposed of in rivers or on land. Pressures by

TABLE III

Seasonal variation in composition of New Zealand and Australian wheys

Component	Whey variety				
	Rennet casein (1)	Sulphuric casein (2)	Lactic casein (1)	HCl casein (2)	Cheddar cheese whey (2)
Total solids[a]	64–67	60–64	62–64	58–61	61–66
True protein[b]	110–140	85–120	84–110	96–124	99–110
NPN:TN ratio[c]	0·24–0·28	0·24–0·29	0·30–0·35	0·18–0·24	0·22–0·30
Ash[b]	74–78	120–132	117–123	116–194	76–91
Lactose[b]	750–810	680–760	620–690		744–810
Calcium[b]	6–8	2–24	24–27		6–7
Inorganic phosphate[b]	10–16	29–34	30–34		8–30
Potassium[b]	18–26	8–24	20–24		19–25
Sodium[b]	6–16	4–14	6–12		6–11
Chloride[b]	14–21	12–18	13–19		19–20

[a]Value expressed as g solids per kg whey.
[b]Values expressed as a g kg^{-1} solids.
[c]Non-protein nitrogen to total nitrogen ratio.
(1) Data extracted from Matthews (1978).
(2) Unpublished data, CSIRO Division of Food Research.

environmentalists, and the realisation that whey contains about half the solids of milk and much of its nutritional value, have substantially changed the attitude of the industry. Whey is now recognised as a valuable raw material, with its potential for economic returns to the dairy industry still largely undeveloped.

Figures available on whey production and utilisation from various sources are not always consistent, and this should be borne in mind in interpretation of such figures. Strobel (1972) pointed out that there were few statistics on whey production and whey solids utilisation, and this remains the case. Still, cheese whey solids production can be assessed at approximately 4,500,000 tonne per annum, based on a world cheese production in 1980 of 11,000,000 tonne (Food and Agriculture Organization, 1981). Of the total cheese whey manufactured, approximately 50 per cent is produced in Europe, 18 per cent in North America, 2·5 per cent in Australia and New Zealand, 15 per cent in other dairy-developed countries, and the remainder in under-developed countries (Allum, 1980). The major European producer of whey is France (0·5 million tonne of solids per annum), followed by West Germany (0·3 million tonne) and Italy (0·2 million tonne). Casein production in the major Western countries amounted to 160,000 tonne in 1982 (Australian Dairy Corporation, 1983), resulting in production of an additional 320,000 tonne of whey solids.

The overwhelming use of whey powder in Europe is as animal feed, with calf milk replacers accounting for 90 per cent of all usage in the EC. Similar consumption patterns are observed in Australia and New Zealand. By contrast, in the USA, approximately 66 per cent of dried whey is incorporated into human food, with major users being the dairy and baking industries (Delaney, 1979). Until recently, detailed information on worldwide utilisation of whey solids has been very sparse, although some early details have been reported by Clark (1979) for the USA.

More recently, a report on the status of whey and whey products in the USA reported that annual whey production had increased from 28·7 billion pounds (13 billion kg) in 1973 to 50·9 billion pounds (23 billion kg) in 1985 (Clark, 1987). The major users of whey concentrates and whey powders in the USA were the dairy and baking industries. Zadow (1987) reported that total whey production totals in Australia and New Zealand in 1985 were 1,436,000 and 945,000 tons respectively. Only 44 per cent of Australian whey was used for manufacturing processes (26 per cent via ultrafiltration), 56 per cent was used for pig feed, or as fertiliser, or was sprayed onto land. In Europe, 50 per cent of total whey produced is utilised, mainly as dried whey. In the Netherlands, about 38 per cent is used for dried lactose-free calf milk replacers, 35 per cent for lactose and 11 per cent for dried demineralised lactose-free whey. Only 8 per cent is used as dried whey, and the remainder is used as liquid whey or in the beverage industry (Hoogstraten, 1987).

Throughout the world there has been a slow trend for whey products to become more important in the economics of the dairy industry. But there is little doubt that the general trends in whey utilisation have changed significantly over the past few years. The adoption of membrane processing by many manufacturers in dairy-developed countries has resulted in the production of significant quantities of whey protein concentrates with functional properties of considerable interest to the food industry. In 1978, modified whey products (demineralised, delactosed wheys and whey protein concentrates) accounted for 23 per cent of the total whey processed in the USA for animal feed, and 11 per cent of the whey processed for human consumption (Allum, 1980). The potential of whey protein concentrates, in particular, as ingredients for foodstuffs remains high, and production increased rapidly in the 1980s. Further increases in production can be expected over the next few years to meet demand. It is to be hoped that, with the increasing economic importance of whey products to the dairy industry in the future, more detailed information will become available on production and utilisation.

Economics

The basic decision by a dairy company on the alternative approaches to whey utilisation should be determined by the relative economics of the various options. However, the interplay of these options is very complex, and there is little detailed information available concerning the relative economics of these process options. Table IV outlines some of the many process and product options available to the whey processor, and Table V lists some of the many chemically modified and fermented products which may be manufactured from whey. Clearly the selection of a process and product range to be adopted by a particular manufacturer is extremely difficult. Ryder (1980) has reported some economic aspects of whey processing, examining in detail the costs of dumping, spraying on land and animal feeding. However, detailed comparative figures for the full range of process and product options is not available.

Aspects of the economic utilisation of permeate have been discussed by Coton (1980), and a useful comparison of the costs of concentration of

TABLE IV
Some process and product options for treatment of whey

Processes	*Products*
Separation	Spray dried whey powders
Clarification	Roller dried whey powders
Removal of lipid fractions	Lactose hydrolysed whey powder
Evaporation	Demineralised whey powders
Spray drying	Demineralised lactose
Roller drying	hydrolysed whey powders
Crystallisation of lactose	Concentrated liquid wheys
Refining of lactose	Crude lactoses
Centrifugal separation of	Refined lactoses
lactose crystals	Whey creams
Ultrafiltration	Hydrolysed whey syrups
Reverse osmosis	Hydrolysed permeate syrups
Demineralisation by UO	Whey protein concentrates
Demineralisation by IE	Demineralised WPC
Demineralisation by ED	Demineralised lactose
Acid hydrolysis of lactose	hydrolysed WPC
Enzymic hydrolysis by	Crude lactose
lactose/soluble enzyme	Refined lactose
Enzymic hydrolysis by	Whey protein isolates
immobilised enzyme	Whey protein fractions
Fermentation	Lactoferrin
Chemical modification	Lactoperoxidase
Selective protein	Liquid whey
precipitation	Fermented products
Chromatographic	see Table V
separations	Chemical modified products
	see Table V

After Zadow (1991).

whey by reverse osmosis (RO) and evaporation has been reported by Pepper (1981). Preconcentration of the whey by reverse osmosis, followed by conventional evaporation proved to be the most economic combination. A case history for a plant processing 100,000 tonne of whey per annum showed that the annual savings of this process were greater than the capital cost of the RO equipment. Muller (1979) has detailed the savings to be made by the RO concentration of whey, based on a case study involving the supply of whey to a central processing depot from three small factories. It was concluded that RO concentration offer satisfactory economies, even if only small quantities of whey are available. A further useful study by Kjaergaard-Jensen and Oxlund (1988) reported

TABLE V

Some options for lactose utilisation by fermentation or chemical modification

Product	Applications
Acetic acid	Foods
Acetone	Various
Alcohol	Foods, energy
Amino-acids	Various
Antibiotics	Medical
Butanol	Various
Citric acid	Foods
Food oils	Animal feeds
Fuel gas	Energy
Galactonic acid	Various
Gibberellic acid	Plant hormones
Glucaric acid	Various
Gluconic acid	Various
Hydrolysed lactose	Sweetener, lactose malabsorbers
Itaconic acid	Various
Lactase	Enzyme applications
Lactic acid	Foods
Lactic polymers	Biodegradable plastics, prosthetics
Lactitol	Non-nutritional sweetener
Lactobionic acid	Chelating
Lactose crystals	Food, tablet binder
Lactose foams	Insulation
Lactose polymers	Surfactants
Lactosyl urea	Ruminant feeding
Lactulose	Infant nutrition
Malic acid	Various
Oligosaccharides	Medical
Polysaccharides	Food gums
Single-cell protein	Various
Vitamins	Food fortification

After Zadow (1991).

on cost comparisons for the concentration of whey by several systems, including conventional evaporators and reverse osmosis.

However, the rate of adoption of much new technology has been particularly disappointing. With the exception of ultrafiltration (UF) technology, there is little indication of widespread adoption by the dairy industry of much of the newer technology discussed later in this article. Such technologies include lactose hydrolysis, reverse osmosis, protein fractionation, fermentation and chemical modification of lactose. In many cases, the technology is well understood and requires little if any further technical development. It seems that the uncertain market demand for many of the new products has resulted in considerable reluctance by the dairy industry worldwide to invest in commercialisation of these processes.

WHEY CHARACTERISTICS

Analytical

Harper (1979) has reviewed recent studies on the analysis of whey and whey products, covering most aspects of the determination of whey components. Much other information is also included in the excellent monograph on whey and whey utilisation by Sienkiewicz and Riedel (1990).

The Harland-Ashworth test for measuring undenatured whey protein nitrogen in non-fat dried milk has been modified to measure undenatured whey protein in dried whey. The modified method produced reproducible estimates of undenatured whey protein in both non-fat dried milk and whey (Mahmoud *et al.*, 1990).

There has been particular interest in the measurement of lactose. Some recommended procedures are the Munson-Walker method involving the formation of cuprous oxide, polarimetry, the Chloramine T method (involving back titration of liberated iodine), the phenol–sulphuric acid method, and the picric acid method involving measurement of a coloured product at 520 nm (Jenness and Patton, 1959). Recently, high performance liquid chromatography techniques have been employed (Euber and Brunner, 1979), as well as an enzymatic technique based on the use of a cryoscope (Zarb and Hourigan, 1979). This method involves treatment of the product with lactase in buffer, then determination of the difference in freezing point of the sample on mixing, and after incubation for 1 h at 37 °C. This difference is directly proportional to the concentration of lactose present. Removal of the protein from the sample before analysis is not required, and the method is specific regardless of the presence of other sugars.

Whey Properties

Physical and chemical properties
There is comparatively little data available on the detailed physical properties of whey. The surface tension of cheese wheys has been shown

to vary between 40,000 and 84,000 N m^{-1}, increasing with increasing total solids, and decreasing with temperature (Zaetz *et al.*, 1982). It is probable that some of the observed variation is due to differences in lipid content of the whey.

The viscosity characteristics of whey and concentrated whey are important, not only in terms of evaporation efficiency, but also for operations such as lactose hydrolysis, which might be expected to function more efficiently if applied to concentrates. Studies on the viscosity of whey and whey concentrates have been reported by Chebotarev *et al.*, 1983; lactose hydrolysis reduces viscosity and the degree of non-Newtonian behaviour. Lipatov and Chebotarev (1981) have also examined the viscosity and density characteristics of whey over a temperature range 20–50 °C. They have suggested that whey is a heterogeneous system in which the fat globule size follows a logarithmic distribution and the casein particles have an irregular shape up to 1 mm in diameter. Their results provide for the use of a 'separability index' as a characteristic of the fat–whey system; the results may be used as the basis of a special separator design.

The compositional and thermal properties (by DSC) of Cheddar cheese whey WPC have been studied by Patel *et al.* (1990). Denaturation enthalpy was positively correlated with β-lactoglobulin and protein content. Denaturation temperature correlated positively with phospholipid content.

Flavour

Studies on the flavour of acid whey, based on eight flavour characteristics, have been reported by McGugan *et al.* (1979). When increasing concentrations of whey were added to skim-milk, 'brothiness' was first noted when 20 per cent whey was added, and diacetyl, bitterness and sweetness at 40 per cent addition. Volatile acidity, non-volatile acidity, saltiness and astringency were only noted in 100 per cent whey. Neutralisation of the whey resulted in a change in all flavour characteristics.

Effect of heat

Hillier *et al.* (1979) have examined protein denaturation, using gel electrophoresis, in wheys heated over the range 70–150 °C. The rate of denaturation of α-lactalbumin appeared to be first order, but was actually considered more likely to be second order displaying pseudo-first order kinetics. The rate of denaturation of β-lactoglobulin A and B followed second order kinetics, whereas that of serum albumin was more complex.

Below 95 °C, the rate of denaturation of β-lactoglobulin A was faster than that of β-lactoglobulin B.

A number of studies have reported the effect of total whey solids on the denaturation of whey proteins. The individual proteins in whey have a wide range of denaturation temperature, about 65–75 °C, as may be seen from Table VI (Ruegg *et al.*, 1977). In general, it has been found that, with higher total solids, the denaturation of β-lactoglobulin A and B slows, but the denaturation of α-lactalbumin increases. Increased lactose concentrations reduced denaturation of both proteins, perhaps as a result of the formation of heat-induced complexes. Increased calcium contents, up to 0·4 mg ml^{-1}, tended to slow denaturation, but above this level, little further effect was observed. The rate of denaturation of both proteins was slower at pH 4 than at pH 9 (Hillier *et al.*, 1979). Other comparisons have shown that the method used for concentration may significantly affect the nature and detailed conformation of whey proteins, with membrane processing having much less severe effects than conventional evaporation.

TABLE VI

Thermodynamic parameters for heat denaturation of whey proteins in simulated milk ultrafiltrate

Protein	Denaturation temp.[a] $TD(°C)/SD$	Denaturation enthalpy $HD(kJ\,mol^{-1})/SD$	Renaturation (per cent of HD)
α-chymotrypsin	55·7/0·3	458/32	0
Xanthine oxidase[b]	61·4/0·8	3230/300	5–10
	77·6/0·5		
Serum albumin	2·2/0·5	939/88	0
Ribonuclease	62·7/0·2	418/18	60–70
Apo–lactoferrin	64·7/0·3	100/95	40–50
α-lactalbumin	65·2/0·2	318/17	80–90
Fe–lactoferrin[c]	69·0/0·5	2020/90	0
	83·5/0·5		
Lysozyme	70·5/0·2	494/10	10–20
β-lactoglobulin	72·8/0·4	227/21	0
γ-globulin	72·9/0·4	4120/260	0

[a]Extrapolated values from heating rate of 0 °C min^{-1}.
[b]Estimated from differential scanning calorimetry thermograms of samples denatured, heated at 5 °C min^{-1} then rescanned.
[c]Overlapping peaks HD calculated from total area of unresolved peaks.
SD = standard deviation.
After Ruegg *et al.* (1977).

Pearce (1983) has described a novel method for the manufacture of enriched α-lactalbumin and β-lactoglobulin fractions from whey, relying on previously unknown solubility characteristics of these proteins, and using temperatures below those at which denaturation is generally considered to start. (This process is described in more detail later in this chapter.) The changes occurring in the proteins at these temperatures may partly explain the improvement in flux commonly obtained by holding whey at 55 °C before ultrafiltration. Clearly, there is still much work to be done to understand the effect of heat, concentration and aqueous environment on the behaviour of whey products.

Nutritional aspects

Many of the major aspects of the nutritional properties of whey-based products are covered in the excellent monograph by Renner (1983) on the topic of milk and dairy products in human nutrition. Whey proteins, with their high cysteine content, are considered good quality foodstuffs for human consumption. For example, studies have shown the benefits of including whey proteins in foodstuffs for low birth weight babies (Berger et al., 1979), while other reports indicate that ingesting high levels of dried whey has little effect on plasma lipids. The main nutritional problem of many whey-based products is the lactose intolerance of certain sectors of the population, particularly non-Caucasian races; the extent of this problem has been reviewed by Garfield (1980). In Japan, for instance, digestive problems from drinking milk affect more than 20–25 per cent of the population, and clinical trials indicate that over 50 per cent of Japanese children have low lactase activity.

One option available to whey processors supplying these markets is to hydrolyse the lactose in their products. However, trials on rats in the early 1970s suggested that a high intake of galactose might result in high blood galactose levels which, in turn, might be responsible for cataract formation. However, a number of studies have now indicated that, when galactose is ingested with equimolar quantities of glucose (as in hydrolysed lactose products), blood galactose levels rise very little (Coton, 1980; Williams and MacDonald, 1982). The incorporation of whey products into domestic animal foods – currently a growth market – is also restricted to some extent by lactose intolerance in many animals; lactose hydrolysed products will also have a large potential application in this area.

A reduction of the serum cholesterol and high density lipoprotein cholesterol levels in swine has been reported (Ritzel *et al.*, 1979) following whey feeds.

PROCESSING OF WHEY

Clarification

For many purposes, clarification is an essential part of whey processing. Clarification is particularly important for the manufacture of whey protein concentrates by ultrafiltration, where the presence of even small amounts of lipid can have an adverse influence on the functional properties of the finished powder. Means for clarification include the modification of whey composition through demineralisation to specific levels, followed by sedimentation of the aggregated material (Wit and Klarenbeek, 1978), the use of anionic precipitating agents (Best *et al.*, 1982) and microfiltration (Merin *et al.*, 1983). In microfiltration, whey was filtered through a membrane with a pore size of 1·2 nm, producing a permeate free of fat, but containing all the other constituents of whey. This product had a 30 per cent higher ultrafiltration flux rate than unfiltered whey. Small fat globules were therefore suggested as major contributors to membrane fouling during the ultrafiltration of cheese whey. Similar studies have been reported by Hanemaaijer (1985).

Storage of Whey

A number of reagents have been suggested as additives to extend the useful storage life of whey through inhibition of bacterial action. The processes reported include the addition of benzoic acid (with pH adjustment to 3·0–4·2), centrifugal clarification followed by the addition of hydrogen peroxide (0·04 per cent), the addition of 0·3 per cent propionic acid (delaying spoilage of whey at 20°C by about 2 days), the addition of orthophosphoric acid and an organic carboxylic acid, and the addition of sodium sorbate, formaldehyde (0·03 per cent), or sodium or magnesium bisulphite (0·5 per cent). It should be noted that some of these approaches may not be legally acceptable treatments for whey in many countries.

Evaporation and Concentration

A number of important developments are occurring in the industry in
the field of evaporation technology. Alternatives to conventional, multistage evaporators are now becoming available, and include mechanical
vapour recompression evaporators (MVR), submerged combustion techniques (Lovell-Smith, 1982), and membrane processing using reverse
osmosis. Current evidence suggests that for large-scale applications,
MVR and RO are closely competitive. For many applications, the most
economic mix appears to be the use of RO to concentrate the whey to
about 25 per cent solids, with further concentration by conventional
methods. For small-scale operations in particular, the RO–evaporator
combination appears to have economic advantages. In practice however, there appears to be a tendency for industry to continue to use
conventional evaporative systems in preference to an RO–evaporator
combination. The use of RO is mainly confined to those applications
where concentration to less than 25 per cent solids is required, for
example to assist in a reduction in transport costs.

Advances in the concentration of whey by evaporation have recently
been reviewed, Sandfort (1987); by reverse osmosis, Pepper and Pain
(1987); and by some newer drying technologies, Kjaergaard-Jensen
(1987).

Dehydration

One problem in drying whey based products with high lactose contents
is caused by the hygroscopic nature of α-lactose. α-Lactose is comparatively insoluble (7 g per 100 ml at 15 °C) compared to its isomer, β-
lactose (50 g per 100 ml). In aqueous equilibrium, a lactose solution
contains about 63 per cent of the β-form, due to mutarotation. But on
crystallisation, the more insoluble α-form will initially precipitate, and
some of the β-lactose will be converted to the α-form; crystallisation
ultimately yields mostly α-hydrate crystals in the product. α-Lactose
exists as a monohydrate, whereas solid β-lactose contains no water of
crystallisation. When solutions containing lactose are dried rapidly,
there may be insufficient time for the α-lactose to form as the monohydrate, and the dried lactose is essentially in the same form as the liquid.
Neither α-lactose hydrate nor β-lactose is hygroscopic, but anhydrous
α-lactose is strongly hygroscopic, and can absorb moisture from the air
forming the hydrate, which occupies a greater volume than the anhy-

drous form. This action is responsible for the caking and lumping observed in many whey based products.

Manufacturing procedures need to be modified to overcome this problem, and such processes involve the conversion of much of the lactose into crystalline α-lactose monohydrate before drying. This change may be effected either by holding the concentrate for sufficient time for crystallisation of the α-hydrate to occur, or by using techniques similar to instantising. Thus, if the surfaces of the particles emerging from the spray drier are partially humidified, or the product is only partly dried, conversion of the α-lactose to the hydrated form can occur before final drying. Recent studies on the crystallisation process have been reported by Hynd (1980), and on the instantising process by Lewicki *et al.* (1981). In these latter studies, the best results were obtained by six treatments (30 s each) involving wetting of the product by air with a moisture content of 3·4 per cent, before final drying. Kjaergaard-Jensen and Oxlund (1988) have recommended the addition of 0·1 per cent fine grained α-lactose to assist with controlled crystallisation of the concentrate. Levels of 0·5–1·0 per cent are recommended by Sottiez (1985). Useful reports on the dehydration of lactose and drying of whey-based products are included in the works of Visser *et al.* (1988) and Kjaergaard-Jensen and Oxlund (1988).

Storage-Induced Changes in Powder

The major changes occurring during storage of whey based products involve changes in the physical state of the lactose, and browning reactions. The rate of moisture uptake by whey based products depends on a number of variables, particularly protein content, relative humidity, temperature and pH (Johns, 1982). Other studies (Saltmarch, 1980) have shown that moisture sorption isotherms for hygroscopic (but not non-hygroscopic) dried wheys stored at 25, 35 and 45°C show discontinuities between water activities of 0·33 and 0·44, coinciding with the initiation of the transition from the amorphous to the crystalline form of α-lactose. Lactose crystallised at water activity 0·40 after about one week at 25°C, and at water activity 0·33 after 1 week at 35 or 45°C (Saltmarch and Labuza, 1980). The maximum rate of browning and loss of protein quality occurred at water activity 0·44. Loss of protein quality at constant water activity, but under fluctuating storage temperatures, was greater than at constant temperature (Labuza and Saltmarch, 1982). This has particular implications for countries exporting whey powders to tropical areas.

The changes in the functionality of cheese whey WPC during storage for 6 months at temperatures ranging from -40 to $40°C$ and water activities from 0.15 to 0.41 have been examined by Hsu and Fennema (1989). Of the product attributes studied, browning was most important; of the storage variables, temperature and time were of primary importance, with water activity of secondary importance. Recommended conditions for storage were temperature not exceeding $20°C$ and water activity not exceeding 0.2.

UNIT PROCESSES AND PRODUCTS

Demineralisation

The influence of demineralisation on overall whey processing costs is considerable, and may be comparable with the cost of spray drying per tonne of solids. Such a high cost factor means that demineralisation is only employed where a high return for the product is guaranteed. Furthermore, the need to handle large quantities of difficult effluents from these processes has restricted their acceptance by industry.

The two main methods most commonly used for the demineralisation of whey – ion exchange (IE) and electrodialysis (ED) – result in products of somewhat different composition. Ion exchange is relatively non-selective and removes both monovalent and polyvalent ions, whereas electrodialysis is more dependent on ionic mobility and tends preferentially to remove monovalent ions. Recent summaries of each of these technologies have been reported by Batchelder (1987) (ED), and Jonsson and Arph (1987) (IE). The relative economics of each process have been examined by Marshall (1979). Economically, only about 90 per cent demineralisation is possible with electrodialysis, with 50 per cent being much more viable. By comparison, ion exchange processes can readily achieve close to 100 per cent demineralisation. The generally lower capital cost of ion exchange gives it a potential advantage in small systems, particularly where high degrees of demineralisation are required. Electrodialysis is likely to have advantages for plants with high hourly utilisation rates, for the manufacture of products which do not require high levels of demineralisation, and where low cost electricity is available. Both systems produce substantial quantities of effluent, generally at least equal in volume to the quantity of whey processed.

A novel approach to demineralisation by ion exchange is the SMR process developed by the Central Laboratory of the Swedish Dairies Association (SMR) in conjunction with Wedholms AB. The cations in whey are exchanged for ammonium ions, and the anions exchanged for bicarbonate ions. The resultant ammonium bicarbonate is then evaporated from the whey in the form of ammonia, carbon dioxide and water, and is recovered for regeneration of the ion exchange resins. High degrees of demineralisation are possible, and the product retains less than 0·25 per cent free ammonia. More than 80 per cent of the ammonium carbonate is recovered for use as regenerant, facilitating effluent treatment, and furthermore, only one solution needs to be passed through the two ion exchange systems. Total costs are estimated to be substantially less than for conventional ion exchange systems, and the mineral content of the effluent from the process is approximately half that of conventional systems (Jonsson and Forsman 1978). A full-scale plant, with a capacity of more than 4 tonne per day, has been installed in Sweden.

Manczak (1979) has indicated that the improvements in the efficiency of demineralisation by electrodialysis may be achieved by processing at 50°C rather than at 20°C; less electrical energy is required. Replacement of the anionic membranes with neutral membranes resulted in a three-fold increase in the electrical energy required for operation, and a less effective system for mineral removal. Nevertheless, this approach may be justified, in some cases, as anionic membranes are very expensive and wear out rapidly.

An alternative approach to improve the efficiency of electrodialysis has been reported by Ennis and Higgins (1981). In these studies, the calcium in permeate was replaced with sodium by ion exchange before electrodialysis. Permeate conductivity, the current carrying capacity of the plant, and the rate of ash removal all increased as the extent of replacement increased. In permeate where nominally 100 per cent of the calcium had been replaced, there was an initial rapid decline in ash removal rate, after which no further decline occurred. In the control samples, the ash removal rate followed an exponential decay pattern.

More recently, 'open' RO membranes have been adopted commercially for partial demineralisation of whey and permeate. These membranes are designed to allow a proportion of the ash content of whey to pass through into the permeate, whilst retaining most of the lactose and protein in the retentate (Eriksson, 1986a, b; Pepper, 1988). In the past decade, advances in this technology have been considerable, with the technology moving from laboratory to commercial applications. Among the installations of this technology are a plant in the USA for treatment of salt whey drippings (Gregory, 1986), two Alfa-Laval plants in Sweden, and a plant in Australia (Marshall, personal communication). Processing conditions for open RO are similar to those for conventional RO, but use somewhat lower operating pressures. As would be expected, fluxes are somewhat higher than for conventional RO. In general, the economic limit for demineralisation with this technology is about 50%. A significant benefit is the significant degree of concentration effected during the process. The availability of this technology, possibly in combination with ion exchange, may result in the ability to produce a wider range of demineralised products more economically.

Two further developments in demineralisation technology have also been described recently. 'Sirotherm', thermally regenerable ion exchange resins, may have potential in this field. These resins may be regenerated by treatment with warm water, rather than through the use of chemical reagents (Jackson, 1980). However, at present, some resins in the Sirotherm range are limited in their ability to process streams with the levels of calcium commonly found in whey, and have a low ion exchange capacity per kilogram of resin. The second technology for demineralisation, 'counter diffusion', has been described by Lee and Johnson (1987). In this methodology, the concentration gradient of the ionic species is the only driving force across a selective membrane. Lactose transport is reduced by the presence of immobilised inorganic crystals on the surface of the counter diffusion membrane. Plants utilising this technology are operating in Australia and Europe.

Hayes (1982b) has described a novel technique for the demineralisation of permeate from the ultrafiltration of whey. The method involves the addition of calcium to the permeate to adjust the Ca/P ratio in such a way as to precipitate calcium phosphate in the apatite form. The precipitate can be removed by mild centrifugation, effectively reducing the calcium content of the permeate by more than 60 per cent, and the phosphorus content by more than 50 per cent. A similar approach has been described by Brothersen et al. (1982). This approach is simple and inexpensive and offers advantages where only partial demineralisation is required, or as a preliminary treatment before conventional processing. The recovered calcium may have pharmaceutical applications.

Whey Protein Recovery and Modification

Lactalbumin
Traditionally, lactalbumin is made by heat precipitation of proteins from whey at an acidic pH (Robinson *et al.*, 1976). The product is comparatively insoluble so its functional properties are limited. As an ingredient of high nutritional quality, it still commands an important sector of the market. Recent modifications suggested for the manufacture of lactalbumin include cation exchange of part of the whey, employing a novel approach to reduce the cost of pH adjustment before heat treatment. A portion of the whey is mixed with a cation exchange resin, resulting both in partial demineralisation and a drop in the pH of this fraction. This portion of the whey is then mixed with the remainder of the whey to obtain the desired pH for lactalbumin production by heat treatment. Process improvements are also claimed by coupling the traditional process with ultrafiltration to improve efficiency and yield (Buhler *et al.*, 1981). The general steps involved in the manufacture of lactalbumin are shown in Fig. 1 (after Matthews, 1984). The application

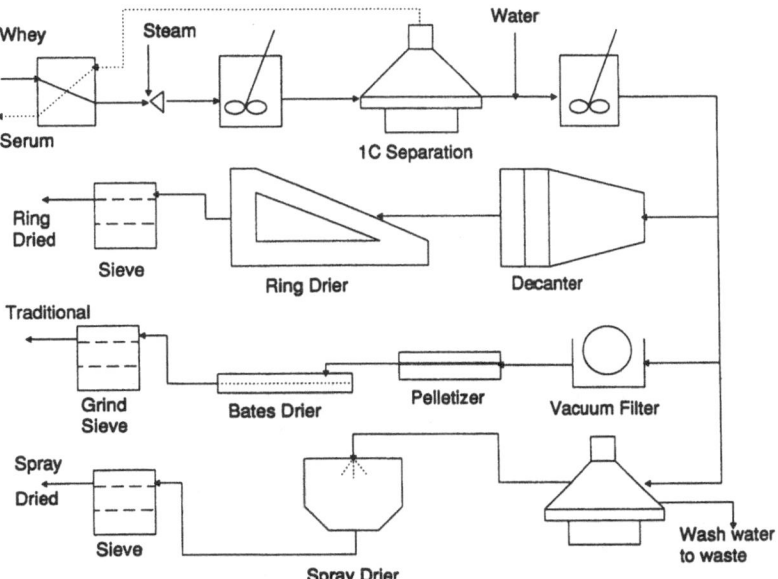

Fig. 1. Options for the production of lactalbumin (after Matthews, 1984).

of UF in the overall process has been studied by Modler and Harwalker (1981). It was reported that substantial improvements could be achieved in yield by the application of this technology.

Protein isolates by selective precipitation

Recently, two techniques have been described for the manufacture of whey fractions rich in α-lactalbumin and β-lactoglobulin (Amundson *et al.*, 1982; Pearce, 1983). Essentially the processes rely on modifying the pH, ionic composition and temperature of the whey to selectively precipitate one of the proteins. In essence, these technologies rely on exploitation of the fact that the solubility of α-lactalbumin changes markedly under specific conditions of pH and temperature. Thus, at pH values near its isoelectric point, α-lactalbumin will undergo aggregation and flocculation. It can thus be separated from whey, leaving a soluble fraction rich in β-lactoglobulin. Development of this methodology (Pearce, 1987) has shown that the sedimented fraction also contained most of the lipid-containing material present in the whey. A process based on a similar approach has been outlined by Maubois *et al.* (1987). The two processes, that of Maubois (1987) and Pearce (1987) have been compared, as shown in Fig. 2, by Modler and Jones (1987). These processes are currently undergoing commercial development. It is anticipated that the fraction rich in β-lactoglobulin will show desirable functional properties in food applications, and the fraction rich in α-lactalbumin may have pharmaceutical and infant food applications. Yoshida (1990) has also described a method for the isolation of β-lactoglobulin and α-lactalbumin, using gel filtration with Sephacryl S-200 and purification by DEAE ion exchange chromatography.

Co-precipitation with soluble polymers or phosphates

Each of these processes is capable of producing an essentially undenatured whey protein fraction, although some end-products (for example those derived using soluble carboxymethyl cellulose, iron complexing and phosphate precipitation) would also contain some precipitant. It appears that there would be considerable difficulty in scaling up many of these processes from laboratory and pilot-scale applications to commercial reality and, in spite of the high functionality of many of these fractions, markets for the products have not yet been developed.

However, there has been considerable research interest in the further development of these techniques for the manufacture of protein concentrates (Marshall, 1979). Humbert and Alais (1982) have reported on the effectiveness of some of these processes. Using carboxymethyl cellulose as the precipitant, protein recovery was 69 per cent, with the product containing about 60 per cent protein and 1·7 per cent ash; with ferric chloride, the corresponding values were 89 per cent, 79 per cent and 18 per cent; with ferric polyphosphate, the values were 92 per cent, 70 per cent and 28 per cent; for polyacrylic acid, the values were 63 per cent, 82 per cent and 13 per cent; and with sodium hexametaphosphate, the values were 76 per cent, 84 per cent and 13 per cent. All of the protein concentrates had excellent functional properties, and were considered to have considerable potential as food ingredients.

Saito *et al.* (1991) have described a new isolation method for caseinoglycopeptide from sweet cheese whey, using alcohol precipitation and ion exchange chromatography after heat coagulation of the whey protein. The most successful method was heating a whey solution (10 per cent w/v) at pH 6·0 for 1 h, followed by precipitation with cold ethyl alcohol (50 per cent v/v). The yield was approximately 1·1 g of peptide from 100 g of cheese whey powder.

More recently, there has been interest in the selective extraction of some of the minor constituents of whey by chromatographic means, utilising either conventional chromatographic supports or affinity chromatography. Two components of particular interest are lactoferrin and lactoperoxidase. Foley and Bates (1987) have described a process for extraction of lactoferrin from human milk whey, achieving a recovery of more than 80 per cent with a simple, rapid process based on adsorption by cellulose phosphate. There appears to be interest in similar applications in a number of research centres throughout the world. For example, Yoshida (1989) has described a process for preparation of lactoferrin by hydrophobic interaction chromatography of whey. It was estimated that approximately 80 mg of lactoferrin could be isolated per litre of whey by this methodology. The extraction of lactoperoxidase and lactoferrins from bovine milk whey using carboxymethyl cation exchange chromatography has also been described by Yoshida and Ye-xiuyun (1991); yields were 41 mg of lactoperoxidase, 21 mg of lactoferrin A and 67 mg lactoferrin B per litre of whey. The isolation of ovine lactoferrin has been outlined by Buchta (1991); an overall yield of 55 per cent was obtained by separation from ovine colostrum on CM-Sephadex C-50 and Blue-Sepharose.

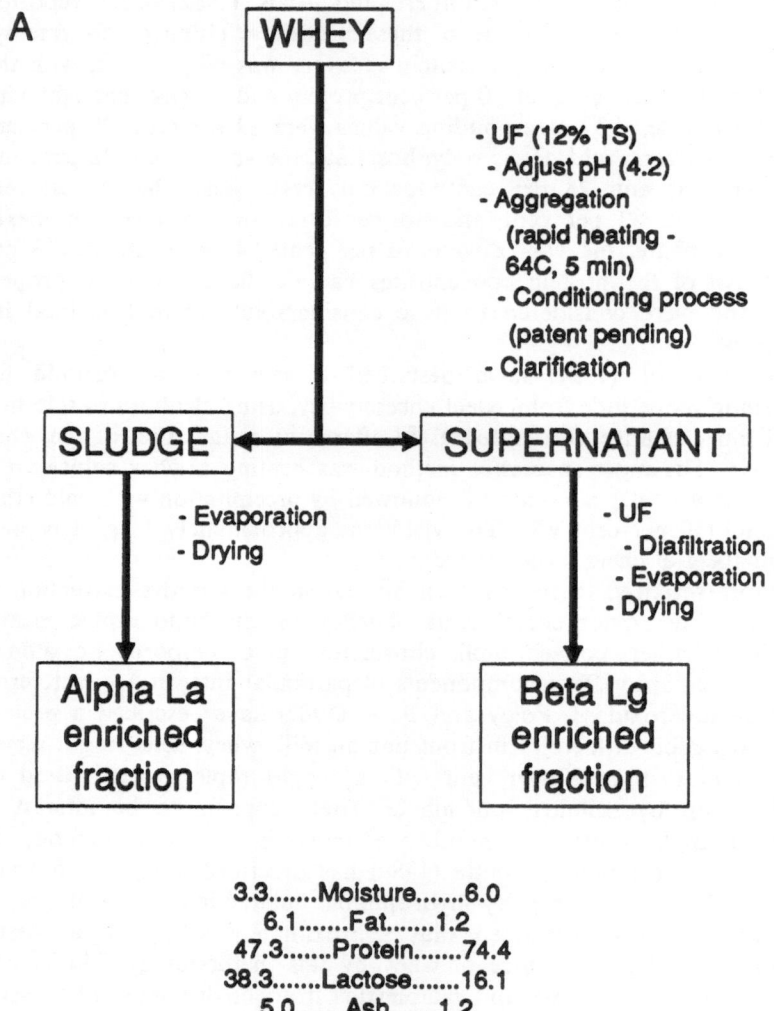

A

WHEY

- UF (12% TS)
- Adjust pH (4.2)
- Aggregation
  (rapid heating -
  64C, 5 min)
- Conditioning process
  (patent pending)
- Clarification

SLUDGE ← → SUPERNATANT

- Evaporation
- Drying

- UF
  - Diafiltration
  - Evaporation
- Drying

Alpha La enriched fraction

Beta Lg enriched fraction

```
3.3.......Moisture.......6.0
6.1.......Fat.......1.2
47.3.......Protein.......74.4
38.3.......Lactose.......16.1
5.0.......Ash.......1.2
```

Fig. 2. Processes for the separation of α-lactalbumin and β-lactoglobulin enriched fractions. Fig. 2a is the Australian process of Pearce (1987), Fig. 2b, c are the French processes for removal of phospholipid and the preparation of β-lactoglobulin and α-lactalbumin (Fauquant *et al.*, 1985a), from Modler and Jones (1987).

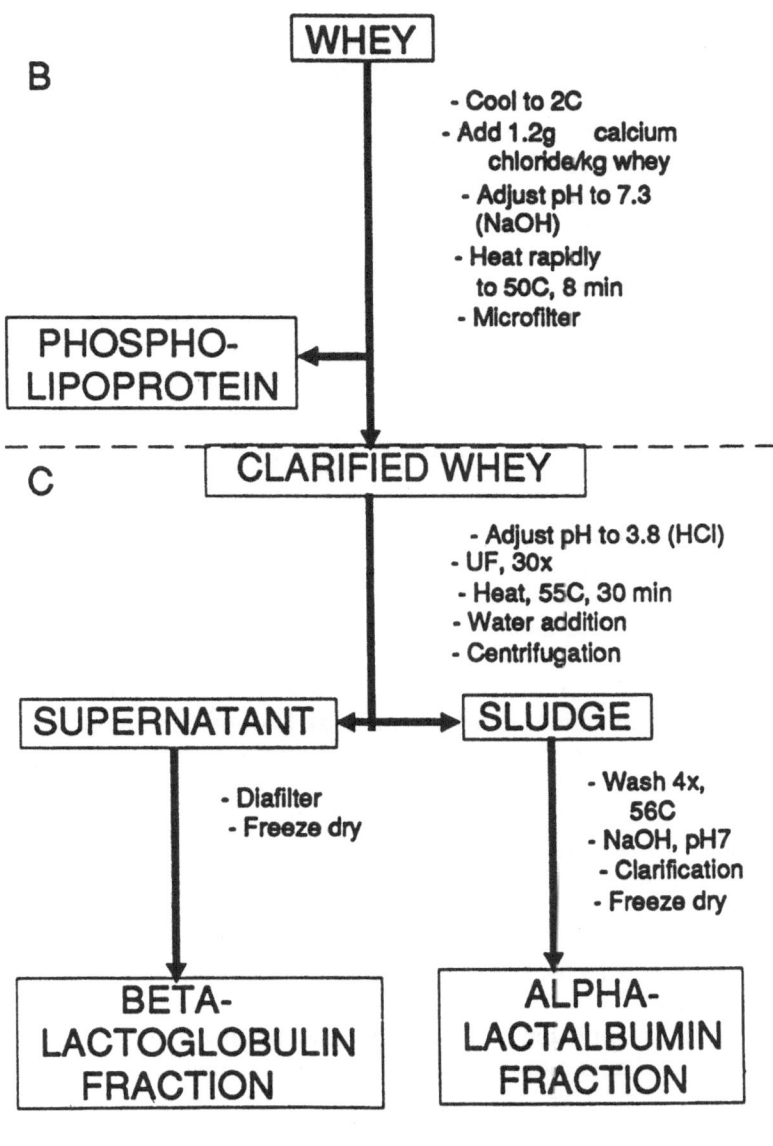

Fig. 2. *Cont.*

However, the uncertain magnitude of the markets and the price elasticity for such products suggests that profitable commercial manufacture of these components is unlikely in the near future.

Adsorption onto insoluble supports – 'Spherosil' and 'Bi-pro'

Rhone-Poulenc Chemie Fine has announced the development of a novel system for whey protein recovery, based on passing whey through a column containing porous, silica microbeads which specifically absorb protein (Mirabel, 1978). The protein is recovered by washing the column with a solution which has a slightly different pH to that of the whey. The process offers substantial advantages in protein recovery from whey, including production of a pure protein product with high functionality, low operational costs, and a protein-free waste material, similar in composition to UF permeate. The product has superior functional properties to 80 per cent whey protein concentrates prepared by UF, reflecting, perhaps, both the higher protein content of the Spherosil product, and the effect of energy input during ultrafiltration. Typically, whey at pH 6·5 is passed initially through a column of Spherosil QMA anion exchange resin, and then through Spherosil XOB cation exchange resin. The columns are washed with water, and eluted with 0·1 M hydrochloric acid and ammonium hydroxide respectively. The columns may be reused after further washing. Spherosil QMA has a protein adsorption capacity of about $100\,\mathrm{mg\,g^{-1}}$, and adsorbs mostly β-lacto-globulin, α-lactalbumin, serum albumin and some immunoglobulins, whereas the Spherosil XOB adsorbs mainly immunoglobulins. The recovered protein fractions are undenatured, and contain less than 3–4 per cent ash. At least two large commercial enterprises in France are now employing this process to manufacture high quality protein fractions from whey. Recent applications of Spherosil for whey protein recovery have been described by Skudder (1985), Nichols and Morr (1985) and Barker and Morr (1986). In general, the procedure suffers from the difficulties associated with the large volumes of regenerant and wash water required.

An alternative approach has been developed by Bio-Isolates, who have used an ion exchange process based on carboxymethyl cellulose for the manufacture of a 97 per cent protein powder (Bi-pro Dairy Albumin) from whey. The product is virtually pure protein, substantially free of fat and lactose, and with a low ash content. It has a digestibility of 99 per cent, a biological value of 94 per cent, net protein utilisation of 93 per cent and a protein efficiency ratio of 3:2. The excellent functional properties of the product are similar to egg white in many respects. Production in 1982 was 30–40 tonne, with a larger-scale plant being planned (Hannigan, 1982). An alternative cellulose-based system has been outlined by Ayers *et al.* (1984), with increased protein binding capacity. The process has been commercially adopted for the recovery of proteins from whey (Ayers and Petersen, 1985). Again, drawbacks of this process relate to high regenerant and wash water volume requirements.

Processes for the modification of whey proteins

Currently, there appears to be little commercial interest in the direct modification of whey proteins by enzymatic means, but this situation may well change in the future. Zadow (1979) has reviewed the status of existing techniques for modifying whey proteins. There has been recent interest in improving the solubility of lactalbumin by treatment with protease, so as to extend its applications (O'Keeffe and Kelly, 1981), and in chemically modifying the functionality of whey proteins, e.g. by succinylation, to improve solubility (Thompson and Reyes, 1980). Succinylation of heat-coagulated cheese whey protein concentrate resulted in improved functionality, allowing it to absorb more than 11 times its weight of water. Possible applications for this product include baked goods and dairy analogues where high water and fat adsorption are required.

Membrane Processes; Ultrafiltration and Reverse Osmosis

Processes

The application of membrane processes to the dairy industry has been reviewed by Matthews (1979). The same year, the IDF published a very useful survey of equipment available for membrane processes, covering IE, UF and RO membranes (Anon, 1979). Over the past decade, there have been continuing developments and improvements in membrane composition and equipment design to increase flux and membrane life, and to reduce cost and improve ease of cleaning and sanitation. Commercially available membrane systems now include tubular, plate and frame and spiral wrap designs; these developments have made membrane processing more economical for the industry. Although ultrafiltration has been most widely applied, the applications of RO are increasing for specific purposes.

Of particular importance to the industry are the sanitation problems posed by membrane processing, particularly with the advent of the spiral wrap system. Beaton (1979) has discussed the hygienic design and operation of membrane systems, particularly considering cleanability and sanitation. The extensive and very successful utilisation of spiral membranes for ultrafiltration of milk and whey in the last decade has now resulted in considerable confidence that such systems may be operated in a fully sanitary manner, and that the cleaning regimes are now very reliable. Thus, sanitation problems in UF systems have been virtually overcome, but some difficulties remain in ensuring sanitary operations of RO plants, particularly in milk processing. Nonetheless, considerable improvements in methodology and technology are occurring, resulting in turn in further improvements in the reliability of RO cleaning and sanitation operations.

Another important factor in achieving commercial acceptance of UF for use on whey was the development of techniques that reduced flux decline during processing. Muller *et al.* (1973) and Hayes *et al.* (1974) showed that, by a combination of pH adjustment and heat treatment before processing, the extent of flux decline could be substantially reduced. The effect of such pretreatments on the processing characteristics of whey was reviewed by Muller and Harper (1979), and on whey processing in general by Zadow (1982). Factors affecting flux included flow velocity, transmembrane pressure, temperature, pH, time of operation, retentate concentration, whey type and whey pretreatments. In studies on the use of chemical pretreatments and fouling on UF of acid cheese whey, Heng and Glatz (1991) reported that the extent of flux decline increased with the level of calcium and phosporus deposition.

Tong, *et al.* (1989) characterised proteinaceous membrane foulants on UF of whey. In general, it was found that low flux during whey UF is associated with adsorbtion of various peptides to the UF membrane. Addition of *Mucor pusillus* protease fragmented many of these peptides which were in whey produced from cheese made using calf rennet. This treatment increased flux levels of the whey to levels similar to those occurring on UF of whey from cheese made using *Mucor pusillus* protease.

Thus, the physical and chemical characteristics of the various types of whey markedly influence their performance during UF on different plants. Some of the whey treatments that improve performance are: clarification, centrifugation (sometimes preceded by calcium addition), heating under conditions determined by the type of whey and pH, demineralisation, pH control and concentration; for RO, only demineralisation and pH adjustment seem to be effective. A further factor which has been implicated is the particular starter strain used in the manufacture of the cheese from which the whey has been prepared. Overall, factors that control performance of whey in membrane processing are not well understood.

Recent studies have focussed on the components in whey involved in membrane fouling, and hence flux decline. Hickey and Hill (1980) have reported that the k-casein macropeptide present in Cheddar cheese whey contributes to fouling during ultrafiltration. Furthermore, the formation of calcium phosphate gels reduced permeation rates when circumstances favoured formation of the gel in membrane pores. The calcium ion concentration and the ionic strength of the whey directly influence permeation rates. Patel and Merson (1978) have examined the fouling characteristics of Cottage cheese whey fractions. Their results suggested that flux decline may be attributed to membrane compaction and fouling by proteinaceous materials which accumulate on the membrane surface; other studies have emphasised the role of whey salts in membrane fouling. Cheryan and Merin (1979) reported from model system studies that membranes may be negatively charged due to salt binding, and such adsorbed salts could act as bridges between the membranes and protein, thus increasing the fouling layer. It has been reported that salts have a detrimental effect on the rate of flux decline (Cheryan and Merin, 1980). α-Lactalbumin has the strongest gel forming tendencies, and β-lactoglobulin the worst long-term fouling effect. Whey showed greater fouling tendencies than did the model systems examined, due to the presence of salts and protein. Removal of the salts appears, therefore, to be beneficial in purely practical terms, but the cost/benefit ratio may not be satisfactory overall. Lee and Merson (1976) showed that flux could be improved very substantially by passing the whey through a 0·4 μm prefilter. Other workers suggesting the removal of lipid fractions from whey by microfiltration include Merin *et al.* (1983) and Hanemaaijer (1985). As indicated above, starter strains have also been implicated in problems of flux decline on ultrafiltration of cheese whey. Interestingly, those starters which produce the most acceptable cheese are often those which result in a rapid decrease in flux on UF. Although flux decline is still a major problem to the industry, existing methods of pretreating whey are generally adequate to make the process commercially viable, although further improvements remain desirable. The use of permeate as a starter medium in cheese production has been studied by Christopherson and Zottola (1989). It was reported that growth in permeate of those starter organisms studied was similar to growth in milk or other whey based media.

The problem of membrane fouling is of less significance in RO processing of whey than for UF. Cheddar cheese whey causes much less fouling on RO or UF than does casein whey. However, the pretreatments described by Hayes *et al.* (1974) for fouling reduction in UF lead to an increase in fouling with RO. This may be caused by different ionic compositions of the membrane surface layer (Smith and MacBean, 1978). In a recent review on the concentration of whey by RO, Pepper and Pain (1987) reported that, in RO plants in the USA, the practical concentration limit was about 25 per cent total solids. Although the plants are capable of producing a concentrate of up to 28 per cent total solids, precipitation of calcium phosphate in the system could not be reliably avoided at these higher concentrations.

Membrane fouling is not only dependent on the nature of the fluid being processed, it is also dependent on the membrane material itself. Currently, virtually all UF membranes are synthetic materials, mostly based on polysulphone. In the past, RO membranes were mostly based on cellulose acetate, but in the past decade a range of membranes based on synthetic materials has become available, with resultant advantages to the dairy processor. Two major problems with cellulose acetate membranes are their limited temperature and pH ranges; below about 30°C and close to neutral. Synthetic membranes, are more resistant to extremes of pH and heat, and may be cleaned more severely. More highly specialised membranes are now becoming available, particularly for UF and microfiltration. Such membranes, for example, may have ion exchange capabilities, may have enzymes bound on their surface, or may be used to retain particular protein fractions through affinity chromatographic techniques.

In 1981, Maubois *et al.*, (1981) reported on a membrane formed dynamically from an inorganic, zirconium oxide supported on a graphite base. Such dynamic membranes are normally formed *in situ* by coating the surface of the support with the active membrane-forming material before operations commence. Unlike the best synthetic UF membranes, based on polysulphone and limited to a maximum operating temperature of about 75°C within a pH range 2–12, these zirconium membranes are resistant to temperatures up to 400°C over the full pH range. Such membranes offer advantages in terms of sanitation, although there is not yet much information concerning their use in commercial operation. Other membranes based on ceramics have also become commercially available. In common with many of the dynamic membranes, ceramic membranes are capable of operation over wide ranges of temperature and pH, and their activity is not affected by harsh cleaning regimes. The major drawback of such systems is their comparative fragility, and their somewhat higher costs compared to polysulphone systems. But the expected life of such ceramic membranes is claimed to be considerably greater than conventional membranes, and this offsets their higher initial cost. Ceramic membrane systems are rapidly gaining increased acceptance, particularly in Europe.

There is little information available on the effect of membrane processing on protein structure, but it is well recognised that ultrafiltration of milk may result in significant changes in product functionality if the energy input during ultrafiltration is excessive. Similar changes are likely to occur during ultrafiltration of whey. Dziuba and Chojnowski (1982) have reported that the UF of whey results in a small increase in protein solubility. Changes in the ultraviolet spectra indicated an increase in hydrophobic groups, but this did not result in the association of whey proteins at the pH of the concentrates (5·8–6·0).

Of particular importance in ultrafiltration processes is the quality of the water used during diafiltration and washing, and very low levels of silica and/or iron will result in severe and permanent loss in flux. These materials appear to be irreversibly adsorbed onto the surface of the membranes. In most parts of the world, it is essential, therefore, to provide either facilities for the deionisation of wash and diafiltration water, or a supply of RO permeate. Care should also be taken in the selection of cleaning reagents employed in UF plants to ensure they have very low levels of iron and silica. Problems due to the contamination of RO membranes by iron and silica are much less severe. However, most RO membranes are extremely sensitive to low levels of chlorine, and action must be taken to ensure that wash waters used in RO operations are chlorine free.

Functionality

Whey protein concentrates (WPCs) may be prepared with a very wide range of functional properties. It is this characteristic that can increase the attractiveness of WPCs as an ingredient, as functionality may be tailor-made to specific requirements. Functionality is affected by a number of factors, such as source of whey, protein, mineral and lipid composition of the WPC, heat treatment given to the whey or retentate during manufacture, and the pH of the system in which the WPC is employed.

Although there has been much work done to develop an understanding of the relationship between compositional characteristics, manufacturing variables and the functional properties of WPC, it has generally been unsuccessful, even in model systems. However, as more detailed information becomes available from model food system studies, a greater understanding of the influence of molecular structure and interaction on the functional properties of WPC should be achieved. This area is particularly complex, and such an understanding is unlikely to be reached in the immediate future.

One complication is the lack of standard methods for the testing of even such straightforward functional properties as gelation strength, emulsification capacity and foaming characteristics. This lack of standardisation is partly the result of the need to assess such properties in widely differing food systems. Further standardisation of functionality tests is currently being considered by the International Dairy Federation.

Some recent reviews covering the functionality of whey proteins are those of Craig (1979), Morr (1979, 1982, 1984), Kamiya and Kaminogawa (1980), Kinsella (1981), Zadow (1985) and Marshall and Harper (1987). In detailed studies, Johns and Ennis (1981) have reported that 33, 67 or 100 per cent replacement of calcium ions with sodium ions in acid casein whey before UF treatment, substantially altered the functional properties of the resultant WPC. Protein solubility increased (especially at pH 4.9), and gel texture parameters improved when at least 67 per cent of the calcium was replaced, with gelation time being directly related to the level of replacement; whipping properties decreased with increasing replacement of calcium ions. Such modified WPCs were, therefore, more suitable for applications requiring good gelling characteristics rather than for those requiring good whipping properties.

Dunkerley and Zadow (1981) have reported that the gelation strength of WPC is decreased by preheating the whey, and, in the main, by demineralisation. Gel strengths comparable to those of egg white were obtained under some circumstances. Other studies have compared the effects of calcium, magnesium and sodium ions on the heat aggregation of WPC, and in general, it has been found that aggregation was increased by addition of these salts, with calcium having the greatest effect. Denaturation temperatures of the proteins decreased on addition of calcium or magnesium, but increased in the presence of sodium (Varunsatian et al., 1983).

Sulphydryl interactions are also involved in the formation of heat gels from whey proteins, and gel formation involves polypeptide chains cross-linked by disulphide bonds. The gels dissolve if a sulphydryl agent is added after heating, and their formation is retarded if compounds that react with sulphydryl groups are added before heating. Gels form more slowly at alkaline pH, perhaps because of the increased electrostatic repulsion between proteins (Hillier *et al.*, 1980). Proteose peptones do not appear to be involved in gelation characteristics (Hillier and Cheeseman, 1979). The rheological properties of heat-induced β-lactoglobulin gels has been studied by Paulsson *et al.* (1990). Shear modulus (a measure of gel stiffness) of such gels continued to increase after 6 h ageing. The gels formed at pHs greater than 6·0 were, in the shorter term, relatively more elastic and less viscous.

Current studies support the concept that gelation is controlled to a large extent by the mineral composition of the product and by sulphydryl interactions. Studies on these variables using multiple regression and response surface analysis have been reported by Schmidt *et al.* (1979).

The foaming and emulsifying properties of WPC have been studied by Patel and Kilara (1990). In general, free fat and bound fat were negatively correlated with foaming and emulsifying properties, whereas ash, calcium and denaturation enthalpy were positively correlated with these properties. A marked improvement in the foaming characteristics of whey protein isolates prepared from acid whey was achieved by centrifugation and microfiltration to remove insoluble foam depressants (Phillips *et al.*, 1991). Such treatments yielded products with greatly improved and highly consistent foaming ability. Typical overruns achieved were 1100 per cent at 20 min, with foam stability of 17 min after 15 min of whipping – comparable to egg white.

Reports on factors affecting the emulsifying properties of WPC are conflicting, and often reflect the differing systems used for examination of this characteristic. In general, conditions affecting emulsifying properties are similar to those controlling foaming and whipping, and are affected by the system used for emulsification, the energy input, the ionic strength of the environment and the type of WPC employed. Functionality of proteins in foams theoretically depends on their ability to reduce interfacial tension, increase liquid phase viscosity and form strong films. These changes may be facilitated by alterations to the tertiary and quaternary structure of the proteins, the extent of which will depend on conditions in the system. Foaming performance of whey protein products has been found to be improved by the addition of calcium ions, hydrogen peroxide and hexametaphosphate, and impaired by the addition of sucrose, lipids and surfactants; interaction effects have also been observed. Reversible improvement of foaming by heat treatment seems to be related to protein–lipid interactions (Richert, 1979).

Many studies on functionality have shown that only minor changes in techniques of manufacture of WPC may have a very considerable impact on functionality. There are very significant functional differences, for example, in WPC prepared from whey preheated at 72°C for 15 s, or 85°C for 15 s. The degree of difference in denaturation indices in such WPC is small, indicating that this parameter is of little value in predicting the impact of processing on functionality.

Because of the sensitivity of functionality to minor changes in processing, there appears to be a strong case for functionality studies to use only samples prepared under strictly controlled conditions. As a minimum, it is essential that data on whey source composition and processing history be included in all reports on WPC functionality. Of particular interest are suggestions that whey protein fractions may show enhanced functionality compared to WPC (Amundson *et al.*, 1982; Slack *et al.*, 1986a, b). The removal of lipid from many of the fractions is also essential if maximum functionality is to be achieved. Furthermore, certain whey fractions may have specific nutritional and therapeutic applications (Maubois, 1984).

Fermentation

Fermentation of whey and whey by-products may be used for the production of a wide range of products. A number of such processes are in commercial operation with, for example, large-scale equipment for the manufacture of alcohol being installed in Eire and New Zealand. A company based in Wisconsin, USA, has been manufacturing single-cell protein for some years, and more recently has commenced production of potable alcohol. Friend and Shahani (1979) have reviewed the fermentation of whey and its constituents, with emphasis on factors influencing the production of alcohol, lactic acid, citric acid and single-cell protein.

Single-cell protein

France manufactures more than 600 tonne per annum of single-cell protein for animal and food use. In one French plant, the process is based on fermentation of deproteinised whey with strains of *Kluyveromyces marxianus* var. *lactis* and *Kluyveromyces marxianus* var. *marxianus* (*Kluy. fragilis*), followed by centrifugation of the liquor to produce a 10–11 per cent concentrate, which is then further concentrated by filtration to 18–22 per cent solids, heated and dried. The yield corresponds to about 53 per cent of the weight of lactose in the initial whey, and the product contains about 50 per cent protein, 30 per cent carbohydrate, 6 per cent lipid, 8 per cent mineral, and moisture content of about 4·5 per cent (La Gueriviere, 1981). Sandhu and Waraich (1983) screened 13 yeast species belonging to nine genera for single-cell protein production from cheese whey. Of the species screened, *Wingea robertsii* showed the highest specific growth rates, could also produce large amounts of β-galactosidase, and gave higher yields for shorter incubation times.

Permeate from the ultrafiltration of whey is a useful substrate for manufacture of single-cell protein. It has been reported that, of 10 strains examined, *Kluy. marxianus* var. *marxianus* ATCC8582 was the most productive over a wide range of pH and lactose concentrations. The most economical permeate concentration from batch fermentation was 5 per cent (Gawel and Kosikowski, 1978). Mahmoud and Kosikowski (1978) examined the use of concentrated whey permeate for single-cell protein and alcohol manufacture. It was found that fermentation of a concentrated permeate (27 per cent solids) with five adapted strains of lactose fermenting *Kluyveromyces* species resulted in single-cell protein yields of 2–8 g l^{-1} dry matter, with about 12 per cent ethanol produced from the process. The use of two-stage fermentation with the initial stage using *Kluy. marxianus* var. *marxianus* and the second *Propionibacterium* species gave a higher yield than did the *Propionibacterium* species alone (Giec *et al.*, 1978). In recent studies, Mahmoud and Kosikowski (1982) have reported that *Kluy. marxianus* var. *marxianus* NRRL Y 2415 produced the highest yield of alcohol (9·1 per cent from concentrated whey permeate) and *Kluy. bulgaricus* ATCC1605 gave the highest yield of biomass (13·5 mg ml^{-1}). High ash contents inhibited both biomass and alcohol production.

The use of hydrolysed whey as a substrate has also been examined, as with this material, non-lactose fermenting organisms, such as *Saccharomyces cerevisiae*, could be used for alcohol production. The metabolism of *Kluy. marxianus* var. *marxianus* was slower with hydrolysed whey than with unhydrolysed whey, but alcohol yields were improved on fermentation with a high alcohol yielding strain of *Saccharomyces cerevisiae*; the galactose present was not utilised. Reviews of the production of biomass from whey include those of Meyrath and Bayer (1979) and Moebus and Tueber (1983).

Alcohol

The Carbery process of Express Dairies is the best-known commercial operation for the production of alcohol from whey streams (Sandbach,

1981a, b). In the batch process, whey permeate (about 4·5 per cent lactose) is fed from a balance tank into either a 110,000 l fermenter or a 26,000 l yeast propagation tank. Six fermenters are used at Carbery, Eire, each approximately 15 m in diameter with a conical base. After 20 h fermentation, the yeast cream is recovered by centrifugation, and the clear liquid, containing about 2·8 per cent alcohol, is passed to a balance tank for distillation. The resultant ethanol (96·5 per cent by volume) is suitable for gin or vodka production. Particular care must be taken during production to avoid yeast and mould contamination from nearby cheese plants. The Carbery plant processes about 600,000 l of whey per day yielding about 22,000 l of alcohol. Efficiency of conversion is about 86 per cent. Waste products from fermentation and distillation have necessitated the installation of a purification plant for the daily treatment of about 900,000 l of waste, with a biological oxygen demand (BOD) of 4000 kg. During winter, when supplies of permeate are low, molasses may be used as a 'make up' for the substrate. If it is assumed that the permeate has no value (alternative whey treatments may, in fact, often operate at a loss), the estimated annual return is about 28 per cent on the £5 million pound (UK) investment. Experiments have shown that concentration of the permeate to 10 per cent solids has little effect on process efficiency.

Plants using this system are considered economic provided that more than 500,000 l of whey per day are available for processing. In Eire, potable alcohol is the most profitable outlet, but in other countries, anhydrous alcohol for industrial or power use may also be attractive. For example, in New Zealand there is little indigenous alcohol production, and until recently it was necessary for that country to import the majority of its alcohol. Under these circumstances, local manufacture of alcohol from permeate was attractive, and recent years have seen the installation of a number of such plants in New Zealand. Alcohol also has a potential use in 'gasohol', and aspects of utilising permeate in this product are described by Lyons and Cunningham (1980).

Some recent research interest in fermentation technology has centred on means of immobilising the organisms involved so as to develop continuous systems. *Kluy. marxianus* var. *marxianus* immobilised in polyacrylamide gel has been used to treat whey permeate for alcohol production; a column half-life of 50 days has been achieved. With a column volume of 0·04 l and a flow rate of 0·34 column volumes per hour, 80 per cent of the lactose was utilised and ethanol production was 45 millimoles per hour. Ethanol production costs were estimated at NZ$ 0·59 1^{-1} (Bartosh and Manderson, 1982). Other approaches have employed the immobilisation of cells, with and without β-galactosidase, and again, considerable success has been achieved using this approach (Linko *et al.*, 1981; Hartmeier, 1982). Developments of such systems to commercially viable processes may offer worthwhile economic gains to the industry. Janssens *et al.* (1984) has described the continuous manufacture of ethanol from whey with cell recycling. The process utilised two steps, the first building biomass in a 100 per cent aerobic operation, followed by a second anaerobic step in which *Kluy. marxianus* var. *marxianus* was used to ferment the sugars to ethanol.

Fuel gas

There is considerable potential for the use of anaerobic fermentation of the permeate for the production of fuel gas. Recent studies (Lang, 1980) have indicated that, on average, biogas production by the anaerobic fermentation of 1 l of cheese whey at 55°C yields 36–48 l of biogas, with about 50 per cent methane and an energy content of 960 kJ l^{-1}. A study on the application of biogas to a dairy complex of 284 farms and one central processing plant, in the USA, estimated that 75 per cent of the cheese plant's liquid fuel and propane requirements could be directly supplied by utilisation of whey permeate for fuel gas manufacture. In practice, however, direct applications of whey fermentation for biogas appear to be rare. Given the increasing cost of energy and the problems of whey disposal, utilisation of whey through such a system is likely to become more attractive in the near future. Research needs to place emphasis on the development of systems that are economic for small-scale manufacturers.

Ammonium lactate/lactic acid

Techniques for the fermentation of whey to lactic acid have been well known for some years, and there appear to have been few recent developments. Commercial interest in developing the technology to commercial application appears minimal, in spite of the high current price of lactic acid, and the moderate demand for this product. In the past, much of the technical work has focussed on development of methodology for removal of the lactic acid from the reaction mixture. The development of membrane processes appears to offer a useful option for removal of lactic acid by employing membranes, such as cellulose acetate, which have very low rejections for this compound. There has also been recent interest in the production of ammonium lactate by continuous fermentation of deproteinised whey to lactic acid, using ammonia to maintain constant pH. In studies on this process using a strain of *Lactobacillus delbrueckii* sub-species *bulgaricus* at pH 5·5 and 44°C, lactate concentration was a maximum (58·7 g l⁻¹) at a cell retention time of 27 h, and cell mass was at a maximum (1·9 g l⁻¹) at 9·3 h (Stieber and Gerhardt, 1979). A mathematical model for the process has been developed, simulating the effect of cell retention time on substrate, product and cell mass concentration in a single-stage system.

The production of lactic acid from whey permeate by continuous fermentation in a membrane reactor has been described by Boyoval *et al.* (1988). Lactate productivity rates of $20 \, g \, l^{-1} \, h^{-1}$ at a dilution rate of $0.77 \, h^{-1}$ were achieved. Rockey and Zaror (1988) have considered the effect of key economic, political and technological factors on the economic feasibility of the production of lactic acid and other fermented products produced from food industry waste. They report that lactic acid production appears feasible on a large scale for the majority of the cases considered. They suggested that further improvements in separation methods will make this option very attractive.

Oil production

The production of oil by the fermentation of whey and permeate using *Candida curvata* and *Trichosporon cutaneum* has been examined by Moon and Hammond (1978). Fermentation for 72 h reduced the chemical oxygen demand (COD) of permeate by 95 per cent, and produced $4.0–5.6 \, g \, l^{-1}$ of oil, $19.6–26.8 \, g \, l^{-1}$ biomass and $2.2–2.3 \, g \, l^{-1}$ protein. *Candida curvata* was more efficient in producing oil and reducing COD. The oil contained approximately 50 per cent oleic, 30 per cent palmitic, 15 per cent stearic and 8 per cent linoleic acids. Extraction of the oil was best done by sequential treatment with methanol and benzene, then ethanol and hexane. The oil mass was a useful base for animal feed.

Citric acid production

Much of the world's supply of citric acid comes from fermentation of molasses, a commodity which may be variable in composition, and is often limited in supply. As there are few manufacturers of citric acid in the world, most countries import their requirements, but for some dairy-developed countries, there may be potential in employing local supplies of permeate as a feedstock for manufacture of this product. Australian research (Tonks and Hawley, personal communication) indicated that permeate has potential in this area, although difficulties may be encountered using permeates with a high ash content. Other studies (Somkuti and Bencivengo, 1981) have reported that acid whey is a suitable substrate for production of citric acid by fermentation with *Aspergillus niger*. Optimal production of citric acid was reached after 8–12 days at 30°C, with a maximum yield of $10 \, g \, l^{-1}$ obtained with a permeate solids content of 15 per cent; throughout the fermentation, galactose was apparently co-metabolised with glucose. Aspects of the process have also been discussed by Hossain *et al.* (1983). A more recent report by Playne (personal communication) reported on the selection of *Candida* yeasts for commercial citric acid production. Two strains from international culture collections were found to produce commercially interesting levels ($> 50 \, g \, l^{-1}$) from galactose. Successful levels of production were not achieved by fermentation of lactose. However, low world prices for citric acid, coupled with oversupply, have meant that further development of a process based on these results is not to proceed.

Butanol
Production of the important solvent, *n*-butanol, by fermentation of permeate has been described by Maddox (1980). Whilst the traditional fermentation of corn mash results in a mixed end-product with a ratio of butanol: acetone: ethanol of 6:3:1, fermentation of permeate results in a product with the favourable ratio of 10:1:1, and a butanol yield of 1·3 per cent. Stevens *et al.* (1988) have described a process in which whey is fermented to butanol and butyric acid using a mixed culture. Ennis and Maddox (1987) report that an appropriate lactose content in the substrate for fermentation to butanol–acetone–ethanol is 5 per cent, with whey UF permeate being particularly suitable.

Polysaccharides
The production of extracellular, microbial polysaccharides by the fermentation of whey, or hydrolysed whey, has also been reported. Test organisms were selected from strains known to produce gum when using glucose as a substrate, such as *Alcaligenes*, *Xanthomonas*, *Arthrobacter* and *Zooglea* species. All the organisms studied produced polymers from hydrolysed whey or glucose/galactose, but only *Alcaligenes visosus* and *Zooglea ramigera* produced a gum from whey under the conditions employed (Stauffer and Leeder, 1978).

Other products
Whey-based products have been suggested as substrates for production of a number of other products, including 2,3-butylene glycol, lysine,

threonine, giberellic acid, β-galactosidase, vitamin B_{12} and β-carotene, and for the growth of mushroom mycelium.

General

A detailed comparison of the advantages of bioprocesses compared to conventional chemical processes has been made by Greenshields and Rothman (1986). The advantages include generally milder reaction temperatures, less hazardous operations, greater specificity, and less expensive and often renewable resources. Disadvantages may include difficulties in downstream processing and control of contamination during the process. Of the various processes and products described above, only those for the manufacture of single-cell protein/biomass, alcohol and fuel oil appear to be potentially economically viable at present. It is possible that many of the remaining processes will be further developed into industrial realities.

Lactose Hydrolysis

In the USA, it appears unlikely that a market for hydrolysed products will develop to any considerable extent (with the possible exception of hydrolysed milk). The competition from corn syrup sweeteners, an established and highly competitive product, is likely to make the economics of lactose hydrolysis unattractive. In parts of the world such as Australia, which have high local sugar prices and no major corn syrup industry, lactose hydrolysis may have potential. In countries such as New Zealand, with no indigenous sugar industry, the potential could be even greater.

Lactose hydrolysis may be carried out either by treatment with acid at high temperatures, or through the use of β-galactosidase. The enzymes used for the latter process are extracted from yeasts, such as *Kluy. marxianus* var. *lactis* (optimum pH 6–7, optimum temperature about 35°C), or *Kluy. marxianus* var. *marxianus* (optimum pH about 6·5, optimum temperature about 40°C), or from fungi such as *Aspergillus niger* (optimum pH about 4·8, optimum temperature about 50°C). More recently, the development of a new lactase from a *Bacillus* species has been announced with a temperature optimum of 65°C and a pH optimum of 6–8. Comparative studies on catalytic and enzymatic methods have indicated that catalysis produces a higher quality product of better keeping quality than does enzymatic hydrolysis. However, such results should be interpreted with care, as the quality of enzymes available for hydrolysis varies widely, and this may significantly affect the quality and storage characteristics of products produced by enzymatic processes.

Acid hydrolysis/cation exchange

This method is only viable for treatment of protein-free streams, and 80 per cent hydrolysis may be readily obtained. Adjustment of the pH may be made by direct addition of acid to the system, or by treatment of the permeate with an ion exchange resin in the cation form. Typically, the pH of the permeate is adjusted to 1·2 (perhaps by cation exchange), and the product heated at temperatures up to 150°C for hydrolysis. During hydrolysis, a brown colour is formed, the extent of which is dependent on the non-protein nitrogen content of the sample. The product may be neutralised by addition of alkali, or by treatment with an anion exchange resin; a final clarification with charcoal may be employed, if desired. A number of modifications to the basic process have been suggested, particularly involving a combination of the steps of pH adjustment and hydrolysis. In this procedure, the permeate is passed directly onto the ion exchange resin at temperatures generally slightly below 100°C, so that both processes proceed virtually simultaneously (MacBean *et al.*, 1979; Haggett, 1976; Demaimay *et al.*, 1978). Boer and Robbertsen (1981) have studied a system for permeate hydrolysis in which the pH of concentrated UF permeate was adjusted to 1·2 through the action of a strong acidic ion exchange resin. The liquid was then processed at 150°C for 3 min to reach a degree of hydrolysis of 80 per cent. The product was purified by further ion exchange processing and concentrated to 62 per cent solids. The cost of the process was estimated to be Fl 0·368 kg^{-1} syrup. The costs of regeneration are significant in the economics of this process. As yet, there is apparently no commercial application of this procedure, although, in terms of simplicity and capital cost, it appears attractive.

Enzymatic hydrolysis

Three techniques may be employed for the enzymatic hydrolysis of lactose streams. These are the use of 'single-use' or 'throw-away' systems, lactase recovery systems (generally based on membrane recovery of β-galactosidase to allow its reuse) and immobilised enzymes. The relative costs of each of these processes are changing rapidly with developments in technology. The cost of single-use enzymes per tonne of hydrolysed lactose continues to drop, with some decrease also occurring in the cost of immobilised enzyme systems. The cost of soluble enzymes is closely related to their purity. However, the use of some of the cheaper soluble enzymes can result in the development of off-flavours in the product, probably from the activity of protease impurities.

Single-use enzymes

There is comparatively little detailed information on single-use enzymes for the hydrolysis of whey. Optimal conditions vary widely depending on the source of the enzyme and its purity. In selection of the enzyme, factors to be considered include the pH of the substrate, maximum temperature of hydrolysis possible, enzyme activity and cost. If economic concentrations of single-use enzymes are used, then several hours may be required for hydrolysis, and there are microbiological benefits in the use of enzymes that operate efficiently at temperatures where microbial growth is retarded. Chiu and Kosikowki (1985) have reported on the hydrolysis of permeate from demineralised whey. During the 5 h hydrolysis at 40°C, the total colony count increased from 50 to 1800 colony-forming units (cfu) ml^{-1}, but with 24, 48 or 72 h hydrolysis at 5°C, the count increased from 40 to 41, 48 and 140 cfu ml^{-1} respectively. Giec et al. (1981) showed that during incubation for 24 h, lactose hydrolysis increased linearly with lactose concentration; at 4°C, 60–70 per cent hydrolysis occurred in 24 h. Studies on the kinetics of soluble enzyme hydrolysis have been reported by Griethuysen et al. (1985) and Flaschel et al. (1982).

Systems using ultrafiltration

Ultrafiltration equipment can be used in three ways in conjunction with enzymes for lactose hydrolysis. In the simplest approach, hydrolysis of protein-free whey is carried out in a batch process, and ultrafiltration is applied to recover the enzyme for further use (Norman et al., 1978). A second approach involves immobilising the enzyme adjacent to the ultrafiltration membrane by a pressure-induced flow regime (Maculan et al., 1978). It has also been suggested by ultrafiltration equipment manufacturers that this equipment can be used in the feed/bleed mode as an enzyme reactor. In studies of this approach (Hayes, personal communication), a high concentration of enzyme was added to permeate being recycled in an ultrafiltration plant. When the desired level of hydrolysis was reached, the system was operated in the feed/bleed mode at the level of 75 per cent hydrolysis; continuous operation at 55°C and pH 4·6 for 30 h resulted in only a small loss of enzyme activity.

A similar system has been studied by Hayes (personal communication), in which a Ladish Triclover UF system was operated at a recirculation flow rate of $54 \, l \, min^{-1}$, at a pressure drop of $1400 \, kg \, m^{-2}$. Hydrolysis rates of 75 per cent were achieved at an operating temperature of 55°C and a pH of 4·6. More than 1400 times the volume of permeate was processed than would have been possible with the same quantity of enzyme on a simple single-use basis.

These approaches can be used only for the treatment of protein-free whey streams. With whole whey, the process involves initial ultrafiltration of the whey, treatment of the permeate by one of the techniques, and blending of the hydrolysed permeate with the UF retentate. The complexity of such a system can make it unattractive for commercial applications.

Immobilised systems

The use of immobilised enzyme systems for lactose hydrolysis appears to have the greatest potential for large volume applications for the treatment of whey or permeate. A number of supports for the immobilisation have been suggested, including silica, ion exchange resins, Duolite cellulose derivatives, phenol-formaldehyde, titanium dioxide, aluminium oxide, chitosan, diazotised glass, silica gel and silochrome. The most frequently used enzyme in immobilisation studies is that obtained from *Aspergillus niger*. Techniques employed for the utilisation of the immobilised systems involve both fixed bed and partially or wholly fluid bed systems.

Pastore and Park (1981) have described immobilisation of lactase from *Scopulariopsis* species on Duolite. The enzyme did not lose activity after 60 h operation at 55°C, but lost 45 per cent after 52 h at 60°C. After 1284 h of suitable operation, the enzyme retained 94 per cent of its activity overall. Marconi *et al.* (1980) employed a radial reactor for hydrolysis, comprising modular units in the form of bobbins prepared by winding fibres containing the immobilised lactase around perforated tubes. In a pilot-plant containing 38 modules, 85–90 per cent hydrolysis was achieved in 130 l of whey over 30 min. In a three month period, 182,000 l were treated with little loss of activity.

A fixed bed system utilising the lactase from *A. niger* and *A. oryzae* immobilised on Duolite has been described by Prenosil *et al.* (1984a, b). Scott *et al.* (1986) reported a useful lifetime of 90 days in a system operating at 30°C. Ottofrickenstein *et al.* (1984) described a novel system involving an upright cylindrical vessel, the upper part of which is filled with immobilised lactase, and the middle contains a transverse rotating mixer. In operation, whey is passed up through the cylinder from below, to achieve a hydrolysis rate of 70–90 per cent.

A novel approach has been directed by Griffiths and Muir (1980), who immobilized whole cells of *Bacillus stearothermophilus* by covalent linkage onto DEAE cellulose treated with glutaraldehyde. There was no apparent loss of lactase activity as a result of immobilisation, and the half-life of enzyme activity was about 15 days at 60°C and pH 7. Lactase activity increased with pH, and decreased with temperature and ionic strength. The bound cells were equally effective for hydrolysis of lactose in permeate, but were four times less efficient with skim-milk as the substrate.

A number of studies using immobilised enzymes in fluid bed reactors have reported reduction in activity with time, generally suggested to be caused by absorption of solid matter. In some cases, sonic cleaning of the enzyme was found to be beneficial; other aspects of sanitation and cleaning of immobilised systems are discussed later.

The Corning process is best known commercially for lactose hydrolysis using immobilised enzyme technology. The process employs lactase from *Aspergillus niger* covalently bound to a controlled pore size, silica carrier. Its particle size is 0·4–0·8 mm, wet bulk density 0·6, activity about 500 U g^{-1} at 50°C, optimal pH of operation between 3·2 and 4·3, and the estimated laboratory life of the resin is 2 years. A concentrated product, from a demineralised permeate (4 per cent solids), contained 61 per cent solids, 0·8 per cent ash, 12 per cent lactose, 23 per cent glucose and 25 per cent galactose. The process has been discussed in detail by Dohan *et al.* (1980). Two plants are in commercial operation, one in the UK, the other in the USA. The system installed in Kentucky, USA, has a 26 million gallon annual capacity. The cottage cheese whey is processed by UF; the permeate is hydrolysed and used as a substrate for the growth of baker's yeast.

The Valio system, utilising the enzyme from *A. niger* is described by Heikonen *et al.* (1985). The system operates at a pH of about 4·0 and a temperature of 40°C. Recently, a system has also been developed which can operate at closer to neutral pH so it can be used for the hydrolysis of sweet wheys and milk.

The Sumitomo system utilises an enzyme from *A. oryzae* immobilised on an ion exchange resin, and operates at about 35–40°C. With a contact time of about 1·5–2 min for whey, and about 1 min for permeate, levels of hydrolysis of about 70 per cent were achieved. The system employs a fluid bed reactor, thus overcoming dead spots and cleaning difficulties.

Specialised cleaning and sanitation regimes were developed to ensure that the system could be thoroughly cleaned without any resultant loss in enzyme activity. A commercial plant capable of producing 400 kg h^{-1} of hydrolysed skim-milk solids was installed in an Australian factory in 1986. Although the plant operated successfully on both whey and skim-milk, the plant has ceased operation because of poor market demand for hydrolysed products (Hayes *et al.*, 1990).

A number of other systems have been reported, including that of Tate and Lyle, utilising enzymes immobilised on charcoal, and the Amerace system (Hausser, 1984) utilising enzyme immobilised on silica particles embedded in a microporous plastic sheet.

Sanitation

Immobilised systems are designed for long-term operation; sanitation is therefore a prime consideration. Three facets must be considered: maintenance of sanitary conditions during hydrolysis, development of adequate cleaning regimes on completion of hydrolysis, and maintenance of sanitary conditions while the plant is in storage between operations. The severity of sanitation and cleaning solutions that may be employed is restricted by the sensitivity of most immobilised enzyme systems to pH and temperature – the majority of common dairy detergents and cleaners cannot be safely employed. Current recommendations for sanitation include treatment of the resins with water, a range of acids, mild alkali and mild detergents with some bacteriological activity. Some resins are sensitive to chlorine, and care must be taken to avoid its contact with the resin. The chlorine may either inactivate the resin or lead to problems with off-flavours in the product.

Oligosaccharide formation

The formation of oligosaccharides during lactose hydrolysis may have nutritional implications, as well as reducing monosaccharide yields. Galactose appears to be more involved in oligosaccharide formation than does glucose. Studies on lactose hydrolysis with a soluble lactase from *Saccharomyces lactis* have indicated a linear rise in the formation of oligosaccharides with initial lactose concentration, up to 13 per cent by weight of total sugar, when 65–70 per cent of the lactose was hydrolysed to monosaccharides; of the six oligosaccharides identified, all were linear. It was concluded that the enzyme had a high *trans*-glycosylation activity, with specificity for formation of β-(1–6) galactosidic bonds (Burvall *et al.*, 1979; Asp *et al.*, 1980). Other evidence suggests the extent of oligosaccharide formation is much more important in systems employing soluble enzymes with long contact times, than in systems using immobilised enzymes with brief contact times. Analytical methodology for the determination of oligosaccharides by HPLC has been outlined by Jeon and Mantha (1985). The effect of oligosaccharide formation on the assessment by cryoscopy of the degree of hydrolysis has been determined by Jeon and Saunders (1986). The extent of oligosaccharide formation depends on the source of enzyme (Mozaffar *et al.*, 1985). There appears to be little information available on the nutritional or toxicological characteristics of such oligosaccharides; there is a need for research in this area.

Determination of degree of hydrolysis

Cryoscopic methods have been recommended for determining the degree of lactose hydrolysis (Zarb and Hourigan, 1979; Hourigan *et al.*, 1983). The method is based on the fact that the freezing point of the sample will be depressed by an amount depending on the molal degree of hydrolysis. Hayes (1982a) compared polarimetric and cryoscopic methods for the assessment of degree of hydrolysis and, between lactose concentrations of 1·15 and 10·29 per cent, the correlation coefficients for each method were better than 0·99. However, where a high level of precision was required, the polarimetric method was preferred. Chen *et al.* (1981) have reported that oligosaccharides are formed during hydrolysis, and the presence of proteinases in the lactase may cause inconsistencies in results. Other techniques for determination of the degree of hydrolysis have been based on high performance liquid chromatography, chemiluminescence and spectrophotometry.

Sweetness of hydrolysed syrups

Sweetness and solubility data for sucrose, lactose, glucose and galactose are shown in Table VII. Studies on the relative sweetness of hydrolysed syrups have been reported by Shah and Nickerson (1978), using a model system based on unflavoured ice cream mix. The simulated hydrolysed syrups were evaluated at 25 and 50 per cent levels of replacement of sucrose in the control mix, with the results showing a synergistic effect with lactose. For example, less sucrose was needed for equi-sweetness when 25 per cent of the lactose was replaced with a glucose/galactose syrup (70 per cent hydrolysed) than when it was replaced with syrup showing total hydrolysis (100 per cent). Other studies have reported that 30, 60 and 90 per cent hydrolysis is equivalent to the addition of 0·5, 0·6 and 0·9 per cent sucrose to milk. Poutanen *et al.* (1978) have examined the conversion of glucose to fructose in glucose/galactose syrups, and found that the relative sweetness of the product was similar to that of sucrose. This approach may offer potential to the dairy industry if the process is developed to commercial reality. Of even greater potential would be the development of a system to convert galactose, with its comparatively low sweetness, to a sweeter monosaccharide.

TABLE VII
Relative solubility and sweetness of saccharides

Saccharide	Relative sweetness	Solubility (g/100 g solution)		
		10° C	30° C	50° C
Sucrose	100	66	69	73
Lactose	16	13	20	30
Galactose	32	28	36	47
Glucose	74	40	54	70
Fructose	174	–	82	87

Data from Shah and Nickerson (1978) and Pazur (1970).

Drying of hydrolysed products

Drying lactose-hydrolysed products is extremely difficult using conventional spray driers; in particular, spray drying of hydrolysed low protein or protein-free products is virtually impossible in conventional systems, because of the formation of sugar glasses on the surface of the spray drier. However, hydrolysed products with protein contents similar to that of skim-milk may be dried with adequate care. For drying low protein, hydrolysed products, the use of a Filtermat dryer has been suggested. It is claimed that this unit is particularly suitable for such products, as the concurrent flow of product and air reduces the risk of deposits in the dryer (Rheinlander, 1982).

Storage and properties of hydrolysed syrups

Storage of hydrolysed syrups poses a number of problems. In general, they cannot be spray dried without the use of special equipment so they must be stored as concentrated liquids. However, problems may be encountered during storage, either due to the growth of micro-organisms, particularly osmophilic yeasts, or to crystallisation of lactose or galactose. Lactose will be crystallised if the degree of hydrolysis is low; galactose will crystallise if the degree of hydrolysis and the total solids of the system are high. Recently, studies on factors affecting the storage characteristics of hydrolysed Cheddar cheese whey permeate syrups have been carried out in Australia (Hayes and Mitchell, 1983). It was found that both crystallisation and microbiological difficulties could be controlled by storage of the syrup at -10 to $-20\,°C$, provided the total solids of the syrups was about 70 per cent. Lack of crystallisation at low storage temperatures was probably caused by high viscosity under these conditions. Storage of syrups at $50\,°C$ to assist in prevention of bacterial growth resulted in severe browning of hydrolysed syrups in a few weeks.

In comparison with these results, Shah and Nickerson (1978) studied the crystallisation characteristics of simulated hydrolysed permeates (based on mixtures of lactose, glucose and galactose) containing 40–80 per cent solids. They reported that at $25\,°C$, crystallisation did not become a problem until the total solids exceeded 70 per cent. Rexroat and Bradley (1986) in studies on the stability of stored permeate reported maximum stability in a product that was decolourised, deionised, had a level of hydrolysis of the order of 85–90 per cent, had a total solids of about 65 per cent, was heated to $75\,°C$ following concentration and was stored at $30\,°C$. Harfouch *et al.* (1985) found that non-enzymatic browning of concentrated hydrolysed whey could be decreased by addition of metabisulphite. In view of the differing results reported above using permeate, it seems that it is not always possible to extrapolate from model systems to predict the behaviour of hydrolysed products.

Regulatory aspects
With the continuing development of a range of hydrolysed products in the dairy industry, regulatory aspects are being considered by a number of countries. French regulations provide for the use of enzymatically hydrolysed products in fine bakery goods, biscuits, desserts and confectionery. In Canada, hydrolysed products are authorised, with no restrictions on quantity; in the Netherlands, they are permitted in frozen confectionery but not yoghurt; in Italy, hydrolysed market milk is permitted, and in Great Britain, a declaration is sufficient (Luquet, 1980). Doubtless, these regulations will be broadened as applications increase, and as further information on toxicology becomes available.

Lactose Production

For many years, the production of lactose from whey typically involved protein removal (perhaps by lime treatment, or by heat and filtration), concentration of whey, refiltration, further concentration, induction of crystallisation and centrifugation to separate the crystals (Nickerson, 1974; Hobman, 1984; Bos, 1987)). In general, about 50 per cent of the lactose was recovered, and the mother liquor was sold as delactosed whey powder. Such processes have generally proved economic on a large scale, and there appears to be little potential in the market for substantial increases in lactose supply. The use of magnesium chloride to assist in the recovery of lactose has been suggested (Kwon and Nickerson, 1978). More recently, permeate has become a useful raw material for production of lactose, and aspects of its utilisation have been described by Madrid Vicente (1979). Carbohydrates may form complexes with alkaline earth metals, and this is the basis of the Steffan process for separation of sucrose from beet sugar. The potential of this process for recovery of lactose from whey has recently been considered by Quickert and Bernhard (1982). Other studies have shown improvements in yield of lactose by demineralisation of the permeate before crystallisation. Visser (1980) has recently reported that pharmaceutical grade lactose contains a substance with an acidic character that crystallises with the lactose and retards the rate of crystal growth. Ion exchange of the lactose removes this contaminant, yielding a non-ionic lactose that crystallises much more rapidly than does the pharmaceutical grade. It is likely that similar crystal growth retardant materials will be present in whey and permeate, and will contribute to the practical problems of lactose crystallisation. The value of demineralisation may be in the removal of such compounds.

Of the two isomers of lactose, β-lactose is slightly sweeter and more soluble. However, lactose produced by conventional processes contains mainly α-lactose. There is only a small demand for β-lactose as such, despite higher solubility and greater sweetness. Manufacture of this isomer involves crystallisation at temperatures above 90 °C, or treatment of α-lactose with alkaline methanol.

LACTOSE UTILISATION AND CHEMICAL DERIVATIVES

Lactosyl Urea

Although ruminants can use urea as a source of non-protein nitrogen for protein synthesis, it is not an ideal material for this purpose. The non-protein source should be palatable, have controlled nitrogen release, and be of low toxicity. Lactosyl urea, prepared by condensation of lactose and urea, appears to meet these requirements. The product may be prepared from whey or whey permeate, and yields in excess of 80 per cent have been claimed. The product is less toxic than a mixture of lactose and urea, and the linkage of lactose and urea results in slower degradation of the lactose, and consequently better utilisation, as well as removing any problems of crystallisation during storage (Widell, 1979). However, in spite of considerable initial interest in this project in both Europe and the USA, commercial development appears to have been slow.

Lactitol

Lactitol may be considered as the lactose equivalent of sorbitol, and has been suggested for use as a non-nutritional sweetener. Details of lactitol manufacture have been outlined by Saijonmaa *et al.* (1978), covering hydrogenation of lactose by either Raney nickel or borohydride; conversion rates of 97 per cent were achieved. Lactitol is hydrolysed as readily as lactose by the lactase from *Aspergillus niger*, but less rapidly by the enzyme from *Kluy. marxianus* var. *marxianus*.

In a detailed article on the applications of lactitol as a new food ingredient, Booy (1987) considered its functional properties, relative sweetness and physiological properties. Booy reports that adaption, and spreading lactitol consumption throughout the day in combination with other foods, makes a large intake possible without excessive laxative effects. Further, he points out that lactitol behaves as dietary fibre, so the physiological side effects of lactitol consumption should prove favourable. Applications for regulatory approval have been made in most European countries. The use of lactitol as a base for lactose-derived surfactants has been suggested by Velthuijsen (1979).

Lactulose

Lactulose, an isomer of lactose, may be formed by molecular rearrangement of lactose under alkaline conditions. It may also be formed as a by-product of the action of some lactase enzymes on lactose. It has received much attention in the field of infant nutrition, and as a remedy for infant constipation. Matvievskii *et al.* (1978) have reported that the best yield of lactulose, using calcium hydroxide as catalyst, was obtained from 15 per cent solutions of lactose held at 70 °C for 15–20 min at pH 11. More recently, a technique based on the use of boric acid, with either triethylamine or sodium hydroxide as catalysts, has been claimed to be a more economical and efficient process (Hicks *et al.*, 1983). Techniques for the conversion of hygroscopic lactulose powder to the non-hygroscopic form by treatment with ethanol have also been developed. A number of methods for determining lactulose in dairy products are available based on techniques, such as spectrophotometry, enzymatic analysis, HPLC and thin layer chromatography.

Lactulose is about half as sweet as sucrose at concentrations up to 15 per cent; the difference decreases at higher concentrations. The water activity of lactulose solutions is significantly lower than that of sucrose syrups of equal concentration; lactulose may therefore be useful as a humectant for intermediate moisture foods.

An extensive review of lactulose in infant nutrition has been prepared by Mendez and Olano (1979). The presence of lactulose in infant foods encourages the development of *Bifidobacterium bifidum* in the intestinal flora, i.e. similar to the flora found in breast-fed infants. Lactulose also has medicinal uses in the treatment of portal systemic encephalopathy and chronic constipation. However, some concern has been expressed over possible laxative effects of lactulose, particularly with regard to consumption by infants. It has been suggested that such problems might be due to a low colonic pH.

Currently, lactulose commands a high price because of its unique properties of high solubility and non-digestibility. It has considerable potential for further market development. In a recent detailed review on lactulose as a sugar with physiological significance, Mizota *et al.* (1987) reported that the amount of lactulose products employed for medical and pharmaceutical purposes is about 6000 tons per year worldwide. The amount of lactose required for this production is about 3 per cent of total lactose manufactured. Currently, the production of lactulose is increasing extensively. Mizota *et al.* (1987) placed emphasis on the use of lactulose as a humanising factor in infant foods, because of its role as a 'Bifidus' factor. Products containing lactulose and *Bifidobacterium* species manufactured by Morinaga Milk Industries of Japan have gained widespread acceptance in Japan, France and West Germany.

Lactobionic Acid

There appears to be little commercial interest in production of this material from lactose. It may have some use as a food acidulant, and might be considered to have interesting chelating properties (Wright and Rand, 1973). Commercial development seems unlikely at present.

Gluconic Acid

A process for the conversion of lactose into gluconic acid and galactose has been described by Dahlgren (1978). In the process, lactose is treated with bromine under acidic conditions. A process for production of gluconic acid by treatment of hydrolysed lactose in sweet whey permeate by use of *Gluconobacter oxydans* at pH 5·0 and 25 °C has been described by Hunyh *et al.* (1986).

Binders

An interesting non-foodstuff application for lactose is its utilisation as a key ingredient of a binder system in processing iron fines captured in pollution control equipment. Commercial application of this process should result in large savings in energy expended in the manufacture of iron or pellets, and permit recycling of iron/steel dust currently discarded (Ferretti and Chambers, 1979). Recently there has been much interest in the use of lactose as a basis for surfactants (Parrish, 1976). In 1959, the industrial manufacture of sucrose based surfactants was commenced in Japan, with current production about 3000 tonne per year. The method employed for sucrose based surfactants is, however, not satisfactory for lactose, as it is necessary to use acid chlorides rather than esters for esterification. The properties of lactose based surfactants are similar to those based on sucrose. Development of the process has been slow because of the requirement for expensive solvents. Work is continuing on the development of solvent-free systems.

CONCLUDING REMARKS

Of the various options discussed above for whey processing, many are unlikely to become commercially viable in the near future as their development is still at an academic level. Currently, therefore, the whey processor remains restricted to existing commercial operations for whey utilisation. However, the existing mix of processes employed will change significantly, particularly under the impact of new membrane technology. The development of a wide commercial range of whey protein isolates with specific functional properties for specific end-uses offers the best opportunity to profitably exploit whey solids. There is already considerable commercial interest in this newly developing technology which will see increased commercial activity in the near future. However, most of these developments will also result in an increase in the volume of lactose streams, such as permeate which will require processing or disposal. For many countries, lactose hydrolysis, possibly in combination with the use of open RO membranes for demineralisation, could become an economic option for permeate processing. In other countries, particularly the USA, disposal of lactose is likely to continue to pose considerable difficulties. For large-scale manufacturers, options such as fermentation may be attractive for production of end-products with a defined market niche. However, given the capital-intensive nature of fermentation, it is probable that small producers will not find this option particularly viable. In some cases, production of lactose-hydrolysed syrups for in-house use in dairy products may provide an alternative solution. The best option in areas where whey or permeate manufacture is concentrated may be for the dairy factories to supply permeate and/or whey to companies specifically organised to process these products. This approach has been shown to be particularly effective in New Zealand and is a model for the rest of the world. Certainly, whey and permeate are no longer seen as waste products, but as valuable raw materials. There are many options for disposal of the protein fractions, but the effective utilisation of the lactose fraction remains difficult with existing technologies and markets.

REFERENCES

Allum, D. (1980). *Journal of the Society of Dairy Technology*, 33, 59.
Amundson, C. H., Watanawanichakorn, S. and Hill, C. G. (1982). *Journal of Food Processing and Preservation*, 6, 55.
Anon. (1979). *IDF Bulletin*, 115.
Asp, N. G., Burvall, A., Dahlqvist, A., Hallgren, P. and Lundblad, A. (1980). *Food Chemistry*, 5, 147.
Australian Dairy Corporation (1983). *Australian Dairy Corporation*, Planning Division, Australian Dairy Corporation, Melbourne.
Ayers, J. S. and Petersen, M. J. (1985). *New Zealand Journal of Dairy Science and Technology*, 20, 129.
Ayers, J. S., Petersen, M. J., Sheerin, B. E. and Bethell, G. S. (1984). *Journal of Chromatography*, 294, 195.
Barker, C. M. and Morr, C. V. (1986). *Journal of Food Science*, 51, 919.
Bartosh J. A. and Manderson, G. J. (1982). In: *Brief Communications, XXI International Dairy Congress*, Volume 1 (2) p. 526.
Batchelder, B. T. (1987). *IDF Bulletin*, 212, 84.

Beaton, N. C. (1979). *Journal of Food Protection*, **42**, 584.
Berger, H. M., Scott, P. H., Kenward, C., Scott, P. and Wharton, B. A. (1979). *Archives of Disease in Childhood*, **54**, 98.
Best, E., Plainer, H. and Sprossler, B. (1982). European Patent Application 0 057 273 A2.
Blanc, B. (1969). In: *Proceedings, IDF Seminar on Whey Processing and Utilisation*, Weihenstephan, Germany, 11-13/11/1969, pp. 51–56.
Boer, R. and Robbertsen, T. (1981). *Netherlands Milk and Dairy Journal*, **35**, 95.
Booy, C. J. (1987). *IDF Bulletin*, **212**, 62.
Bos, M. J. van den (1987). *IDF Bulletin*, **212**, 99.
Brothersen, C. F., Olson, N. F. and Richardson, T. (1982). *Journal of Dairy Science*, **65**, 17.
Buchta, W. J. (1991). *Journal of Dairy Research*, **58**, 211.
Buhler, M., Olofsson, M. and Fosseux, P. Y. (1981). US Patent 4 265 924.
Burvall, A., Asp. N.-G. and Dahlqvist, A. (1969). *Food Chemistry*, **4**, 243.
Chebotarev, E. A., Nesterenko, P. G., Davydyants, L. E., Mikhailova, N. T. and Chebotareva, N. G. (1983). *Molochnaya Promyshlennost*, **2**, 26.
Chen, S.-L. Y., Frank, J. F. and Loewenstein, M. (1981). *Journal of the Association of Official Analytical Chemists*, **64**, 1414.
Cheryan, M. and Merin, U. (1979). *Abstracts of Papers, National Meeting, American Chemical Society*, **178** (1), COLL 128.
Cheryan, M. and Merin, U. (1980). *Polymer Science and Technology*, **13**, 619.
Christopherson, A. T. and Zottola, E. A. (1989). *Journal of Dairy Science*, **72**, 2862.
Chiu, C. P. and Kosikowski, F. V. (1985). *Journal of Dairy Science*, **68**, 16.
Clark, W. S. (1979). *Journal of Dairy Science*, **62**, 96.
Clark, W. S. (1987). *IDF Bulletin*, **212**, 6.
Coton, S. G. (1980). *Journal of the Society of Dairy Technology*, **33**, 89.
Craig, T. W. (1979). *Journal of Dairy Science*, **62**, 1695.
Dahlgren, S. A. (1978). UK Patent 1 526 903.
Delaney, R. A. M. (1979). In: *Proceedings, Whey Products Conference*, Minneapolis, 1978, p. 111.
Demaimay, M., Le Henaff, Y. and Printemps, P. (1978). *Process Biochemistry*, **13** (4), 3.
Demmler, G. (1968). In: *Handbuch der Lebensmittelchemie*, Volume III/1 863, Springer-Verlag, Berlin, Heidelberg, New York.
Dohan, L. A., Baret, J. L., Pain, S. and Delalande, P. (1980). *Enzyme Engineering*, **5**, 279.
Dunkerley, J. A. and Zadow, J. G. (1981). *New Zealand Journal of Dairy Science and Technology*, **16**, 243.
Dziuba, J. and Chojnowski, W. (1982). In: *Brief Communications, XXI International Dairy Congress*, Volume 1 (2), p. 173.
Ennis, B. M. and Higgins, J. J. (1981). *New Zealand Journal of Dairy Science and Technology*, **16**, 167.
Ennis, B. M. and Maddox, I. S. (1987). *Biotechnology and Bioengineering*, **29**, 329.
Eriksson P. (1986a) AICHE Annual Meeting, Poster. Miami, 2–7 November.
Eriksson, P. (1986b) AICHE Annual Meeting, Poster. Miami, 2–7 November.
Euber, J. R. and Brunner, J. R. (1979). *Journal of Dairy Science*, **62**, 685.

Fauquant, J., Vieco, E., Brule, G. and Maubois, J. L. (1985). *Lait*, **65**, 1, 20.

Ferretti, A. and Chambers, J. V. (1979). *Journal of Agricultural and Food Chemistry*, **27**, 687.

Flaschel, E., Raetz, E. and Renken, A. (1982). *Biotechnology and Bioengineering*, **24**, 2499.

Foley, A. A. and Bates, G. W. (1987). *Analytical Biochemistry*, **162**, 296.

Food and Agriculture Organisation (1981). *FAO Production Yearbook*, Volume 34, p. 234.

Friend, B. A. and Shahani, K. M. (1979). *New Zealand Journal of Dairy Science and Technology*, **14**, 143.

Garfield, E. (1980). *Current Contents: Life Sciences*, **23**, 5.

Gawel, J. and Kosikowski, F. V. (1978). *Journal of Food Science*, **43**, 1717.

Giec, A. and Kosikowski, F. V. (1982). *Journal of Food Science*, **47**, 188.

Giec, J., Czarnecka, I., Baraniecka, B. and Skupin J. (1978). *Acta Alimentaria Polonica*, **4**, 247.

Greenshields, R. and Rothman, H. (1986). In: *Cambridge Studies in Biotechnology*, Volume 3, Cambridge University Press, Cambridge.

Gregory, A. G. (1986). In: *Proceedings of International Whey Conference*, Chicago, 27–29 October.

Griethuysen, E. van, Flaschel, E. and Renken, A. (1985). *Journal of Chemical Biotechnology B*, **35**, 129.

Griffiths, M. W. and Muir, D. D. (1980). *Journal of the Science of Food and Agriculture*, **31**, 397.

Haggett, T. O. R. (1976). *New Zealand Journal of Dairy Science and Technology*, **11**, 176.

Hanemaaijer, J. H. (1985). *Desalination*, **53**, 143.

Hannigan, K. J. (1982). *Food Engineering*, **54** (3), 96.

Harfouch, M., Collard-Bovy, C. and Ramet, J. P. (1985). *Sciences des Aliments*, **5**, 181.

Hargrove, R. E., McDonough, F. E., La Croix, D. E. and Alford, J. A. (1976). *Journal of Dairy Science*, **59**, 25.

Harper, W. J. (1979). *New Zealand Journal of Dairy Science and Technology*, **14**, 156.

Hartmeier, W. (1982). In: *Use of Enzymes in Food Technology, International Symposium*, Versailles, 1982 (Ed. P. Dupuy), Technique et Documentation Lavoisier, Paris, p. 205.

Hausser, A. G. (1984). In: *Proceedings of Whey Production Conference*, Chicago, October 25–26, Philadelphia, p. 63.

Hayes, J. F. (1982a). In: *Working Papers, 2nd Dairy Technology Review Conference*, Glenormiston, Victoria, 1982, p. 211.

Hayes, J. F. (1982b). In: *Working Papers, 2nd Dairy Technology Review Conference*, Glenormiston, Victoria, 1982, p. 213.

Hayes, J. F. and Mitchell, I. R. (1982). In: *Proceedings, 2nd Whey Protein Collaborative Research Group Conference*, Highett, Victoria, 1983, p. 379.

Hayes, J. F., Dunkerley, J. A., Muller, L. L. and Griffin, A. T. (1974). *Australian Journal of Dairy Technology*, **29**, 132.

Hayes, J. F., Mitchell, I. R. and Zadow, J. G. (1990). Summary of lactose hydrolysis project, Dairy Research Laboratory Report, CSIRO, Australia.

Heikonen, M., Harju, U. M. and Tykklainen, P. (1985). *Nordisk Mejeriindustrie* **12**, 76.

Heng, M. H. and Glatz, C. E. (1991). *Journal of Dairy Science*, **74**, 11.

Hickey, M. E. and Hill, R. D. (1980). *New Zealand Journal of Dairy Science and Technology*, **15**, 123.

Hicks, K. B., Raupp, D. L. and Smith, P. W. (1983). *Abstracts of Papers, National Meeting, American Chemical Society*, **186**, AGFD 159.

Hillier, R. M. and Cheeseman, G. C. (1979). *Journal of Dairy Research*, **46**, 113.

Hillier, R. M., Lyster, R. L. J. and Cheeseman, G. C. (1979). *Journal of Dairy Research*, **46**, 103.

Hillier, R. M., Lyster, R. L. J. and Cheeseman, G. C. (1980). *Journal of the Science of Food and Agriculture*, **31**, 1152.

Hobman, P. G. (1984). *Journal of Dairy Science*, **67**, 2630.

Hoogstraten, J. J. (1987). *IDF Bulletin*, **212**, 16.

Hossain, M., Brooks, J. D. and Maddox, I. S. (1983). *New Zealand Journal of Dairy Science and Technology*, **18**, 161.

Hourigan, J. A., Zarb, J. M. and Kissell, W. (1983). In: *Proceedings, 2nd Whey Protein Research Conference*, Highett, Victoria, 1983.

Hsu, K. H. and Fennema, O. (1989). *Journal of Dairy Science*, **72**, 829.

Humbert, G. and Alais, C. (1982). *La Technique Laitiere*, **952**, 41.

Hunyh, N. V., Decleire, M., Voels, A. M., Molte, J. C. and Monseur, X. (1986). *Process Biochemistry*, **21**, 31.

Hynd, J. (1980). *Journal of the Society of Dairy Technology*, **33**, 52.

Jackson, M. B. (1980). *Research Review*, CSIRO Division of Chemical Technology, p. 45.

Janssens, J. H., Bernard, A. and Bailey, R. B. (1984). *Biotechnology and Bioengineering*, **26**, 1.

Jenness, R. and Patton, S. (1959). *Principles of Dairy Chemistry*, Wiley, New York, p. 89.

Jeon, I. J. and Mantha, V. R. (1985). *Journal of Dairy Science*, **68**, 581.

Jeon, I. J. and Saunders, S. R. (1986). *Journal of Food Science*, **51**, 245.

Johns, J. E. M. (1982). *Brief Communications, XXI International Dairy Congress*, Volume 1 (2), p. 196.

Johns, J. E. M. and Ennis, B. M. (1981). *New Zealand Journal of Dairy Science and Technology*, **16**, 79.

Jonsson, H. and Arph, S. O. (1987). *IDF Bulletin*, **212**, 91.

Jonsson, H. and Forsman, B. (1978). *Nordisk Mejeriindustri*, **5**, 636.

Kamiya, T. and Kaminogawa, S. (1980). *Japanese Journal of Dairy and Food Science*, **29**, A-15, A-27.

Kinsella, J. E. (1981). *Food Chemistry*, **7**, 273.

Kjaergaard-Jensen, G. and Oxlund, J. K. (1988). *IDF Bulletin*, **238**, 4.

Kjaergaard-Jensen, G. (1987). *IDF Bulletin*, **212**, 27.

Kosikowski, F. V. (1977). *Cheese and Fermented Milk Foods*, 2nd edn, Edward Bros., Ann Arbor, Michigan, p. 448.

Kwon, S. Y. and Nickerson, T. A. (1978). *Journal of Dairy Science*, **61** Supp. 1, 112.

La Gueriviere, J.-F. de. (1981). *La Technique Laitiere*, **952**, 89.

Labuza, T. P. and Saltmarch, M. (1982). *Journal of Food Science*, **47**, 92.

Lang, F. (1980). *Milk Industry*, **82** (2), 30.

Lee, C. H. and Johnson, R. A. (1987). In: *Proceedings, International Congress on Membranes and Membrane Processes*, Tokyo, 8–12 June.

Lee, D. N. and Merson, R. L. (1976). *Journal of Food Science*, **41**, 778.

Lewicki, P. P., Galoch, I. and Slesinka, K. (1981). *Przemysl Spozywczy*, **35**, 102.

Linko, Y. Y., Jalanka, H. and Linko, P. (1981). *Biotechnology Letters*, **3**, 263.

Lipatov, N. N. and Chebotarev, E. A. (1982). *Izvestiya Vysshikh Uchebbnykh Zave denii, Pischchevaya Tekhnologiya*, **2**, 41.

Lovell-Smith, J. E. R. (1982). *New Zealand Journal of Dairy Science and Technology*, **17**, 161.

Luquet, F.-M. (1980). *La Technique Laitiere*, **940**, 23.

Lyons, T. P. and Cunningham, J. D. (1980). *American Dairy Review*, **42** (11), 42A.

MacBean, R. D., Hall, R. J. and Willman, N. J. (1979). *Australian Journal of Dairy Technology*, **34**, 53.

McGugan, W. A., Larmond, E. and Emmons, D. B. (1979). *Canadian Institute of Food Science and Technology Journal*, **12**, 32.

Maculan, T. P., Hourigan, J. A. and Rand, A. G. (1978). *Journal of Dairy Science*, **61** Supp 1, 114.

Maddox, I. S. (1980). *Biotechnology Letters*, **2**, 493.

Madrid Vicente, A. (1979). *Industrie Alimentari*, **18**, 811.

Mahmoud, M. M. and Kosikowski, F. V. (1978). *Journal of Dairy Science*, **61** Supp. 1.1, 114.

Mahmoud, M. M. and Kosikowski, F. V. (1982). *Journal of Dairy Science*, **65**, 2082.

Mahmoud, R., Brown, R. J. and Ernstrom, C. A. (1990). *Journal of Dairy Science*, **73**, 1694.

Manczak, M. (1979). *Przemsyl Spozywczy*, **33**, 382.

Marconi, W., Bartoli, F., Morisi, F. and Marani, A. (1980). *Enzyme Engineering*, **5**, 269.

Marshall, S. C. (1979). *New Zealand Journal of Dairy Science and Technology*, **14**, 103.

Marshall, K. R. and Harper, W. J. (1987). IDF Document No. 388/B 21.

Matthews (1984). *Journal of Dairy Science*, **67**, 2680.

Matthews, M. E. (1978). *New Zealand Journal of Dairy Science and Technology*, **13**, 149.

Matthews, M. E. (1979). *New Zealand Journal of Dairy Science and Technology*, **14**, 86.

Matvievskii, V., Ya., Kravchenko, E. F. and Khramtsov, A. G. (1978). *Trudy, Vsesoyuznyi Nauchno-issledovatel'skii Institut Maslodel'noi i Syrodel'noi Promyshlennosti Nauchno-proizvodstvennogo Ob"edinenya 'Uglich'*, **26**, 13.

Maubois, J.-L. (1982). 'Les proteins lactoserum extraites pour ultrafiltration'. In: *Proteines animales* (ed. C. M. Bourgeois and P. Roux), Technique and Documentation, Lavosier, p. 172.

Maubois, J.-L., Brule, G. and Gourdon, P. (1981). *Technique Laitiere*, **14**, 86.

Maubois, J.-L., Pierre, A., Fauquant, J. and Piot, M. (1987). *IDF Bulletin*, **212**, 154.

Mendez, A. and Olano, A. (1979). *Dairy Science Abstracts*, **41**, 531.

Merin, U., Gordin, S. and Tanny, G. B. (1983). *New Zealand Journal of Dairy Science and Technology*, **18**, 153.

Meyrath, J. and Bayer, K. (1979). *Economic Microbiology*, **4** (microbial biomass), 207.

Mirabel, B. (1978). *Informations Chimie*, **175**, 105.

Mizota, T., Tamura, Y., Tomita, M. and Okonogi, S. (1987). *IDF Bulletin*, **212**, 69.

Modler, H. W. and Harwalker, V. R. (1981). *Milchwissenschaft*, **36**, 537.

Modler, H. W. and Jones, J. D. (1987). *Food Technology*, **41** (10), 114.

Moebus, O. and Tueber, M. (1983). In: *Production and feeding of single cell proteins. Proceedings, COST workshop, Zurich, Switzerland*, (Ed. M. P. Ferranti and A. Fletcher), Applied Science, London p. 124.

Moon, N. J. and Hammond, E. G. (1978). *Journal of the American Oil Chemists Society*, **55**, 683.

Morr, C. V. (1979). *New Zealand Journal of Dairy Science and Technology*, **14**, 185.

Morr, C. V. (1982). In: *Developments in Dairy Chemistry – 1 – Proteins* (Ed. P. F. Fox), Applied Science, New York, pp. 375–99.

Morr, C. V. (1984). *Food Technology*, **38**, 39.

Moulin, G. and Galzy, P. (1981). *Biotechn. Gen. Eng. Rev.* **1**, 374.

Mozaffar, Z., Nakanishi, K., and Matsuno, R. (1985). *Journal of Food Science*, **50**, 1602.

Muller, L. L. (1979). *New Zealand Journal of Dairy Science and Technology*, **14**, 121.

Muller, L. L. and Harper, W. J. (1979). *Journal of Agricultural and Food Chemistry*, **28**, 70.

Muller, L. L., Hayes, J. F. and Griffin, A. T. (1973). *Australian Journal of Dairy Technology*, **28**, 70.

Nichols, J. A. and Morr, C. V. (1985). *Journal of Food Science*, **50**, 610.

Nickerson, T. A. (1974). In: *Fundamentals of Dairy Chemistry* (Ed. B. H. Webb, A. H. Johnson and J. A. Alford), AVI Publishing Co., Westport, Connecticut, p. 273.

Norman, B. E., Severinsen, S. G., Nielsen, T. and Wagner, J. (1978). *Nordeuropaeisk Mejeri Tidsskrift*, **44**, 129.

O'Keefe, A. M. and Kelly, J. (1981). *Netherlands Milk and Dairy Journal*, **35**, 292.

Ottofrickenstein, H., Plainer, H., Sprossler, B. and White, H. (1984). German Patent Application 33 10 430 A1.

Parrish, F. W. (1976). In: *Proceedings, Whey Products Conference*, Atlantic City 1976, p. 104.

Pastore, G. M. and Park, Y. K. (1981). *Ciencia e Tecnologia de Alimentos*, **1**, 65.

Patel, P. C. and Merson, R. L. (1978). *Journal of Food Science and Technology, India*, **15**, 56.

Patel, M. T. and Kilara, A. (1990). *Journal of Dairy Science*, **73**, 2731.

Patel, M. T., Kilara, A., Huffman, L. M., Hewitt, S. A. and Houlihan, A. V. (1990). *Journal of Dairy Science*, **73**, 1439.

Paulsson, M., Dejmek, P. and Vilet, T. van (1990). *Journal of Dairy Science*, **73**, 45.

Pazur, J. H. (1970). In: *Carbohydrates: Chemistry and Biochemistry*, Volume IIA, (Ed. W. Pigman, D. Horton and A. Herp), 2nd edn, Academic Press, New York, p. 69.

Pearce, R. J. (1983). *Australian Journal of Dairy Technology*, **38**, 144.

Pearce, R. J. (1987). *Australian Journal of Dairy Technology*, **42**, 75.
Pepper, D. (1981). *Dairy Industries International*, **46**, 24.
Pepper, D. (1988). *Proceedings, IMTEC Meeting*, Sydney 15–17 November.
Pepper, D. and Pain, L. H. (1987). *IDF Bulletin*, **212**, 25.
Phillips, L. G., Hawkes, S., Rosmussen, R. A., Barbano, D. M. and Kinsella, J. E. (1991). *Journal of Dairy Science*, **74**, 101.
Pirhonen (1987). *Finnish Journal of Dairy Science*, **45**, 62.
Posati, L. P. and Orr, M. L. (1976). *Agriculture Handbook*, US Department of Agriculture, No. 8–1.
Poutanen, K., Linko, Y.-Y. and Linko, P. (1978). *Nordeuropaeisk Mejeri Tidsskrift*, **44**, 90.
Prenosil, J. E., Stuker, E., Hediger, T. and Bourne, J. R. (1984a). *Annals of the New York Academy of Science*, **434**, 136.
Prenosil, J. E., Stuker, E., Hediger, T. and Bourne, J. R. (1984b). *Biotechnology*, **2**, 441.
Quickert, S. C. and Bernhard, R. A. (1982). *Journal of Food Science*, **47**, 1705.
Regester, G. O. and Smitters (1991). *Journal of Dairy Science*, **74**, 796.
Renner, E. (1983). *Milk and Dairy Products in Human Nutrition*, Volkswirtschaftlicher Verlag, Munich.
Rexroat, T. M. and Bradley, R. L. (1986). *Journal of Dairy Science*, **69**, 1762.
Rheinlander, P. M. (1982). *North European Dairy Journal*, **48**, 121.
Richert, S. H. (1979). *Journal of Agricultural and Food Chemistry*, **27**, 665.
Ritzel, G., Stahelin, H. B., Schneeberger, H., Wanner, M. and Jost, M. (1979). *International Journal for Vitamin Nutrition Research*, **49**, 419.
Robinson, B. P., Short, J. L. and Marshall, K. R. (1976). *New Zealand Journal of Dairy Science and Technology*, **11**, 114.
Rockey, J. and Zaror, C. (1988). In: *Developments in food microbiology – 4* (Ed. R. K. Robinson), Elsevier Applied Science, London, p. 188.
Roeper, J. (1971). *New Zealand Journal of Dairy Science and Technology*, **6**, 112.
Ron, I. (1986). *Meieriposten*, **75**, 356.
Ruegg, M., Moor, U. and Blanc, B. (1977). *Journal of Dairy Research*, **44**, 509.
Ryder, D. N. (1980). *Journal of the Society of Dairy Technology*, **33**, 73.
Saijonmaa, T., Heikonen, M., Kreula, M. and Linko, P. (1978). *Milchwissenschaft*, **33**, 733.
Saito, T., Yamaji, A. and Itoh, T. (1991). *Journal of Dairy Science*, **74**, 2831.
Saltmarch, M. (1980). *Abstracts International B*, **41**, 879.
Saltmarch, M. and Labuza, T. P. (1980). *Scanning Electron Microscopy*, **3**, 659.
Sandbach, D. M. L. (1981a). *Cultured Dairy Products Journal*, **16** (4), 17.
Sandbach, D. M. L. (1981b). *Proceedings, Second Bi-ennial Marschall International Cheese Conference*, Madison, 1981, p. 17.
Sandfort, P. H. (1987). *IDF Bulletin*, **212**, 21.
Sandhu, D. K. and Waraich, M. K. (1983). *Biotechnology and Bioengineering*, **25**, 797.
Schmidt, R. H., Illingworth, B. L., Deng, J. C. and Cornell, J. A. (1979). *Journal of Agricultural and Food Chemistry*, **27**, 529.
Scott, T. C., Hill, C. G. and Amundson, C. H. (1986). *Biotechnology and Bioengineering Symposium*, **17**, 585.
Shah, N. O. and Nickerson, T. A. (1978). *Journal of Food Science*, **43**, 1575.

Sienkiewicz, T. and Riedel, C.-L. (1990). *Whey and Whey Utilization*, Verlag. Th. Mann, Gelsenkirchen-Buer, Germany.

Skudder, P. J. (1985). *Journal of Dairy Research*, **52**, 167.

Slack, A. W., Amundson, C. H. and Hill, C. G. (1986a). *Journal of Food Processing and Preservation*, **10**, 31.

Slack, A. W., Amundson, C. H. and Hill, C. G. (1986b) *Journal of Food Processing and Preservation*, **10**, 81.

Sloth Hansen, P. and Kjaergaard-Jensen, K. G. (1977). *Investigations concerning variations in the composition of whey. 224*, Beretning, States Forsogmejerie, Hillerood, Denmark, pp. 1–77.

Smith, B. R. and MacBean, R. D. (1978). *Australian Journal of Dairy Technology*, **33**, 57.

Somkuti, G. A. and Bencivengo, M. M. (1981). *Developments in Industrial Microbiology*, **22**, 557.

Sottiez, P. (1985). *Produits derives des fabrications Fromages. Lactis et produits lactierie V2. 347* (Ed. F. M. Luquet), Lavoisier, Paris.

Stauffer, K. R. and Leeder, J. G. (1978). *Journal of Food Science*, **43**, 756.

Stevens, D., Alam, S. and Bajpai, R. (1988). *Journal of Industrial Microbiology*, **3**, 15.

Stieber, R. W. and Gerhardt, P. (1979). *Journal of Dairy Science*, **62**, 1558.

Strobel, D. R. (1972). *Foreign Agriculture*, **10** (46), 9.

Thompson, L. U. and Reyes, E. S. (1980). *Journal of Dairy Science*, **63**, 715.

Tong, P. S., Barbano, D. M. and Jordan, W. K. (1989). *Journal of Dairy Science*, **72**, 1435.

Varunsatian, S., Watanabe, K., Hayakawa, S. and Nakamura, R. (1983). *Journal of Food Science*, **48**, 42.

Velthuijsen, J. A. van (1979). *Journal of Agricultural and Food Chemistry*, **27**, 680.

Visser, R. A. (1980). *Netherlands Milk and Dairy Journal*, **34**, 255.

Visser, R. J., Bos, van den, M. P. and Fergusson, W. P. (1988). *IDF Bulletin*, **233**, 33.

Webb, B. H. (1972). *American Dairy Review*, **34**, 6.

Widell, S. (1979). In: *Proceedings, Whey Products Conference*, Minneapolis, 1978, p. 53.

Williams, C. A. and MacDonald, I. (1982). *Annals of Nutrition and Metabolism*, **26**, 374.

Wilson, L. G., Bottomley, R. C., Sutton, P. M. and Sisk, C. H. (1988). *Journal of the Society of Dairy Technology*, **41**, 10.

Wit, J. N. de and Klarenbeek, G. (1978). British Patent 1 519 897.

Wright, D. G. and Rand, A. G. (1973). *Journal of Food Science*, **38**, 1132.

Yoshida, S. (1989). *Journal of Dairy Science*, **72**, 1446.

Yoshida, S. (1990). *Journal of Dairy Science*, **73**, 2292.

Yoshida, S. and Ye-xiuyun (1991). *Journal of Dairy Science*, **74**, 1439.

Zadow, J. G. (1979). *New Zealand Journal of Dairy Science and Technology*, **14**, 131.

Zadow, J. G. (1982). *Proceedings, 2nd Australian Dairy Technology Review Conference*, Glenormiston, Victoria, 1982, p. 276.

Zadow, J. G. (1985). *Australian Journal of Dairy Technology*, **41**, 96.

Zadow, J. G. (1987). *IDF Bulletin*, **212**, 12.

Zadow, J. G. (1991). *CSIRO Food Research Quarterly*, **51**, (1/2), 99.
Zaetz, N. E., Kubanskaja, D. M. and Kocherov, N. I. (1982). In: *Proceedings, XXI International Dairy Congress*, Volume 1 (2), p. 657.
Zarb, J. M. and Hourigan, J. A. (1979). *Australian Journal of Dairy Technology*, **34**, 184.

Chapter 8

Utilisation of Milk Components: Casein

C. R. Southward
New Zealand Dairy Research Institute, Private Bag, Palmerston North, New Zealand

Casein, the principal protein in cow's milk, has been extracted commercially for most of the twentieth century. The casein content of whole milk varies according to the breed of cow, stage of lactation and time of milking, but is generally in the range 24–29 gl^{-1} (Jenness and Patton, 1959a; Fox, 1989). The skim-milk produced from the separation of whole milk to a 40 per cent fat cream (the first stage in the manufacture of casein) consequently contains 25–31 gl^{-1} casein (Jenness and Patton, 1959b; McDowall, 1971), which accounts for 75–80 per cent of the nitrogen in skim-milk (Dolby *et al.*, 1969; Gordon and Kalan, 1974; Fox, 1989). With a content of 0·7–0·9 per cent phosphorus, covalently bound to the casein by a serine ester linkage, casein as a phosphoprotein is a member of a relatively rare class of proteins (Gordon and Kalan, 1974). The amino-acid content of casein has been well established and includes high proportions of all amino-acids essential to man (Gordon and Kalan, 1974) with the possible exception of cysteine.

Casein exists in milk in complex micelles that consist of casein molecules, calcium, inorganic phosphate and citrate (Farrell and Thompson, 1974). It has been established that whole casein is a heterogeneous mixture of several individual casein components ($\alpha_{s1}, \alpha_{s2}, \beta, \kappa$) each of which has slightly different properties (Farrell and Thompson, 1974; Gordon and Kalan, 1974; Fox, 1989; Swartz *et al.*, 1991). For the purpose of commercial production, however, it is only necessary to consider whole casein, which contains all of these components.

EXTRACTION OF CASEIN FROM MILK

Casein is extracted from milk by the process outlined in Figs. 1 and 2. After separation of the whole milk to produce cream and skim-milk, the casein is precipitated, washed and dried (Kastens and Baldauski, 1952; Spellacy, 1953). The precipitation state is outlined in Fig. 1. When the casein curd has been precipitated and the mixture heated, the curd is separated from the whey (Figs. 2 and 3) and is subsequently washed several times with water in vats (Jordan, 1983) or towers (Boullé, 1986; Fig. 4) prior to mechanical dewatering by pressing (Munro and Vu, 1983) or centrifuging (Munro *et al.*, 1983) and drying. In New Zealand, fluid bed dryers are typically used in this operation (Fig. 5). The warm, granular casein is then cooled in a 'tempering' process (Fig. 2) in which casein is continuously circulated through several bins to attain equilibra-

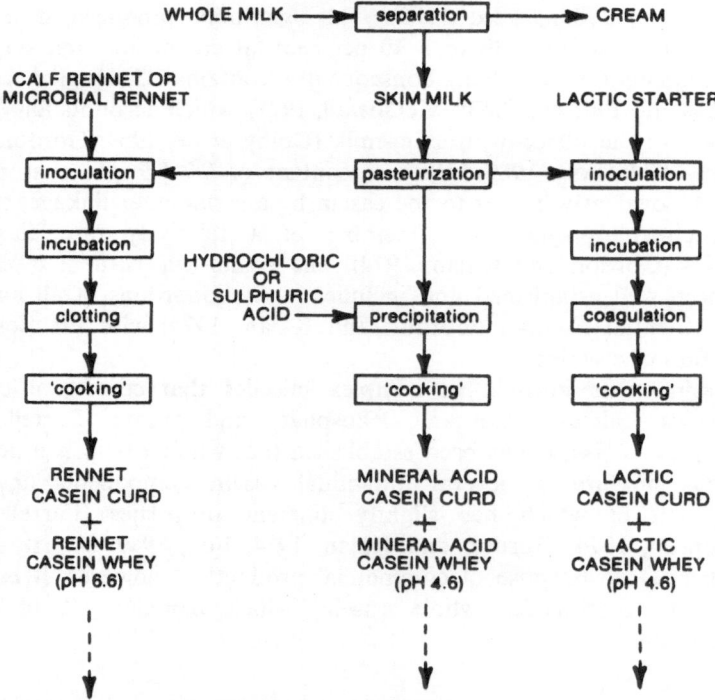

Fig. 1.   Processing steps involved in the precipitation of acid and rennet caseins from milk. (Reproduced from Southward and Walker (1980) by permission of the publishers.)

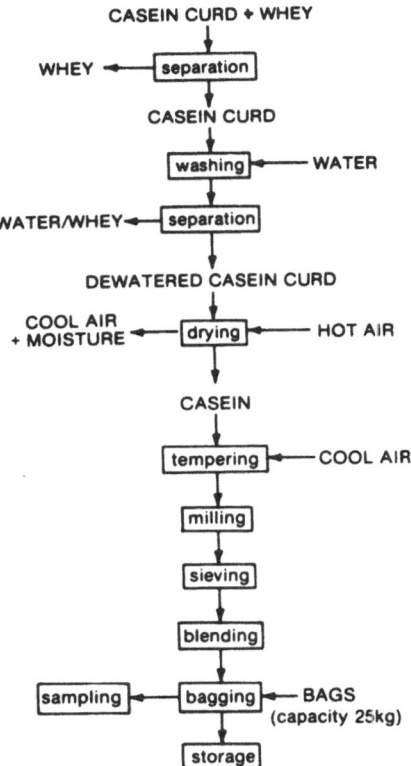

CASEIN CURD + WHEY

WHEY ← separation

CASEIN CURD

washing ← WATER

WATER/WHEY ← separation

DEWATERED CASEIN CURD

COOL AIR + MOISTURE ← drying ← HOT AIR

CASEIN

tempering ← COOL AIR

milling

sieving

blending

sampling ← bagging ← BAGS (capacity 25kg)

storage

Fig. 2. Processing operations in the washing and drying of caseins. (Reproduced from Southward and Walker (1980) by permission of the publishers.)

tion of moisture throughout the different sized particles. The fully tempered casein is passed over a magnetic separator before grinding in roller mills. The milled casein is then sieved to various particle sizes, depending upon market requirements, blended (King and Jebson, 1970) and finally bagged, ready for storage or shipment.

Two basic types of casein are produced, and these are named in accordance with the coagulating agent employed. They are acid casein and rennet casein. Three types of acid casein are made commercially; lactic, hydrochloric and sulphuric acid caseins have been produced for most of the twentieth century. In New Zealand, lactic casein is by far the most common product, although a small quantity of sulphuric casein is also produced for particular requirements. In Australia and Europe,

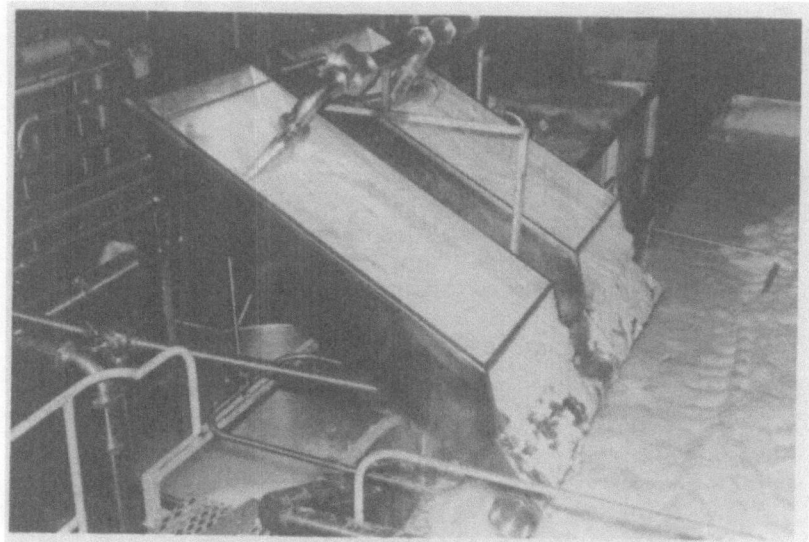

Fig. 3. Separation of casein curd and whey by inclined stationary screens.

however, hydrochloric acid is commonly employed as a precipitant in casein manufacture since this acid is available as a relatively inexpensive by-product of the chemical industry. More recently, other techniques have been proposed for the production of acid casein. Acidification of milk by ion exchange plus acid (Salmon, 1983) or by ion exchange alone (Anon, 1984) has been developed in France. A proposed alternative involves electrodialysis of skim-milk to pH 5·0 followed by acidification to pH 4·6 with acid (Laiteries Triballat, 1979). A specific advantage of these methods is they produce acid whey with reduced mineral content. This acid whey is more readily utilised than acid whey produced by the normal acidification process. Other studies have evaluated the use of high pressure carbon dioxide for precipitation of casein from milk (Jordan *et al.*, 1987; Caron *et al.*, 1991).

The other type of casein manufactured commercially is rennet casein, produced from skim milk clotted by the action of calf chymosin (rennin).

Manufacturing processes for casein have been reviewed by Muller (1971, 1982a, b), Gwozdz (1978), Segalen *et al.* (1985) and Mulvihill (1989).

Fig. 4. Washing towers for casein curd (after Boullé, 1986).

Acid Casein Manufacture

Figure 1 shows the processing operations in the manufacture of acid casein.

Mineral acid casein
Skim-milk (pH 6·6) is pasteurised (72 °C/15 s) and mixed with dilute (0·5 N) acid at a temperature of about 20 °C to a pH of approximately 4·6. The mixture is heated, for instance by injection of steam, to a temperature of 50–55 °C in order to aid agglomeration of the casein particles.

Fig. 5. Fluid bed drier for casein.

Following a short period of residence in a 'cooking' line and 'acidulation' vat, the resultant curd is separated from the whey, washed and dried.

Lactic acid casein

Pasteurised skim-milk for lactic casein manufacture is inoculated with strains of lactic acid-producing bacteria, known as 'starters' (e.g. *Lactococcus lactis* sub-species *cremoris*, 0·1–0·5 per cent of milk volume) at a temperature of 22–26 °C (Thomas and Lowrie, 1975a, b; Heap and Lawrence, 1984). The milk is incubated in several silos, each with a capacity of up to 250,000 l, for 14–16 h. During this period, some of the lactose in the milk is fermented to lactic acid by the starter and the pH of the milk is reduced to about 4·6 causing coagulation of the casein. The lactic coagulum is then 'cooked' and processed further in a manner similar to that described for mineral acid casein.

Process schematics for lactic, hydrochloric and acid casein manufacture in New Zealand, Australia and Europe, respectively, have been published (King, 1970; Muller, 1971, 1982a; Segalen *et al.*, 1985). Skim-milk throughputs in commercial plants in New Zealand range from 20,000 to 90,000 l h^{-1}, representing casein production of 0·6–2·7 tonne h^{-1}. During the seasonal peak (October and November) some units produce 25–45 tonne casein per day.

Rennet Casein Manufacture

In the manufacture of rennet casein (Fig. 1), calf rennet or microbial rennet (Southward and Elston, 1976) is added to skim-milk (ratio 1:7000) at a temperature of about 29 °C (Weal and Southward, 1974). Where rennet casein is intended for industrial purposes, it is important that the skim-milk should not be pasteurised, for it has been observed (Munro *et al.*, 1880) that rennet casein plastic produced from pasteurised milk has a darker colour than casein plastic from unpasteurised milk.

During the renneting of the skim milk, the enzyme specifically cleaves one of the bonds in κ-casein, releasing glycomacropeptide (Ernstrom and Wong, 1974; Fox, 1989). This action destabilises the casein micelles and they form a three-dimensional clot with calcium ions. This process normally takes place in about 30 min at pH 6·6. The clotted milk may then be 'cooked' and the casein processed in a manner similar to that described for acid casein.

Yield

The yield of commercial casein from 100 kg skim-milk is approximately 3 kg.

Co-Precipitates

The action of heat on skim-milk helps to induce complex interactions between casein and the whey proteins (mainly β-lactoglobulin). This phenomenon forms a basis for the commercial preparation of a third general group of casein products, namely the milk protein co-precipitates. These products are manufactured by heating skim-milk (generally to between 85 and 95 °C) to induce casein–whey protein interactions, subsequently causing co-precipitation and coagulation of protein either by reduction of the pH of heated milk or by addition of inorganic calcium salts to the hot skim-milk (Muller *et al.*, 1967).

All casein and co-precipitate products produced by the three processes described above are insoluble in water. However, addition of alkali (such as sodium, potassium, calcium or ammonium hydroxide) will cause the undried curd or the rehydrated casein product to dissolve. The soluble proteins formed by reaction of acid casein with alkali are known as caseinates.

MANUFACTURE OF CASEINATES

Caseinates may be produced from acid casein curd or from dry acid casein by reaction under aqueous conditions with any one of several different dilute alkalis, as outlined in Fig. 6. The resulting homogeneous caseinate solution may be dried by the spray or roller process to produce a caseinate powder having a moisture content of 3–8 per cent, depending on the manufacturer and customer requirements.

Spray Dried Sodium Caseinate

Aspects of the manufacture of spray dried sodium caseinate have been reviewed by Fox (1968) and Muller (1971, 1982a) and described by Burston *et al.*, (1967), Australian Society of Dairy Technology (1972), Bergmann (1972), Towler (1976a) and Segalen (1982).

Raw materials
In casein manufacturing countries, it is common practice to prepare sodium caseinate from fresh acid casein curd since it is considered to be blander in flavour than sodium caseinate made from dried casein (Cayen and Baker, 1963; Burston *et al.*, 1967). Sodium caseinate prepared from dry casein will also incur the additional manufacturing costs associated with drying, dry processing, bagging and storage of the casein before its conversion to sodium caseinate.

The most common alkali used in the production of sodium caseinate is sodium hydroxide solution, with a strength of 2·5 M (Australian Society of Dairy Technology, 1972; Towler, 1976a). The amount of sodium hydroxide required is generally 1·7–2·2 per cent by weight of the casein solids (Australian Society of Dairy Technology, 1972). Other alkalis

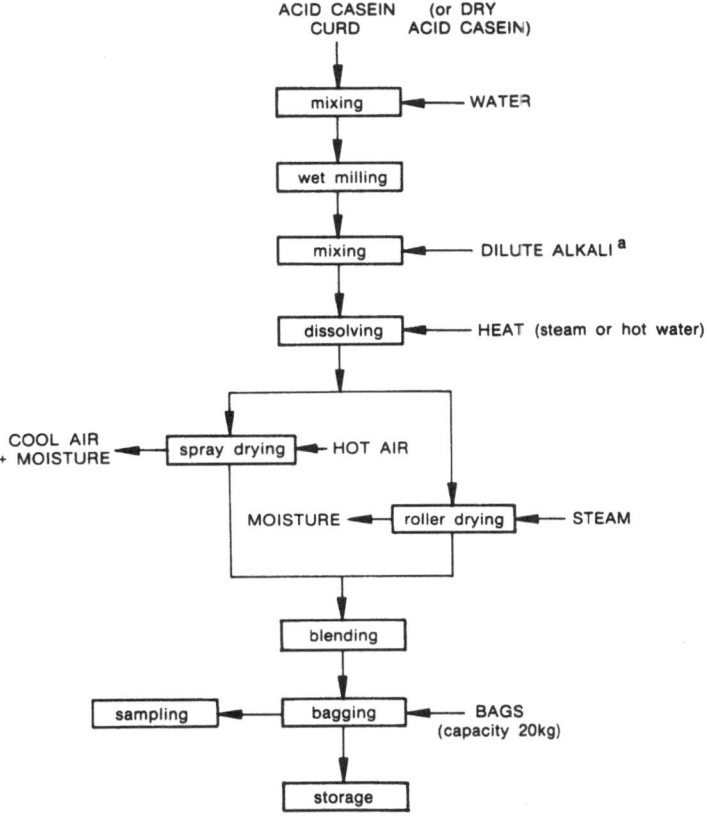

Fig. 6. Manufacture of spray dried or roller dried caseinate from acid casein curd or dry acid casein.

may be used such as sodium bicarbonate or sodium phosphates (Segalen, 1982) but the amount required and their cost are greater than those of sodium hydroxide.

Equipment

Milling
One or more colloid mills are required to reduce the particle size of curd for rapid dissolving. When dry casein is used, 30 mesh casein is preferred to the finer 80–90 mesh product which tends to form lumps when added

to water (Burston *et al.*, 1967; Fox, 1968). The use of a funnel in a recirculation line (as used in reconstituting milk powder) may help to overcome this problem.

Alkali dosing
Suitable equipment for metering the alkali into the casein slurry is necessary.

Dissolving vats
One or more dissolving vats, equipped with powerful agitators and recirculating pumps, are needed. The capacity and number of vats may vary, depending on the throughput of the drier and whether the dissolving process is batch or continuous. It is preferable to keep the size of any dissolving vat below 5000 l as it is difficult and expensive to achieve good mixing of the viscous caseinate solution in larger vessels.

Heating
The caseinate solution must be heated during the dissolving process. Indirect heating may be carried out by means of a tubular heat exchanger in the recirculation line or through the walls of a jacketed dissolving vat. Alternatively, it is possible to inject steam directly into the caseinate solution, provided that the steam is of good quality. However, this results in dilution of the caseinate, and has the potential for producing scorched particles.

pH and viscosity measurement
It is essential for pH and viscosity of the caseinate solution to be controlled, and suitable instruments are necessary for the measurement of these properties. It is possible to use a measurement of refractive index of a solution of caseinate to estimate its solids content and hence use this information to assist in the control of the dissolving operation (Towler, 1976b).

Spray drying
Sodium caseinate can be dried in spray dryers equipped with either disc or pressure nozzle atomisation (Segalen, 1982) from solutions with a concentration of up to 20 per cent solids.

Roller Dried Sodium Caseinate

By comparison with the literature on spray dried caseinates, very little information has been published on roller drying these products, though it is known that the technique is used quite extensively in Europe (Towler, 1977). Roller dried caseinate is used mainly in the meat industry (Gwozdz, 1978). Solutions of sodium caseinate for roller drying may be prepared in a similar manner to those intended for spray drying (Fox, 1968; Muller, 1971). A recent patent (Lippe *et al.*, 1983) describes using a drum dryer for drying sodium caseinate solutions with a solids content of 44–47 per cent. Alternatively, casein curd with a moisture content of 50–65 per cent may be mixed with sodium carbonate or bicarbonate and fed on to the hot drum of a roller dryer to produce a completely reacted sodium caseinate product before drying is completed (Towler, 1976c).

Granular Sodium Caseinate

The high costs of spray drying solutions of sodium caseinate, together with the problems associated with the highly viscous nature of the solutions during dissolving and the costs of packaging and transport of the relatively low bulk-density product, have prompted investigations into the production of granular sodium caseinate (Towler, 1977, 1978; Gwozdz, 1978, 1983; Roeper, 1982; Segalen, 1982). The manufacture of this product, which may be either partly or fully reacted, may include the partial drying of casein curd, mixing with an alkali (such as sodium carbonate) and, finally, drying of the caseinate (Towler, 1977, 1978). Where partial drying of the curd is not carried out before addition of alkali, it is possible to produce an extremely sticky and rubbery curd-like mass which is difficult to subsequently process. Methods of overcoming this have included low initial drying temperatures for the caseinate, mixing of curd with the sodium salt of an organic acid (e.g. sodium citrate) before addition of the alkali (Wakodo Co., 1969), and cooling either the reaction/mixing vessel (Towler, 1978) or the curd itself (Gwozdz, 1983). Alternatively, the solids content of the curd can be increased by addition of dry casein (Gwozdz, 1983).

Preparation of Concentrated Caseinate Solutions

One further technique for the manufacture of sodium caseinate has been reported recently (Lippe *et al.*, 1983). A mixture of casein (curd or powder), water and a solubiliser such as NaOH is pumped into a steam-heated reaction chamber. This concentrated solution (typically 33–47 per cent solids) is prepared in two steps, first to pH 5·0–5·3 (where the viscosity is at a minimum) and then to pH 6·5–7·0. The solution may then be spray (at up to 33 per cent solids) or roller dried. Alternatively, the solution may be forced from the chamber through orifices to produce fine jets of caseinate which can be dried in hot air to form threads of sodium caseinate.

Extruded Sodium Caseinate

As noted by Towler (1977), it is possible to produce sodium caseinate from casein in the presence of a limited amount of water by use of extrusion techniques. In most of the published information, dry casein is used as the starting material and water and alkali are added to form a mixture for extrusion (e.g. Wiedmann and Millauer, 1986). The moisture content (10–30 per cent) is generally lower than that used for the granular caseinates discussed above, and fairly high pressures and temperatures may be used during extrusion to facilitate the formation of sodium caseinate. Newer developments in the process for extrusion of caseinates have been reviewed recently (Boullé, 1987).

Ammonium Caseinate/Potassium Caseinate

As observed by Gwozdz (1978) ammonium caseinate, potassium caseinate and sodium caseinate have similar physical properties. Thus, it is possible to add ammonium hydroxide to a slurry of casein and water to produce a solution of ammonium caseinate which may be subsequently spray dried in the same manner as sodium caseinate. Alternatively, a granular ammonium caseinate may be produced by exposing either acid casein curd (Krause, 1953) or dry acid casein (Girdhar and Hansen, 1974) to gaseous ammonia. Excess ammonia is removed during drying of the converted curd or on air flushing of the converted casein.

It has also been reported that small amounts of ammonium hydroxide may be used to increase the pH of calcium caseinate solutions, either when there is an upper limit placed on the calcium content of the dried product (Muller, 1971) or to decrease the amount of sedimentable matter (Roeper, 1977a; Lippe et al., 1983). Most of the added ammonia is evaporated during the subsequent drying process.

Potassium caseinate may be prepared in a similar manner to sodium caseinate by substituting potassium hydroxide, for example, in place of sodium hydroxide (e.g. Towler, 1977; Lippe et al., 1983).

Calcium Caseinate

In contrast to the translucent, viscous, straw-coloured sodium, potassium and ammonium caseinates, calcium caseinate forms micelles in water, producing an intensely white, opaque, 'milky' solution of relatively low viscosity (Roeper, 1977a, b). Its preparation follows the same general process as that used for sodium caseinate with one or two important exceptions. Calcium caseinate solutions may be destabilised on heating (Roeper, 1977a, b), especially at pH values below 6 (Hayes *et al.*, 1968). This sensitivity decreases with increase in pH and decrease in concentration, and is manifested as a reversible heat gelation, first reported by Zittle *et al.* (1956).

During the dissolving process, it has been found that the reaction between acid casein curd and calcium hydroxide (the alkali most commonly used in the production of calcium caseinate) proceeds at a much slower rate than that between curd and sodium hydroxide. The temperature of conversion is a particularly important factor in determining the completeness of solubility (amount of sedimentable matter) of the calcium caseinate (Roeper, 1977a, b). Therefore, the dissolving process must be closely monitored to ensure production of calcium caseinate of good solubility.

Since calcium hydroxide itself has very limited solubility in water, it is common practice to add it to the mixture of casein in water as an aqueous slurry.

Other Caseinates

Magnesium caseinate has also been briefly mentioned in the literature (Towler, 1977; Muller, 1982a), being prepared from casein and a magnesium base or basic salt, such as magnesium oxide (Zavagli and Kasik, 1978), magnesium hydroxide, carbonate or phosphate (Segalen, 1982), or by ion exchange (Hidalgo *et al.*, 1981).

Compounds of casein with aluminium have been prepared for medicinal use (Schuette, 1939; Paterson, 1955) or for use as an emulsifier in meat products (Barrat, 1978).

Heavy metal derivatives of casein which have been used principally for therapeutic purposes (Schuette, 1939) include those containing silver, mercury, iron and bismuth. Iron and copper caseinates have also been prepared by ion exchange for use in infant and dietetic products (Hidalgo *et al.*, 1981).

MANUFACTURE OF CO-PRECIPITATES

Following addition of small quantities of calcium chloride or acid to skim-milk, the mixture is heated to 85–95 °C and held at that temperature for a period of 1–20 min, to allow interaction between the caseins and the whey proteins. Precipitation of the proteins from the heated milk is then effected by controlled addition of either calcium chloride solution (to produce high calcium co-precipitate) or dilute acid (to produce medium calcium or low calcium co-precipitate depending upon the amount of acid added and the pH of the resultant whey). The curd is subsequently washed and either dried to produce granular insoluble co-precipitates, or dissolved in alkali as described in the methods for manufacture of caseinates to produce soluble or 'dispersible' co-precipitates. While co-precipitate manufacture was first reported in the USSR in the early 1950s (e.g. D'yachenko *et al.*, 1953), the commercial potential of co-precipitates was developed in the 1960s and 70s by the Australian Commonwealth Scientific and Industrial Research Organisation (CSIRO). This, and subsequent work, has been extensively reviewed (Muller, 1971, 1982a, b; Southward and Goldman, 1975).

The processes described above cause heat denaturation of the whey proteins. As a consequence, the solubility of the resultant co-precipitates in alkali at pH 6·6–7·2 is not complete (Southward and Aird, 1978). By application of a lower heat treatment to skim-milk under alkaline conditions, however, it has been found possible to alter the whey proteins so they precipitate with the casein at pH 4·6 but dissolve completely with the casein when alkali is added to pH 6·6–7·0 (Connolly, 1983). The product is referred to as 'total milk protein'. Another, similar, product has also recently been developed in the Netherlands (Stiching Nederlands Instituut voor Zuivelonderzoek, 1982).

COMPOSITION OF CASEIN

The typical composition of adequately washed acid casein is shown in Table I. When the same manufacturing operations are employed, the caseins produced from lactic, sulphuric or hydrochloric acid precipitation are indistinguishable from one another (Roeper, 1974). As indicated in Table I, rennet casein differs from acid casein particularly in ash content and in the pH of a water extract. During the acidification process in the manufacture of acid casein the calcium and inorganic phosphate, associated with the casein micelle in milk, are dissolved and leached from

TABLE I

Composition of casein and granular co-precipitates

	Acid casein[a]	*Rennet casein*[a]	*Co-precipitate*[b]		
			High-calcium	*Medium-calcium*	*Acid*
Moisture (%)	11·4	11·4	9·5	9·5	9·5
Protein (%)	85·4	79·9	81·7	85·6	86·7
Ash (%)	1·8	7·8	7·7	3·7	2·4
Lactose (%)	0·1	0·1	0·5	0·5	0·5
Fat (%)	1·3	0·8	0·6	0·7	0·9
pH	4·6–5·4	7·3–7·7	6·5–7·2	5·6–6·2	5·4–5·8
pH of whey after separation of curd	4·3–4·6	6·5–6·7	5·8–5·9	5·1–5·3	4·9–5·1

[a] Caseins dried to a maximum moisture content of 12%.
[b] Co-precipitates washed twice in water and dried to a nominal maximum moisture content of 10% (data of Southward and Aird, 1978).

the curd leaving only the organic phosphorus and a small residue of calcium. Rennet casein contains about 3 per cent calcium and approximately 1·4 per cent phosphorus (McDowell, 1968).

PROPERTIES OF CASEIN

Acid and rennet caseins are not soluble in water but when the dry powders are wetted, the particles absorb water and swell. Both types of casein may be dissolved in alkali and, for virtually all industrial applications, acid casein is dissolved before use. Acid casein is soluble at pH 7 or above in most alkalis: sodium and potassium hydroxides, carbonates and bicarbonates, trisodium phosphate, borax and lime. Rennet casein may be rendered soluble in solutions of complex phosphates, such as sodium tripolyphosphate, at pH 7–8, or may be dissolved in sodium hydroxide at pH 9·5 or higher. Solutions of all caseins (except rennet casein in sodium tripolyphosphate) are very sticky and make very good adhesives and films. The solutions are also very viscous, especially at concentrations above $150 \, \mathrm{g \, l^{-1}}$.

COMPOSITION OF CASEINATES

The typical composition of sodium and calcium caseinate, produced from well-washed acid casein, is shown in Table II. The ash content of the

TABLE II

Composition and physical properties of spray dried soluble co-precipitates and caseinates.

| | Sodium caseinate | Calcium caseinate | Soluble co-precipitate[a] | | |
			High-calcium	Medium-calcium	Acid
Moisture (%)	3·8	3·8	(4·0)[b]	(4·0)	(4·0)
Protein (N × 6·38) (%)	91·4	91·2	81·4	88·1	90·5
Ash (%)	3·6	3·8	13·5	6·7	4·1
Lactose (%)	0·1	0·1	(0·5)	(0·5)	(0·5)
Fat (%)	1·1	1·1	0·6	0·7	0·9
Sodium (%)	1·2–1·4	<0·1	1·9	–	1·1
Calcium (%)	0·1	1·3–1·6	2·9	1·2	0·5
Iron (mg/kg)	3–20	10–40			
Copper (mg/kg)	1–2	1–2			
Lead (mg/kg)	<1	<1			
pH	6·5–6·9	6·8–7·0	7·1–7·2	6·6–7·2	6·6–7·2
Solubility (%)	100	90–98	92	95–98	97–98
Farinograph water absorption[c] (%)	271	143	278	282	292

[a]Data of Southward and Aird (1978).
[b]Assumed values in parentheses.
[c]Data of Knightbridge and Goldman (1975).

caseinates includes approximately 1·8 per cent derived from organic phosphorus, an integral part of the casein molecule. With a pH generally in the range 6·5–7·0, sodium caseinate will usually contain 1·2–1·4 per cent sodium, whereas the calcium content of calcium caseinate is generally in the range 1·3–1·6 per cent.

COMPOSITION OF CO-PRECIPITATES

Table I shows the proximate compositional analysis of granular (insoluble) co-precipitates, prepared from curd washed twice in water. The data show that the calcium content of co-precipitates decreases steadily with a reduction in the ash content, both of which are caused by a decrease in the pH of precipitation. The fat content of the co-precipitates increases from 0·6 to 0·9 per cent as the pH of precipitation is reduced from 5·9 to 4·9. The high ash content of high calcium co-precipitate leads to a consequential reduction of 3–5 per cent in the protein content of the product relative to lactic and acid caseins and acid co-precipitate.

Proximate compositional analysis and physical properties of soluble co-precipitates are presented in Table II. Soluble high-calcium co-precipit-ate has a very high ash content (13·5 per cent), due in part to the presence of sodium tripolyphosphate, and contains almost 2 per cent sodium. Soluble acid co-precipitate has a composition similar to that of sodium caseinate (Table II). In order to render medium calcium co-precipitate substantially (>90 per cent) soluble at pH 7·5, it is necessary to use both sodium hydroxide and sodium tripolyphosphate (or another complex phosphate).

PROPERTIES OF CASEINATES AND CO-PRECIPITATES

Solubility in Water

The granular co-precipitates shown in Table I are, like acid and rennet caseins, insoluble in water. Of the 'soluble' casein products shown in Table II, sodium caseinate is the only one that is completely soluble, i.e. can be centrifuged without depositing any insoluble particles, and pro-duces a pale straw-coloured translucent solution. Calcium caseinate and the various soluble co-precipitates are not completely soluble in water, but vary in solubility from just over 90 per cent to about 98 per cent. Calcium caseinate produces a very white colloidal dispersion. The heat denaturation of the whey proteins during the manufacture of co-precipit-ates causes these products to exhibit some measure of insolubility, though the amount of insoluble material is only a fraction of the whey protein present in the co-precipitate. Variables that affect the appearance of solutions of co-precipitates and their solubility have been discussed by Smith and Snow (1968) and Hayes *et al.* (1969).

Functional Properties

The terms 'functional property' and 'functionality' are generally used to describe the physical effects of, in this case casein, products on foods. 'Functionality' may be used to include almost all properties (e.g. colour, flavour, nutrition and solubility as well as the functional properties listed below) (American Chemical Society, 1981). However, casein products provide functional properties of more specific interest: water absorption, fat emulsification, viscosity and gelation, whipping and foaming, and texturisation (Morr, 1982). A number of reviews of milk protein function-ality have recently been published (de Moor and Huyghebaert, 1983; Fox and Mulvihill, 1983; Kinsella and Whitehead, 1987; Mulvihill and Fox, 1989; Lorient et al., 1991).

Water Absorption

The Farinograph water absorption characteristics of the dispersible and soluble casein products are shown in Table II. These values vary from about 130 to approximately 290 per cent, i.e. 1·3–2·9 g water absorbed (g product)$^{-1}$. By contrast, acid and rennet caseins and the insoluble co-precipitates have much lower absorptions (80–110 per cent) while calcium caseinate shows an intermediate absorption (143 per cent) (Knightbridge and Goldman, 1975). The water absorption properties give an indication of the suitability of casein products for different bakery applications; products with high water absorption (>200 per cent), such as sodium caseinate and the soluble co-precipitates, function as water binders in intermediate and high moisture foods, whereas products of medium (100–200 per cent) (e.g. calcium caseinate, dispersible and insol-uble co-precipitates) and low water absorption (<100 per cent) (e.g. casein and insoluble co-precipitates) find a variety of applications ranging from use in bread and baked goods to confections and frozen desserts (Knightbridge and Goldman, 1975). Water absorption by milk proteins has been extensively reviewed in recent literature (Kinsella and Fox, 1986; Mulvihill and Fox, 1989).

Viscosity and Gelation

The viscosity characteristics of solutions of sodium caseinate, rennet casein (dissolved in sodium tripolyphosphate) and soluble co-precipitates have been studied (Towler, 1974; Southward and Goldman, 1978; Roeper and Winter, 1982) over a range of concentration and temperature. The products with the highest viscosity, soluble rennet casein and soluble high calcium co-precipitate, also have the highest calcium content. Calcium caseinate also has a relatively high calcium content but a comparatively low viscosity, since it exists in water as a white colloidal dispersion (Roeper, 1977a).

The viscosity of casein solutions has an important bearing on their manufacturing costs and on their applications. Thus, a number of investigations have been carried out to decrease the viscosity of casein solutions by enzymatic hydrolysis either during manufacture (Hooker *et al.*, 1982) or for paper coating (Muller and Hayes, 1963; Salzberg and Simonds, 1965). On the other hand, an increase in viscosity, leading to the formation of casein gels, can also be of value in various food applications. Thus, acidified casein gels may have application in the preparation of milk protein/fruit juice drinks and jellies (Korolczuk, 1982).

Fat Emulsification

Sodium caseinate has been used for many years as an emulsifier of oils and fats in such food products as comminuted meats, whipped toppings and coffee whiteners. Many studies have been carried out to determine the relative emulsifying power and emulsion stability of milk proteins (Tornberg, 1980; Fox and Mulvihill, 1983; Reimerdes and Lorenzen, 1983), but it is apparent that a standardised procedure is needed to enable true comparisons to be made among the different groups of research workers.

This subject continues to be of very great interest to the food industry, as indicated in the recent literature (Dickinson, 1986; Mitchell, 1986; Jost *et al.*, 1991; Mulvihill and Murphy, 1991).

Whipping and Foaming

The surfactant properties of milk proteins make them particularly useful in whipping and foaming applications such as in whipped toppings. A review on foam formation and stability is given by Mulvihill and Fox (1989). Measurements of whipping and foaming properties of casein products have been made by Southward and Goldman (1978), Goldman and Toepfer (1978) and Kitabatake and Doi (1982). Recently, collaborative studies for measuring the foaming properties of proteins were made in order to establish standardised procedures for those properties (Patel *et al.*, 1988; Phillips *et al.*, 1990).

Texturisation

The ability of caseins to yield products of different texture is widely known (Burgess and Coton, 1982; Tuohy *et al.*, 1983; Tolstoguzov *et al.*, 1985). This property has been used to produce meat-like products either by spinning from a caseinate solution (Amiot *et al.*, 1979; Kelly *et al.*,

1983) using techniques similar to those used in the late 1930s for producing textile fibres (Ferretti, 1938), or by production of a 'curd' (Goldman, 1974; Suchkov et al., 1988). Casein based textured products may be made by extrusion (Szpendowski et al., 1983). Casein has also been used extensively as the basic texture-forming matrix in imitation cheese products (Vernon, 1972).

Flavour of Casein Products

The flavour of casein products has assumed a much greater importance during the past 20 years as their use in foods has increased (Southward et al., 1978). Various workers have studied the effects of manufacturing technique and treatment of the final casein on its flavour (Cayen and Baker, 1963; Hansen et al., 1970), and considerable research has been undertaken in an attempt to isolate the gluey (Ramshaw and Dunstone, 1969a, b) or musty (Walker, 1972, 1973; Walker and Manning, 1976) storage off-flavours in acid caseins and their derivatives. It has generally been found, however, that fresh curd caseinates (Cayen and Baker, 1963) and rennet casein (Walker, 1972) have the blandest and most stable flavour of the casein products. The flavour of co-precipitates tends to follow a similar pattern to that of the caseins; high calcium co-precipitates are more stable than low calcium (acid) co-precipitates, and fresh curd soluble co-precipitates are better than those reconstituted from dry granular insoluble co-precipitates (Southward and Goldman, 1978). As a class, the co-precipitates may also tend to exhibit 'cooked' flavour overtones as a result of the high heat treatment given to the milk during their manufacture. Particle size of granular caseins and co-precipitates also appears to affect their flavour; finely ground products tend to exhibit stronger off-flavours than coarser fractions.

Nutritional Properties

Amino-acid analysis
The nutritive quality of a protein is determined primarily by its essential amino-acid composition. For adult man, eight amino-acids are essential, i.e. they must be supplied in the diet. These are isoleucine, leucine, lysine, methionine, phenylalanine, threonine, tryptophan and valine; the infant requires histidine as well. In 1973, an 'ideal' protein composition was published (Food and Agriculture Organization, 1973) derived from the probable pattern of amino-acid requirements in man. By comparing

the essential amino-acid composition of dietary proteins with that of the reference protein, a preliminary evaluation of their nutritive quality can be made. In comparison with the reference protein, caseins are only deficient (or limiting) in the sulphur-containing amino-acids methionine and cystine.

Biological evaluation

Biological availability of the amino-acids cannot always be determined by chemical methods and, for this and other reasons, biological tests of protein quality continue to be necessary. Protein efficiency ratio (PER), biological value (BV) and net protein utilisation (NPU) are all used to assess protein quality. Rats are the most widely used test animals and, of the methods based on the growth of rats, PER is the most widely used. A standardised version of this test, officially recommended for the biological estimation of protein quality (Association of Official Analytical Chemists, 1970), uses ANRC casein as the reference protein, with a PER of 2·5. PER is defined as the ratio of weight gained to protein eaten.

Chemical analysis of amino-acids present in proteins has been regarded as inadequate as the sole measure of their quality and nutritive value, but this view has been challenged recently (e.g. Young and Pellett, 1991). An alternative procedure to PER, labelled protein digestibility-corrected amino-acid score (PDC-AAS), has been proposed for evaluating the protein quality of foods and infant formulae.

Milk proteins are regarded as the main source of biologically active peptides (Schlimme *et al.*, 1989), so the 'nutritive value' of food proteins consequently needs a much broader consideration than that described for intact milk proteins (e.g. Hambraeus, 1982, 1985). Bioactive peptides of interest include casein phosphopeptide (CPP) (Meisel and Frister, 1989; Saito, 1990), opioid peptides such as β-casomorphin (Hamosh *et al.*, 1989; Meisel, 1991) and peptides involved in immunomodulation (Migliore-Samour *et al.*, 1989).

CASEIN PRODUCTION AND TRADE

From the data in Table III, it can be seen that the mean annual production of casein during 1984–1990 has been approximately 300,000 tonne, substantially higher than the 225,000 tonne per annum estimated for the period 1977–1983 (Southward, 1986). That earlier figure had been substantiated by Morr (1987), who estimated production of casein during

TABLE III

World production of casein 1984–90 (tonnes)[a]

	1984	1985	1986	1987	1988	1989	1990
Austria	1 800[b]	2 000[b]	500[b]	500[b]			
Denmark	11 800	15 000	18 500	16 400	21 300	18 500	13 200
France	44 891	48 655	44 133	54 439	61 529	47 076	26 897
Germany[c]	20 434	20 496	20 622	23 604	24 978	21 236	13 247
Irish Republic	19 100	30 000	20 300	39 200	41 000	36 500	30 000[d,e]
Netherlands	20 000[b]	20 000[b]	20 000[b]	20 000[b]	20 000[d]	20 000[d,f]	20 000[d,e]
Norway	3 800	4 000[b]	4 000[g]	4 600[g]	4 900[g]		
Poland	43 000[h]	34 000[h]	28 800[i]	21 900[i]	23 800[i]	32 700[i]	40 000[i,f]
USSR	46 000[h]	46 000[h]	49 000[h]				
United Kingdom	3 800[b]	2 700[b]	2 000[j]	3 000[j]	1 300[j]	1 000[d,f]	1 000[d,e]
Argentina	1 791[k]	1 914[k]	1 847[k]	1 900[b]			
Uruguay	2 587[k]	4 214[k]	3 068[k]				
Australia[l]	8 043[m]	8 680[m]	8 236[m]	8 954[m]	6 870[m]	5 328[m]	2 295[m,f]
New Zealand[n]	64 349[o]	75 382[o]	61 840[o]	65 759[o]	54 540[p]	62 907[p]	65 657[p]

[a]Source: Agra Europe (London) (1987–91a) unless shown otherwise.
[b]Source: Commonwealth Secretariat (1987–88).
[c]Data refer to Federal Republic of Germany before re-unification.
[d]Source: United States Department of Agriculture (1989a–1990).
[e]Forecast.
[f]Preliminary.
[g]Source: Food and Agriculture Organisation (1990).
[h]Source: Nienhaus (1988).
[i]Source: Agra Europe (London) (1991b).
[j]Source: Milk Marketing Board (1986–91).
[k]Source: Food and Agriculture Organisation (1987).
[l]Twelve months ending 30 June in following year.
[m]Source: Australian Dairy Corporation (1991).
[n]Twelve months ending 31 May in following year.
[o]Source: New Zealand Dairy Board (1988).
[p]Source: New Zealand Dairy Board (1991).

the early to mid 1980s at 220,000–226,000 tonne, while Gulayev-Zaitsev (1987) assessed world casein production in 1986 at 240,000 tonne. Major producers are New Zealand (mean production 64,000 tonne), France (46,000 tonne), Poland (32,000 tonne) and the Irish Republic (31,000 tonne). The USSR also appears to produce some 25,000–30,000 tonne of casein each year, but complete statistical data for this period are not available.

According to the figures for exports and imports of casein, shown in Tables IV and V, respectively, approximately 195,000 tonne of casein entered international trade each year during this period. New Zealand contributed 64,000 tonne of this total (33 per cent) while French and Irish exports amounted to an average of 33,000 and 31,000 tonne, respectively, during this period. The major casein importing countries are the USA (90,000 tonne per year), Japan (24,000 tonne) and Germany (23,000 tonne). Data from 1920 to 1983 have been reported earlier (Southward and Walker, 1980, 1982; Southward, 1986).

INDUSTRIAL USES OF CASEIN

The principal industrial uses of acid casein are shown in Table VI.

Casein as an Adhesive

Wood glues
Casein glues have been marketed in two forms: prepared glues and wet-mix glues. Prepared glues are sold in powder form and contain all the necessary ingredients except water. The user simply mixes the powder with water (usually in the proportions 1 part powder to 2 parts water by weight). Once mixed, the glue must generally be used within one day or less. Wet-mix glues are made as required from ground casein, water and the additional chemicals required by the formula (Browne and Brouse, 1939).

Besides casein and water, an alkali must be used to dissolve the casein. This is commonly sodium hydroxide and provides the third ingredient in the production of a simple glue. Where a water resistant glue is required, it is also necessary to add lime to this formulation. The lime promotes cross-linking of the casein molecules and, over a period of several hours, will cause a casein glue to irreversibly form a jelly that is insoluble in water. Many different additives may be used to promote

TABLE IV
World trade in casein. Exports 1984–1990 (tonne)[a]

	1984	1985	1986	1987	1988	1989	1990
Belgium/Luxembourg	843	970	45	6	125	218	413
Denmark	11 790	13 042	15 756	4 417	18 829	17 889	14 701
France	32 096	32 313	31 151	26 902	40 735	31 661	30 830
Germany[b]	14 619	17 297	18 125	13 639	19 687	20 251	25 044
Irish Republic	22 988	32 982	30 055	38 050	37 919	30 428	26 607
Italy	361	373	630	244	1 082	810	544
Netherlands[c]	6 122	5 873	7 100[d]	6 841	4 339	4 728	7 126
Poland	17 000[d]	19 800[e]	15 600[e]	16 400[e]	12 300[e]	7 100[e]	20 000[e,f]
United Kingdom	5 012	3 914	4 848	3 935	3 255	2 710	3 526
Argentina	1 000[d]	1 000[d]	1 000[d]				
Australia[g]	8 300[d]	7 130[h]	8 275[h]	8 343[h]	6 115[h]	4 257[h]	3 190[h]
New Zealand[g]	70 900[d]	71 200[i]	73 700[i]	66 700[i]	44 400[i]	56 100[i]	66 400[i]

[a]Source: Milk Marketing Board (1986–91) unless shown otherwise.
[b]Data refer to Federal Republic of Germany before re-unification.
[c]Excluding caseinates.
[d]Source: Commonwealth Secretariat (1987–88).
[e]Source: Agra Europe (London) (1991c).
[f]Preliminary.
[g]Twelve months ending 30 June in following year.
[h]Source: Australian Dairy Corporation (1991).
[i]Source: New Zealand Department of Statistics (1989–91).

TABLE V
World trade in casein. Imports 1984–1990 (tonne)[a]

	1984	1985	1986	1987	1988	1989	1990
Belgium/Luxembourg	5 011	3 437	3 771	2 584	4 075	3 808	3 288
Denmark	2 048	2 164	2 381	336	2 494	2 226	2 817
France	7 448	8 383	8 420	6 403	12 979	15 115	26 393
Germany[b]	18 385	22 246	21 028	13 588	26 801	22 856	34 716
Greece	682	1 094	1 155	537	952	1 010	1 015
Irish Republic	704	1 590	1 232	420	849	1 476	1 703
Italy	10 751	11 638	12 670	10 153	16 242	13 412	14 635
Netherlands	13 009	14 436	12 177	11 421	13 872	10 850	17 145
Portugal			1 587	1 025	1 543	1 239	1 279
Spain			13 911	5 124	16 593	11 549	10 674
United Kingdom	5 562	7 480	8 602	3 832	10 475	8 609	7 577
United States	87 357[c]	87 321[c]	107 938[c]	108 136[c]	73 676[c]	81 800[d,e]	85 100[d,e]
Japan	22 649[f]	24 439[f]	23 630[f]	26 848[f]	29 482[f]	21 948[g]	21 146[g]

[a]Source: Milk Marketing Board (1986–91) unless shown otherwise.
[b]Data refer to Federal Republic of Germany before re-unification.
[c]Source: United States Department of Agriculture (1989b).
[d]Source: Agra Europe (London) (1991c).
[e]Preliminary.
[f]Source: Japan Ministry of Agriculture, Forestry and Fisheries (1989–90).
[g]Source: Japan Finance Ministry (1990–91).

TABLE VI
Principal industrial uses of acid casein

Coatings for paper and board
Adhesive for wood, e.g. plywood
Paints
Joint cements
Textile sizing
Synthetic fibres
Horticultural spreaders
Leather tanning
Stock foods

various properties of casein glues. For instance, addition of sodium silicate may be employed to prolong the working life of a glue, whereas use of copper chloride will increase the glue's water resistance.

Casein glues are often employed in interior woodwork, for example in the production of interior doors for houses, wooden cabinets, plastic-topped counters and general edge glueing where fast setting, water resistance and gap filling are needed (National Casein Co., 1989). In 1967, approximately 5000 tonne of casein were used for glues in the USA (Fox, 1970). However, the increase in the price of casein since then has caused a significant decrease in its use for this purpose.

Paper coating
In a development parallel to that of adhesives for wood, casein has also been employed as an adhesive in the coating of board and paper (Salzberg et al., 1961), especially in the production of high quality art paper. During the period 1940–1962, paper coating accounted for the major use of casein – generally 50 per cent (Robinson, 1964) as shown in Fig. 7. Even in 1967, paper coating accounted for about one third of the approximately 50,000 tonne of casein products imported into the USA (Fox, 1970).

When used in coatings with a high solids content, casein forms highly viscous solutions (Salzberg et al., 1961). These limit the solids content of the coating mix and hence the speed of the coating process. The viscosity of casein solutions has been reduced in commercial applications by the addition of urea or dicyandiamide (Salzberg and Marino, 1975), considered to be flow modifiers. There have been a number of successful attempts to lower the viscosity of casein solutions by alkaline or enzymatic hydrolysis (Muller and Hayes, 1963; Salzberg and Simonds, 1965). Disulphide bond-reducing agents, such as

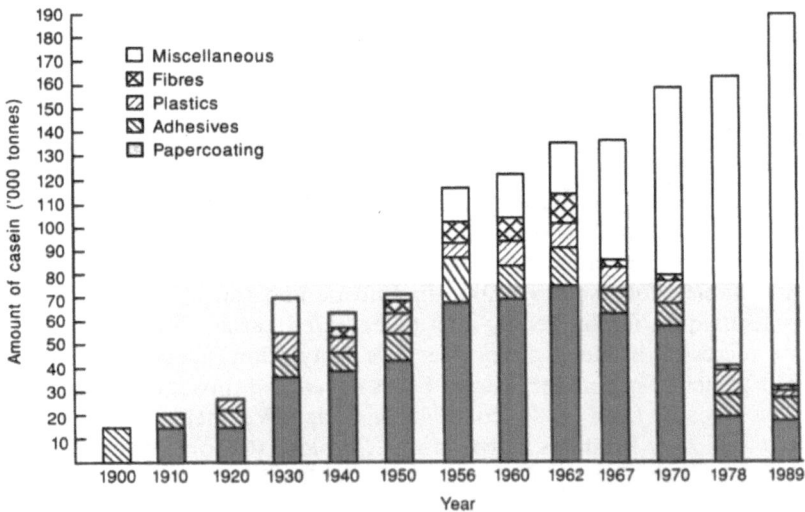

Fig. 7. Probable utilisation of casein 1900–1978. Sources: for period 1900–1962: Robinson (1964); for period 1967–78: Southward and Walker (1980); for 1989: calculated from data supplied by A. Kirby-Lewis, (1990, personal communication).

mercaptoacetic acid and its ammonium and sodium salts, and 2-mercaptoethanol, have also been examined for this purpose (Towler *et al.*, 1981).

Sizing

The film-forming ability of casein is retained even when it is deposited from a very dilute solution. Thus, casein may be applied to wool to reduce its felting properties (Jackson and Backwell, 1955) and for textile sizing generally. Paper surfaces may be made repellent to solvents, oils and greases by applying a film of casein to them ('sizing'). This casein film may also be given a high degree of water resistance by the inclusion of 'hardeners' in the solution or by post-application to the film. Casein then becomes a permanent finish applied to paper to enhance its lustre or stiffness (Salzberg, 1965).

Casein may also be employed as an adhesive for other paper bonds such as laminates, cigarette side-seaming, bottle and can labelling (Salzberg *et al.*, 1974; Salzberg, 1977), and in bonding the striking heads to matches.

Casein Fibre

The process of producing casein fibres includes the following steps. Casein is dissolved in an alkali, such as sodium hydroxide, at a concentration of about $200 \, gl^{-1}$; the solution is then forced through a spinneret into a coagulating bath. The bath usually contains acid, inorganic salts and often heavy metal salts. The fibres thus formed resemble wool except they have a lower tensile strength and do not 'felt' (i.e. shrink on washing) like wool.

Following commercial development of casein fibre by Ferretti (1938), considerable effort was expended in Europe and the USA in perfecting new techniques for hardening this regenerated casein fibre. Among the more successful hardening processes was acetylation (Brown et al., 1944).

The principal proprietary casein fibres developed throughout the world in the decade from 1936 to 1945 include (Wormell, 1954): Aralac (National Dairy Products Corp., USA), Casolana (Co-op Condensfabriek Friesland, Netherlands), Fibrolane (Courtaulds Ltd., UK), Lanital and Merinova (Snia Viscosa, Italy). Of these, only Fibrolane and Merinova were still in production by 1971 (Roff and Scott, 1971).

Estimates of world production suggest that around 5,000 tonne per annum of casein fibre were made during the decade 1940–50. Manufacture increased during the 1950s to about 10,000 tonne by 1960 (Robinson, 1964) (see also Fig. 7).

Casein fibres were employed during and after the war years, usually in combination with wool and other fibres, such as cotton, viscose and rayon, in a variety of products such as flannel, woollen spun cloth (overcoats, blankets), felt hats (up to 25 per cent casein fibre with wool), filling materials, such as artificial horsehair and carpets and rugs (Wormell, 1954). Bristles were also produced from casein fibres (McMeekin et al., 1945) for use in brushes of various types.

The importance of casein fibre in the USA and Europe has now declined in the face of competition from other fibres. However, copolymer fibres containing casein have been prepared in Japan as a substitute for silk (Fox, 1970), and the author has a number of ties and other articles recently manufactured in Japan from casein and casein copolymer fibres (Fig. 8).

Casein in Paints

Casein has been used for many years as a binder and pigment vehicle in water based paints on walls and easel paintings (Scholz, 1953; O'Neil,

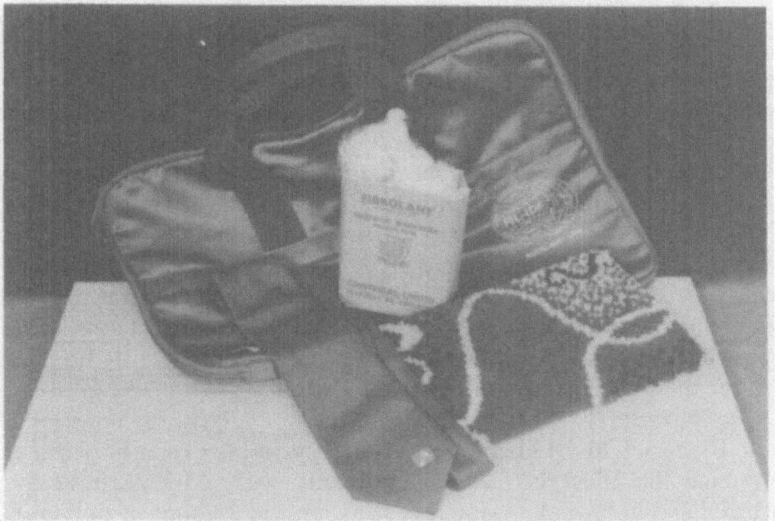

Fig. 8. Articles made from casein fibre. (Reproduced from Southward (1989) by permission of the publishers.)

1971) and as a stabiliser/emulsifier for oil paints (Mummery, 1949). More recently, it has come to be used as a pigment dispersing agent, stabiliser and bodying agent in latex paints (Melsheimer and Hoback, 1953; Salzberg, 1964) usually at a concentration of 1–2 per cent (w/w) of the paint. O'Neil (1971) has described in some detail techniques for preparing finishes on various articles, particularly period furniture, by application of casein paint mixtures.

Casein in the Leather Industry

The use of casein in the leather industry is confined almost entirely to the last of the finishing operations: coating the leather with certain preparations and then subjecting it to mechanical operations such as glazing, plating, brushing and ironing. After finishing in this way, leather is said to have been seasoned (Cavett, 1939; Landmann, 1962). Casein is generally used in these operations in combination (as graft copolymers) with other substances such as acrylates (Langmaier, 1967; Lakshminarayana *et al.*, 1986).

Casein in Animal Feeds

Formaldehyde-treated casein for ruminants

In the 1960s and 70s extensive nutritional studies were carried out on the feeding of casein to ruminants. During normal digestion in the rumen, proteins such as casein are almost completely broken down by the rumen microflora to ammonia, and therefore become unavailable for nutrition of the animal. This problem may be overcome by 'protecting' the casein by first treating it with formaldehyde. Under such circumstances, the treated casein passes undigested through the rumen (neutral pH) and is digested under acid conditions in the abomasum to provide a nutritional supplement. Formaldehyde treated caseins have consequently been examined for their effectiveness in promoting wool growth in sheep (Reis and Tunks, 1969; Barry, 1972) and milk production in lactating cows (Wilson, 1970; Broderick and Lane, 1978). Studies have also been undertaken to determine the effect of feeding ruminants with polyunsaturated oils, encapsulated in formalin-treated caseins, on the amount of polyunsaturated fat in meat and milk (Scott *et al.*, 1970; McGilliard, 1972).

Pet foods

Casein, generally in the form of sodium caseinate, is also finding application as a nutritional supplement and binder in calf milk replacers (Battelle Memorial Institute, 1974) and various pet foods (Burkwall *et al.*, 1976), where it may comprise 3–30 per cent of the food. Figure 9 shows an example of a pet food that contains casein.

Miscellaneous Uses of Acid Casein

Among the many other industrial applications for casein (or at least claimed applications), is in the production of concrete, particularly in eastern Europe, in joint cement for plaster board (Robinson, 1964), as an emulsifying agent in asphalt and bitumen, in photolithography (Salzberg, 1965), as a reinforcing agent for rubber (Genin, 1961), in smoking mixtures, in soaps and dishwashing liquids and in cosmetics such as cold wave lotions (Genin, 1966), hair-sprays, shampoos (Rhone-Electra, 1990) and hand creams (Salzberg, 1967). Casein has also been used as a spreader in horticultural sprays, as an insect attractant (Sharp, 1987) and even as a fertiliser. In beverages, casein has been employed as a clarifying agent for wines (Watts *et al.*, 1981) and beer and in the removal of colour from apple juice (Lodge and Heatherbell, 1976), at a concentration

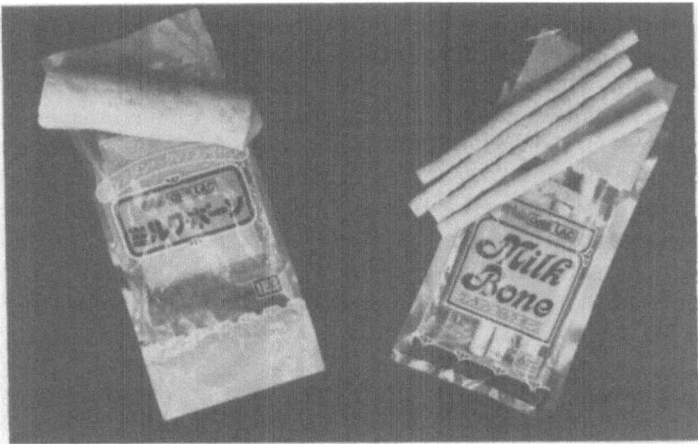

Fig. 9. Pet food – 'Milk Bone', made from casein.

usually less than 0·1 per cent. A later application claimed for casein is in the extraction and recovery of heavy metals from mining, tanning and electroplating wastes (Davey *et al.*, 1980).

USES OF RENNET CASEIN

The principal industrial uses of rennet casein are in the production of plastics such as buttons, buckles and imitation ivory knife handles. Casein plastics were first produced before the turn of the century and commercial production of galalith (Greek: 'milk stone') was commenced in France and Germany by the International Galalith Gesellschaft Hoff & Co. in 1904. After the First World War, the manufacture of casein plastics increased all over the world, producing such trade names as *Erinoid* (UK), *Aladdinite* (USA), *Casolith* (Netherlands) and *Lactoloid* (Japan) (Brother, 1939). Since that time, it is estimated that about 10,000 tonne of casein plastic have been made throughout the world each year (Southward, 1974).

Rennet casein produces a plastic far superior to that from acid casein. In the production of casein plastic, the casein (if unground) is milled to pass a 40 mesh (350 μm) sieve and mixed with dyes and fillers, as required, and water to a final moisture content between 20 and 35 per cent. After a period of equilibration, the moistened casein mixture is extruded through heated nozzles at relatively high pressures (15–75 MPa). The plastic rod that emerges is subsequently cooled in water, trimmed and cut into button blanks. The button blanks are cured for a period of several days or weeks in a dilute solution of formaldehyde before final machining and polishing (McDowell *et al.*, 1976; Munro *et al.*, 1980). Alternatively, if sheet plastic is required, the freshly extruded rods may be placed side by side and formed into sheets in a heated hydraulic press. Samples of casein plastics are shown in Fig. 10. Today, casein plastics are generally fabricated into buttons and buckles. Only a small proportion of this production is channelled into other items, such as imitation ivory knife handles and piano keys.

Fig. 10. Sheets, rods, buttons and button blanks made from rennet casein plastic. (Reproduced from Southward and Walker (1980) by permission of the publishers.)

EDIBLE USES OF CASEIN PRODUCTS

Although casein and its derivatives were used in foods well before 1940, mainly for nutritional purposes (Schuette, 1939), it was not until the 1960s that commercially significant amounts of casein products were employed for this purpose (Poarch, 1967; Fox, 1970), particularly in the developed countries such as the USA (Centre National du Commerce Exterieur, 1970; Hammonds and Call, 1970) and in Europe (van Kreveld, 1969). The significant increase in the 'Miscellaneous' uses category for casein products in 1967 and the 1970s (Fig. 7) is due mainly to the use of casein and its derivatives in food and feed applications.

The uses of milk proteins in formulated foods have been reviewed recently (Koide, 1985; Morr, 1987; Southward, 1989; Mulvihill, 1991), and the principal uses of casein products in foods are shown in Table VII.

TABE VII
Uses of casein products in foods

Meat products	Ice cream and frozen desserts
Coffee whiteners and creamers	Confectionery
Whipped toppings	Cultured milk products, soups and gravies
Bakery	High-fat powders, shortenings and spreads
Instant breakfasts and beverages	Infant foods
Pasta	Sports drinks
Cheese products	Nutritional food bars
Pharmaceuticals	

Casein in Meat Products

The main properties imparted by casein products to comminuted meats are protein enrichment, fat emulsification, water binding and general improvement of consistency (Visser, 1982, 1985). Usually, the amount of added casein is fairly low (less than 5 per cent of the weight of the meat product), though higher amounts of up to 20 per cent of the meat may be employed for specific purposes such as lower cost or an increase in nutritional quality. Some countries have promulgated regulations that restrict the amount of non-meat proteins in meat products (Pfaff, 1974; Fogh, 1975), leading to the development of methods for the detection of these 'foreign' proteins in meats (Kas et al., 1984; Janssen et al., 1987). Synthetic or artificial meats (meat analogues) that possess the fibrous and chewy texture of meat muscle have also been prepared from casein using the following two basic techniques.

(1) The spinning of protein fibre by methods developed in the 1930s for producing textile fibres in which a protein solution is forced through spinnerets into a coagulating bath (Burgess and Coton, 1982).
(2) The production of chewy meat-like gels either by precipitation from concentrated casein solutions, or from moistened casein products by such techniques as extrusion cooking (Skurray and Osborne, 1976).

One of the main problems encountered in the production of fibrous casein for foods has been its relative sensitivity to heat; when such fibres are cooked in a moist environment, they rapidly lose their strength and fibrous nature. Similar problems were encountered in the manufacture and use of casein textile fibres. Future prospects for casein texturisation have been reviewed recently (English, 1981).

Casein in Baked Products and Cereals

Early uses of casein derivatives in baked products were in various dough and biscuit formulations for the principal purpose of nutritional fortification of the wheat flour. One of the most important functional characteristics of the casein derivatives in this application was found to be water binding. Flavour and other sensory properties of baked products are claimed to be influenced by the addition of milk proteins, at least in Europe (Bunnies and Timm, 1983).

While sodium caseinate has apparently found acceptance in controlling texture and uniformity of such baked products as doughnuts, biscuits, waffles and yeast raised doughs (Craig and Colmey, 1971; Cocup and Sanderson, 1987; Schär et al., 1987), it has also been found to produce doughs with an unacceptable 'slimy' mouth feel in some farinaceous products. However, when it is employed in a mixture with whey, its suitability for use in baked foods is greatly extended (Kirk, 1973).

Considerable interest in co-precipitates as milk protein ingredients for baked products was generated some 20 years ago (Craig and Colmey, 1971). Co-precipitates having a range of controlled water absorption characteristics were prepared and used in such foods as the Australian milk biscuit (Townsend and Buchanan, 1967), cake mixes for diabetics (Buchanan and Henderson, 1971), bread, biscuits (Cooper et al., 1984) and in a pastry glaze (Bready, 1966).

Casein in Whipped Toppings

The main constituents of whipped toppings are water, vegetable fat, sugar(s), protein (such as sodium caseinate), emulsifiers and stabilisers. Since sodium caseinate is not defined as a dairy product in the USA whipped toppings that incorporate this protein ingredient are considered to be 'non-dairy' substitutes for cream (Webb, 1970), and may often possess superior whipping properties to cream (McIntire, 1971).

In the preparation of a whipped topping, the water-soluble ingredients (i.e. sugar, sodium caseinate, stabiliser) are usually blended together and then dissolved in the water (Knightly, 1968; Hedrick, 1969). The fat and emulsifiers are melted together, generally at a temperature of 30–46 °C, and added to the aqueous solution which is heated to the same temperature. Following pasteurisation and two-stage homogenisation, generally at pressures of about 7 and 3·5 MPa for the first and second stage, respectively, the whipped topping mixture may be either cooled rapidly to below freezing point or spray dried. The effect of processing variations on the properties of whipped toppings has been reviewed by Knightly (1968).

In whipped toppings, sodium caseinate functions as a film former to entrap aerating gases (Knightly, 1968), a fat encapsulating agent (Cameron *et al.*, 1959), a stabiliser (Pader and Gershon, 1965) and a bodying agent (Clarke and Love, 1974). The amount of caseinate employed for this purpose is usually 5–10 per cent of the dry weight of the ingredients (Centre National du Commerce Exterieur, 1970). Other casein derivatives, such as casein hydrolysates and calcium and potassium caseinates, are also occasionally employed in whipped topping formulations.

Casein in Ice Cream and Frozen Desserts

Although the effect of sodium caseinate in products such as ice cream was studied as early as 1935 (Bird *et al.*, 1935), it has not had an extensive commercial use in this application because of various legal restrictions (Wade, 1977). For example, US standards of identity for ice cream do not permit the use of casein derivatives as part of the minimum requirement for 10 per cent milk solids-not-fat.

Ice cream substitutes have appeared, however, on the market in the USA (Hedrick, 1969; Centre National du Commerce Exterieur, 1970) and a number of them have contained sodium caseinate. The function of sodium caseinate in frozen desserts and ice cream is to impart body (Webb, 1970) and to act as a stabiliser (Little, 1966). It serves a similar function in milk-shakes and instant puddings where foam stability is also important. The amount of caseinate commonly used in ice cream and other desserts varies from 1 to 10 per cent according to the manufacturer's formula (Centre National du Commerce Exterieur, 1970). Figure 11 illustrates samples of desserts and puddings that contain casein products.

Fig. 11.   Samples of desserts and puddings that contain casein products.

Casein in Coffee Whiteners and Creamers

The greatest single use of sodium caseinate in food products in the USA during the 1960s was in coffee whiteners and creamers (Centre National du Commerce Exterieur, 1970). In 1968, approximately 4000 tonne of sodium caseinate, representing 22·7 per cent of the total food use of caseinates for that year, was employed in coffee whiteners. Later data suggest that approximately 6000 tonne of caseinate was used in coffee whiteners in 1978 (United States International Trade Commission, 1979).

Surprisingly little information has been published on the production of 'non-dairy' coffee whiteners and creamers, presumably because the major manufacturers have preferred to keep the details confidential. They are based, however, on vegetable fat and are formulated in a similar manner to whipped toppings (McIntire, 1971). Three retail packs of coffee whiteners are shown in Fig. 12.

The function of sodium caseinate (or other protein) in coffee whiteners is multipurpose: it provides emulsification, provides some whitening, imparts body, improves flavour and promotes resistance to 'feathering' (Clarke and Love, 1974; Buchheim, 1983). It is usually incorporated into the formulation at a level of 7–10 per cent of the dry ingredients. Coffee whiteners may be used in the fluid or frozen state or as a spray dried powder (Knightly, 1969). The importance of manufacturing variables has been reviewed by Knightly (1969), who describes the various stages in the preparation of coffee whiteners. The process is similar to that described

Fig. 12. Coffee creamers and whiteners from the USA, Korea and New Zealand.

for the production of whipped toppings, but one important difference is in the first stage homogenisation pressure. This is recommended to be 14–17 MPa (Hedrick, 1969; Knightly, 1969) in order to reduce the diameter of the fat globules to 0·7–1 μm, whereas a lower first stage pressure of about 7 MPa is employed in the homogenisation of whipped toppings.

Casein in Instant Breakfast, Imitation Milk and Other Beverages

Instant breakfast formulations are a feature of the domestic food scene in the USA (Centre National du Commerce Exterieur, 1970). They usually contain skim-milk powder, sucrose, sodium caseinate, vitamins and minerals, and are sold in different flavours. One sachet of an instant breakfast may be mixed with a glass of milk to provide adequate nutrition for an adult's breakfast. Instant breakfast formulations usually contain 2–4 per cent sodium caseinate (Centre National du Commerce Exterieur, 1970), though some variants may contain considerably more. One product, formulated specifically for nutritional purposes as a diet beverage (Sobotka, 1971), is believed to contain about 30 per cent sodium caseinate. A fruit-type beverage, fortified with protein, has been proposed for this application as well (Appleman, 1973). Since many intact proteins (such as casein) are likely to be precipitated in an acidic beverage, the author of this invention proposed that the proteins first be hydrolysed.

The incorporation of casein products (mostly in the form of sodium caseinate) in imitation milk beverages, based on vegetable fat and a carbohydrate source such as corn syrup solids, also appears to have originated mainly in the USA (Hedrick, 1969; Rusch, 1971; Filer, 1972), although it has been observed that some artificial milk has also been produced in the UK (Waite, 1972) and in the USSR (Voropalva and Ionkina, 1979). Probably the main reasons for the appearance of imitation milks are the lower ingredient cost and the absence of lactose, for which some people show intolerance, compared with cow's milk.

Formulations of imitation milks generally show them to consist of 3–4 per cent vegetable fat, 1–5 per cent protein (usually 1–2 per cent of either sodium caseinate or soy protein), 6–10 per cent carbohydrate (usually corn syrup solids and possibly sucrose), and various stabilisers, emulsifiers etc. Vitamins and minerals are other optional ingredients (Hedrick, 1969; Rusch, 1971; Filer, 1972; Waite, 1972). Processing of these products is carried out in a manner similar to those methods used for coffee whiteners and whipped toppings (Rusch, 1971). Generally, imitation milks are nutritionally inferior to cow's milk and to filled milk, especially in terms of protein, vitamin and mineral content (National Dairy Council, 1968). The low protein content is mainly a consequence of problems encountered by the formulations in providing a beverage of acceptable flavour and consistency (Filer, 1972).

The nutritional properties of casein may be used to good advantage in the fortification of various milk products, including fluid milk (Solms-Baruth, 1972; Downes and van der Merwe, 1982).

A recent new application for sodium caseinate is as a stabiliser in cream based liqueurs, one of the fastest growing markets for cream (Muir and Banks, 1987). The main source of cream based liqueurs has been the Irish Republic, although recent work on these products has been carried out in Scotland (Banks et al., 1983; Muir and Banks, 1984; Muir, 1988).

Casein in Cultured Milk Products, Soups and Gravies

In the USA, the use of sodium caseinate in sour cream products based on vegetable fat, i.e. imitation sour cream, has been reported (Centre National du Commerce Exterieur, 1970). In such products, the sodium caseinate acts as a stabiliser and as an emulsifier for the fat. The amount of sodium caseinate incorporated into sour cream has been estimated to be 2–3 per cent of the weight of all ingredients (Centre National du Commerce Exterieur, 1970). Mayonnaise is another product in which sodium caseinate has been used; its function is to act as an emulsifier for the fat (Cameron *et al.*, 1959; Zalewska and Swiderski, 1984).

In cultured dairy foods, sodium caseinate has been used as a stabiliser in yoghurt in the USA (Centre National du Commerce Exterieur, 1970; Modler *et al.*, 1983) while, in Europe, potassium caseinate has been evaluated for this purpose (Renner and Eiselt-Lomb, 1985). In the USSR, Bogdanova *et al.* (1978) used sodium caseinate for the purpose of increasing the viscosity of cultured cream products.

Casein products have also been used for nutritional fortification and the thickening of soups and gravies (Powell, 1974), while flavour enhancement of these other foods has been achieved by using hydrolysates of casein (Connell, 1966). Several substitute foods based on caseins, such as vegetable cutlets, synthetic caviar (Nesmeyanov *et al.*, 1971) and a nut substitute, have also been prepared in Poland and the USSR.

Casein in High-fat Powders, Shortenings and Spreads

The emulsifying properties of casein derivatives have been used to good effect in a number of high-fat foods, e.g. butter substitutes (Roberts, 1959) and shortening products for baking purposes (Cameron *et al.*, 1959). Casein, usually in the form of sodium caseinate, is generally incorporated at a level of up to 10 per cent by weight of the product and is used in combination with other emulsifiers. It has consequently been possible to manufacture spray dried butter powders with a fat content of up to 80 per cent (Hansen, 1963; Snow *et al.*, 1967).

A number of butter-like foods (Rasic *et al.*, 1978; Stavrova and Mochalova, 1978) and cheese spreads (Elenbogen and Baron, 1968) have been made by the incorporation of sodium and/or calcium caseinate or co-precipitate in emulsions of edible fat, sometimes based on vegetable oils (Anon, 1976, 1987). Whipping fat and whipping cream containing sodium caseinate have also been produced (Cooper and Peacock, 1979).

Casein in Infant Foods

A number of infant food formulations that contain casein derivatives have been reported in the literature. In some cases, the formulations have been prepared specifically for infants with particular dietary problems (Knights, 1985). For instance, Lofenalac[®][a] is a casein hydrolysate

[a]*Lofenalac* is a registered trademark of Mead Johnson & Co.

specially treated to remove almost all the phenylalanine and is used for feeding infants suffering from phenylketonuria (Owen, 1969; Acosta *et al.*, 1977). For those children who have problems in digesting lactose or other sugars, low lactose or carbohydrate-free infant foods have been formulated (Henderson and Buchanan, 1973; Gupta and Rao, 1984). Other infant formulae containing casein have been produced in Japan (Morinaga Milk Industry Co., 1975), the USSR (Korobkina *et al.*, 1974), Yugoslavia (Caric *et al.*, 1981) and the USA (Nichols *et al.*, 1983). A recent review has described the biological effects of bovine casein in human diets with specific reference to digestion in infants (Miller *et al.*, 1990).

Casein in Pasta Products and Snack Foods

The reason for using casein derivatives in pasta and snack foods is mainly to enhance their nutritional quality. Casein has been used to enrich macaroni (Tolstogusow *et al.*, 1975) and spaghetti (Durr and Neukom, 1972), co-precipitate has been formed into imitation rice (Markh *et al.*, 1982) and caseinates have been used in the fortification of rice, pasta and bread (Humphries and Roeper, 1974). There also appears to be some interest in Japan in the fortification of noodle products with casein (Minami *et al.*, 1979). While the amount of casein used to fortify these products is usually between 5 and 20 per cent of the total weight, some applications suggest that the pasta is a synthetic or imitation product that contains casein as one of the major ingredients. In such cases, the casein obviously has a major effect on texture.

Casein products, such as casein, caseinates and co-precipitate curd, can be used to form relatively high protein (30–75 per cent expressed on the basis of the dry food product) snack foods (Wong and Parks, 1970). While these milk proteins form a source of nutrition for the consumer, they are also formed, during the extrusion or other manufacturing process, into the basic matrix that provides the texture necessary for their appeal as a snack food. One such product, Pro-Teens®, has undergone market evaluation in the USA (McCormick, 1973).

Casein in Confectionery

Casein products are used in candy in order to develop a chewy firm body, neither sticky nor tough (Kinsella, 1970; Webb, 1970; Hugunin and Nishikawa, 1977). They also contribute to the colour and flavour of confectionery. During the cooking of products such as caramel and fudge, the casein apparently coagulates (Webb, 1970). It is considered that the fineness of the coagulum contributes towards the final texture of the candy. An extensive review on this topic has been published by Lim (1980). Some studies have been undertaken recently to evaluate the effect of casein supplementation of chocolate confectionery on tooth decay (Reynolds and Black, 1987, 1989).

Casein in Cheese Products

Although casein products have been used in cheese-like foods, such as cheese spreads (Elenbogen and Baron, 1968), it was not until the early 1970s that so-called 'non-dairy' or imitation cheese products, based entirely on vegetable oil and casein, were produced (Vernon, 1972). These were developed initially in the USA and considerable research and development was applied in attempts to produce 'non-dairy' cheese with functional and nutritional properties similar to traditional cheese products such as Mozzarella and processed cheese. Initially, at least, it appears that imitation cheese products were used for 'industrial' purposes by the manufacturers of foods such as pizza and lasagna, by institutional food service companies, and by suppliers to fast-food franchises for use in cheeseburgers and sandwiches.

The volume of cheese substitutes produced in 1978 was 43,000 tonne and represented 2·7 per cent of the natural cheese produced in the USA (Shaw, 1984). Within two years, the production of cheese substitutes had increased to more than twice the 1978 level and, at 90,000–95,000 tonne, was equivalent to about 5 per cent of the 1·8 million tonne of natural cheese produced in the USA in 1980 (Shaw, 1984). By 1990, the market share of imitation cheese in the USA had stabilised at about 7 per cent of total cheese, compared with <3 per cent in Europe (Anon, 1990).

The growth of imitation cheese (real and potential) in the USA has led to considerable study of the subject (Vakaleris, 1980) and an expression of concern that imports of casein should be curtailed (Graf, 1979). However, both Vakaleris (1980) and Shaw (1984) note that it is generally believed that the introduction of imitation cheese products has increased the total amount of 'cheese' consumed, rather than displaced any natural cheese. Further, the price of imported casein into the market in the USA obviously has an effect on the sales of imitation cheese products (Graf, 1986; Herner, 1988).

Interest in imitation cheese has also been shown in other countries. In 1980 production of imitation cheese, based on casein and vegetable fat, commenced in the UK (Shaw, 1984) and in Japan (Akino and Yoshioka, 1982). The international situation with regard to imitation cheese products was reviewed recently (McCarthy, 1990).

Casein derivatives are used in cheese products primarily for their functional properties, e.g. binding of fat and water, texture, melting properties, stringiness and shredding properties (Petka, 1976). The list of casein derivatives used in cheese products includes mainly acid casein and caseinates, but it is evidently a field where other types of casein, such as rennet casein and co-precipitates, may also find application (Roeper, 1976; L. D. Schreiber Cheese Co., 1978). Figure 13 shows a number of imitation cheese products derived from casein.

Fig. 13.   Imitation cheese products derived from casein.

Uses of Casein in Pharmaceutical Products

Casein and its derivatives have been employed for many years in a variety of pharmaceutical applications such as tonic foods, in the treatment of convalescent and undernourished patients, as a therapeutic agent (in combination with some heavy metals), in dressing wounds and in ointments (Schuette, 1939).

During and just after the Second World War, a number of studies were undertaken to determine the nutritional effect of feeding various casein hydrolysate preparations to hospital patients following surgery (Bell,

1947; Robertson and Smaill, 1947). The status of more recent studies has been described and reviewed with respect to the relative advantages of enteral (via nasal tube or by oral supplementation) and parenteral (via a catheter) nutrition (Grimble and Silk, 1989).

Manson (1980) has reviewed the use of milk protein and casein hydrolysates in pharmaceutical applications.

Various casein products have been used in dietary foods and drinks (Fig. 14), for meal replacement, for weight reduction (Sobotka, 1971), in candy for space feeding (Dymsza *et al.*, 1966) and in high protein supplements (Fig. 15). Applications for casein products in the treatment of medical conditions such as anaemia, digestive problems, cancer and disorders of the pancreas have been claimed.

Fig. 14.   Sports nutrition supplements containing casein products.

Other pharmaceutical applications include the use of casein derivatives in dentifrice and dental paste (Galinsky, 1959; University of Melbourne and Victorian Dairy Industry Authority, 1987; Societe des Produits Nestlé, 1990), as disintegrating agents in pharmaceutical tablets, in a protective hand cream and in stabilisation of vitamins A, B_1 and C as listed by Southward and Walker (1982).

Arising from some of the studies noted above, numerous proprietary pharmaceutical compounds that contain casein have been marketed in recent years. A selection of these is shown in Table VIII.

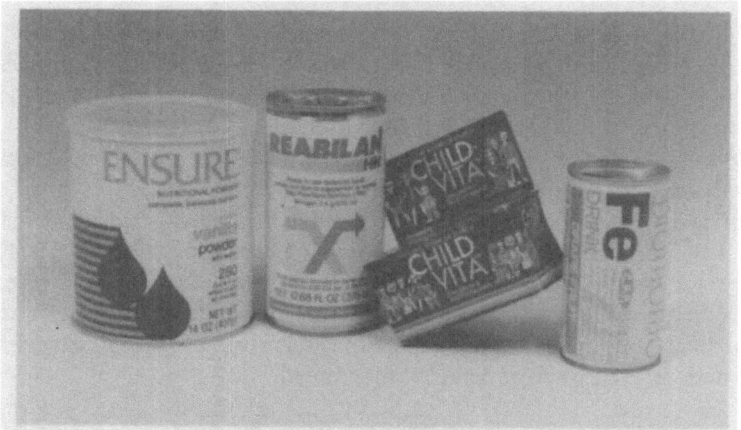

Fig. 15. Pharmaceutical products containing casein and casein hydrolysates.

TABLE VIII

Proprietary[a] pharmaceutical products containing casein

Casein derivative	Pharmaceutical product	Manufacturer	Reference
Sodium caseinate	Meritene	Doyle	Lagua et al. (1974)
	Nutrament	Drackett	Lagua et al. (1974)
	Portagen	Mead Johnson	Lagua et al. (1974)
	Sustacal	Mead Johnson	Lagua et al. (1974)
Calcium caseinate	Casec	Mead Johnson	Robertson and Smaill (1947)
	Casilan	Glaxo	Bender (1975)
	Complan	Glaxo	Bender (1968)
	Metrecal	Mead Johnson	Bender (1968)
	Sustagen	Mead Johnson	Lagua et al. (1974)
Casein hydrolysate (either acid or enzymatic)	Amigen	Mead Johnson	Bell (1947)
	Aminsol	Vitrum	Jorpes et al. (1946)
	Lofenalac	Mead Johnson	Lagua et al. (1974)
	Nutramigen	Mead Johnson	Lagua et al. (1974)
	Parenamine	Frederick Stearns	Bell (1947)
	Pronutrin	Herts Pharmaceuticals	Bell (1947)

[a] Proprietary product names are registered trademarks of the appropriate manufacturer.

NEW DEVELOPMENTS

Intensive research in membrane processing in the dairy industry has previously focussed on whey with the principal aim being enrichment of the whey protein to make whey protein concentrate (Morr, 1989). More recent research has been devoted to membrane filtration (both microfiltration and ultrafiltration) of milk to produce two new casein products, micellar whole casein and β-casein (Smithers and Bradford, 1991). Micellar whole casein is claimed to have improved dispersibility and whitening ability and anticipated increased heat stability and hydration capacity when compared with acid and rennet casein. β-Casein is produced by membrane filtration of skim-milk after cooling and slight acidification; the β-casein dissociates from the micelles and passes through the membrane. The β-casein so produced is claimed to have the potential for use in infant formulae and, in the pharmaceutical industry, as the source material in the preparation of biologically active peptides such as β-casomorphins (Maubois and Leonil, 1989) (see also the section on nutritional properties).

Edible films from casein are strictly not a new development since the property has been referred to in numerous patents, especially for whipped toppings, over the past 30 years or so (e.g. see Southward and Walker, 1982). Furthermore, in technical or non-food applications, casein has been used as a film former in paints and leather finishing for considerably longer. Nevertheless, recent research is being focussed on edible film materials that can be used to cover fresh fruit and vegetables, frozen foods etc., to act as a barrier to permeability by water vapour, oxygen and carbon dioxide (Kester and Fennema, 1986). Casein films are among those being studied.

Casein has been used previously to make a variety of products with different texture. The newly publicised technique of microparticulation to make fat substitutes, e.g. Simplesse[R], from egg and milk (whey) proteins (Kretchmer, 1991) shows potential for a similar type of product from casein based raw materials.

Casein has had a long history of use in industrial applications. During the past 30 years, however, there has been a significant decline in its non-food uses in some countries because of lower cost alternatives. In the same period, edible applications for casein products have increased significantly, particularly in the USA, Europe and Japan. Although casein does have some application in nutritional foods as a high quality protein, its major value appears to be as a specialised food ingredient, included for its functional properties such as water binding, fat emulsification, whipping, foaming and texturisation characteristics. Future trends for its production and world trade are difficult to predict since government trade and dairy support policies will continue to influence the situation. The dairy support programmes in the EC and the USA have, for instance, tended to favour the production of skim-milk powder rather than casein. On the other hand, major casein producers, such as New Zealand, are to some extent limited in the expansion of their production because of trade policies which restrict access to world markets for the remaining butterfat portion of the milk.

REFERENCES

Acosta, P. B., Wenz, E. and Williamson, M. (1977). *American Journal of Clinical Nutrition*, **30**, 198.

Agra Europe (London) (1987–91a). *Preserved Milk*, a monthly supplement to Agra Europe. No. 89, February 1987; No. 106, February 1989; No. 109 June 1989; No. 117, April 1990; No. 130, August 1991.

Agra Europe (London) (1991b). *East Europe and U.S.S.R Agriculture and Food Monthly*, No. 111, p. 34.

Agra Europe (London) (1991c). *Agra Europe Special Report No. 59, 1991 Dairy Review*, pp. 66–67.

Akino, Y. and Yoshioka, S. (1982). *Food Industry (Japan)*, **25** (14), 20.

American Chemical Society (1981). *Protein Functionality in Foods*, (Ed. J. P. Cherry), ACS Symposium Series 147, American Chemical Society, Washington, D.C.

Amiot, J., Brisson, G. J., Castaigne, F., Goulet, G. and Boulet, M. (1979). *Canadian Institute of Food Science and Technology Journal*, **12**, 23.

Anon (1976). *Zuivelzicht*, **68** (11), 258.

Anon (1984). *Technique Laitiere*, No. 986, 23.

Anon (1987). *North European Food and Dairy Journal*, **53** (6), 173.

Anon (1990). *Process (Rennes)*, No. 1046, 18. (*Dairy Science Abstracts* (1991). **53**, 1572.)

Appleman, M. D. (1973). British Patent 1 308 690,

Association of Official Analytical Chemists (1970). *Official Methods of Analysis of the Association of Official Analytical Chemists* (Ed. W. Horwitz), 11th edn, AOAC, Washington, D.C., p. 800.

Australian Dairy Corporation (1991). *Dairy Compendium*, Australian Dairy Corporation, Parkville, Victoria, Australia, pp. 19, 60.

Australian Society of Dairy Technology (1972). *Casein Manual*, Australian Society of Dairy Technology, Parkville, Victoria, Australia, pp. 25–27.

Banks, W., Muir, D. D. and Wilson, A. G. (1983). In: *Proceedings of IDF Symposium on Physico-Chemical Aspects of Dehydrated Protein-Rich Milk Products*, 17–19 May, 1983, Statens Forsogsmejeri, Hillerod, Denmark, pp. 331–8.

Barrat, J.-F. (1978). French Patent 2 370 441.
Barry, T. N. (1972). *New Zealand Journal of Agricultural Research*, **15**, 107.
Battelle Memorial Institute (1974). British Patent 1 350 647.
Bell, M. E. (1947). *New Zealand Medical Journal*, **46**, 255.
Bender, A. E. (1968). *Dictionary of Nutrition and Food Technology*, 3rd edn, Butterworths, London.
Bender, A. E. (1975). *Dictionary of Nutrition and Food Technology*, 4th edn, Newnes–Butterworth, London, pp. 48, 198.
Bergmann, A. (1972). *Journal of the Society of Dairy Technology*, **25**, 89.
Bird, E. W., Sadler, H. W. and Iverson, C. A. (1935). *The Preparation of Non-Desiccated Sodium Caseinate Sol and its Use in Icecream*, Research Bulletin No. 187, Agricultural Experiment Station, Iowa State College of Agriculture and Mechanic Arts (Dairy Industry Section), USA.
Bogdanova, E. A., Padaryan, I. M., Lavrenova, G. S. and Inozemtseva, V. F. (1978). In: *XX International Dairy Congress, Brief Communications*, Paris, p. 845.
Boullé, M. (1986). Australian Patent 84 34989.
Boullé, M. (1987). In: *Extrusion Technology for the Food Industry* (Ed. C. O'Connor), Elsevier Applied Science, London, pp. 96–108.
Bready, P. J. S. (1966). *Australian Journal of Dairy Technology*, **21**, 153.
Broderick, G. A. and Lane, G. T. (1978). *Journal of Dairy Science*, **61**, 932.
Brother, G. H. (1939). In: *Casein and Its Industrial Applications*, (Ed. E. Sutermeister and F. L. Browne), 2nd edn, Reinhold Publishing Co., New York, pp. 181–232.
Brown, A. E., Gordon, W. G., Gall, E. C. and Jackson, R. W. (1944). *Industrial and Engineering Chemistry*, **36**, 1171.
Browne, F. L. and Brouse, D. (1939). In: *Casein and Its Industrial Applications* (Ed. E. Sutermeister and F. L. Browne), 2nd edn, Reinhold Publishing Corporation, New York, pp. 233–92.
Buchanan, R. A. and Henderson, J. O. (1971). *Journal of the Dietetic Association of Victoria*, **22**, 7. (*Dairy Science Abstracts* (1971). **33**, 5048).
Buchheim, W. (1983). In: *Proceedings of IDF Symposium on Physico-Chemical Aspects of Dehydrated Protein-Rich Milk Products*, 17–19 May 1983, Statens Forsogsmejeri, Hillerod, Denmark, pp. 319–30.
Bunnies, H. and Timm, T. (1983). *Gordian*, **83**, 248.
Burgess, K. J. and Coton, G. (1982). In: *Food Proteins* (Ed. P. F. Fox and J. J. Condon), Applied Science Publishers, London, pp. 211–24.
Burkwall, M. P., Jr., Leyh, J. C., Jr. and Reagen, J. G. (1976). US Patent 3 984 576.
Burston, D. O., Muller, L. L. and Hayes, J. F. (1967). In: *International Dairy Federation Seminar on Caseins and Caseinates*, 31 May–2 June 1967, Paris, France, PARIS-SEM Subject 4.
Cameron, D. E., Chilson, W. H., Elsesser, C. C. and Windmuller, R. (1959). US Patent 2 913 342; British Patent 822 614.
Caric, M., Gavaric, D., Milanovic, S., Jakimov, N., Karic, A. and Markovic, D. (1981). *Mljekarstvo*, **31**, 79. (*Dairy Science Abstracts* (1982). **44**, 2732.)
Caron, J.-P., Pean, J.-L., Flamant, H., Roussel, H., Cuq, J.-L. and Alliotte, T. (1991). World Patent 91/18516.
Cavett, E. S. (1939). In: *Casein and Its Industrial Applications* (Ed. E. Sutermeister and F. L. Browne), 2nd edn, Reinhold Publishing Corporation, New York, pp. 354–65.

Cayen, M. N. and Baker, B. E. (1963). *Journal of Agricultural and Food Chemistry*, **11**, 12.

Centre National du Commerce Exterieur (1970). *The United States Market for Edible Caseins and Caseinates*, Centre National du Commerce Exterieur, Paris.

Clarke, R. J. and Love, G. (1974). *Chemistry and Industry*, No. 4 (16 Feb 1974), 151.

Cocup, R. O. and Sanderson, W. B. (1987). *Food Technology, Chicago*, **41** (10), 86.

Commonwealth Secretariat (1987–88). *Meat and Dairy Products*, Commonwealth Secretariat Publications, London, pp. 11, 20 (May 1987), p. 10 (Nov 1988).

Connell, J. E. (1966). *Canadian Food Industries*, **37** (2), 23.

Connolly, P. B. (1983). US Patent 4 376 072.

Cooper, H. R. and Peacock, I. C. (1979). *New Zealand Journal of Dairy Science and Technology*, **14**, 291.

Cooper, H. R., Patten, J.D. and Fletcher, R. H. (1984). *Journal of Food Science*, **49**, 376.

Craig, T. W. and Colmey, J. C. (1971). *Bakers Digest*, **45** (1), 36.

de Moor, H. and Huyghebaert, A. (1983). In: *Proceedings of IDF Symposium on Physico-Chemical Aspects of Dehydrated Protein-Rich Milk Products*, 17–19 May 1983, Statens Forsogsmejeri, Hillerod, Denmark, pp. 276–301.

Davey, P. T., Williams, D. R. and Winter, G. (1980). *Journal of Applied Biochemistry*, **2**, 60.

Dickinson, E. (1986). *Food Hydrocolloids*, **1**, 3.

Dolby, R. M., Creamer, L. K. and Elley, E. R. (1969). *New Zealand Journal of Dairy Science and Technology*, **4**, 46.

Downes, T. E. H. and van der Merwe, N. L. (1982). *South African Journal of Dairy Technology*, **14**, 59.

Durr, P. and Neukom, H. (1972). *Lebensmittel-Wissenschaft und-Technologie*, **5**, 132.

D'yachenko, P., Vlodavets, I. and Bogomolova, E. (1953). *Molochnaya Promyshlennost'*, **14** (6), 33. (*Dairy Science Abstracts* (1953). **15**, 962.)

Dymsza, H. A., Stoewsand, G. S., Donovan, P., Barrett, F. F. and Lachance, P. A. (1966). *Food Technology, Chicago*, **20**, 1349.

Elenbogen, G. D. and Baron, M. (1968). US Patent 3 397 994.

English, A. (1981). *Journal of the Society of Dairy Technology*, **34**, 70.

Ernstrom, C. A. and Wong, N. P. (1974). In: *Fundamentals of Dairy Chemistry*, (Ed. B. H. Webb, A. H. Johnson and J. A. Alford), 2nd edn, AVI Publishing Co., Westport, Connecticut, pp. 662–771.

Farrell, H. M., Jr. and Thompson, M. P. (1974). In: *Fundamentals of Dairy Chemistry*, (Ed. B. H. Webb, A. H. Johnson and J. A. Alford), 2nd edn, AVI Publishing Co., Westport, Connecticut, pp. 442–73.

Ferretti, A. (1938). British Patent 483 731.

Filer, L. J., Jr. (Chairman, American Academy of Pediatrics, Committee on Nutrition) (1972). *Pediatrics*, **49**, 770.

Fogh, A. W. (1975). *Maelkeritidende*, **88**, 352.

Food and Agriculture Organization (1973). *Energy and Protein Requirements*, FAO Nutrition Meetings Report Series No. 52; WHO Technical Report Series No. 522. FAO and WHO, Rome, p. 63.

Food and Agriculture Organization (1987). *FAO Monthly Bulletin of Statistics*, **10** (9), 26.

Food and Agriculture Organization (1990). *Agricultural Review for Europe*, No. 32, 1988 and 1989. Volume V. The Milk and Dairy Products Market. FAO, United Nations, New York.

Fox, K. K. (1970). In: *By Products from Milk* (Ed. B. H. Webb and E. O. Whittier), 2nd edn, AVI Publishing Co., Westport, Connecticut, pp. 331–55.

Fox, P. F. (1968). *Casein, Caseinates and Casein Co-precipitates*, Dairy Research Review Series No. 4, An Foras Taluntais, Fermoy, Cork, Irish Republic.

Fox, P. F. (1989). In: *Developments in Dairy Chemistry – 4. Functional Milk Proteins* (Ed. P. F. Fox), Elsevier Applied Science, London, pp. 1–53.

Fox, P. F. & Mulvihill, D. M. (1983). In: *Proceedings of IDF Symposium on Physico-Chemical Aspects of Dehydrated Protein-Rich Milk Products*, 17–19 May 1983, Statens Forsogsmejeri, Hillerod, Denmark, pp. 188–259.

Galinsky, M. (1959). British Patent 825 115.

Genin, G. (1961). *Le Lait*, **41**, 44.

Genin, G. (1966). *Le Lait*, **46**, 283.

Girdhar, B. K. and Hansen, P. M. T. (1974). *Journal of Food Science*, **39**, 1237.

Goldman, A. (1974). *Food Technology in New Zealand*, **9** (7), 25.

Goldman, A. and Toepfer, N. G. (1978). In: *XX International Dairy Congress, Brief Communications, Paris*, pp. 418–19.

Gordon, W. G. & Kalan, E. B. (1974). In: *Fundamentals of Dairy Chemistry* (Ed. B. H. Webb, A. H. Johnson and J. A. Alford), 2nd edn, AVI Publishing Co., Westport, Connecticut, pp. 87–124.

Graf, T. F. (1979). *Hoard's Dairyman*, **124**, 1531.

Graf, T. F. (1986). *Journal of Dairy Science*, **69**, 1462.

Grimble, G. K. and Silk, D. B. A. (1989). In: *Milk Proteins – Nutritional, Clinical, Functional and Technological Aspects* (Ed. C. A. Barth and E. Schlimme), Steinkopff Verlag, Darmstadt, Germany, pp. 270–82.

Gulayev-Zaitsev, S. S. (1987). In: *Milk the Vital Force. Proceedings of the XXII International Dairy Congress, The Hague, 29 September–3 October 1986*. D. Reidel Publishing Co., Dordrecht, Netherlands, pp. 769–78.

Gupta, S. K. and Rao, Y. V. (1984). *Journal of Food Science and Technology, India* **21**, 143. (*Food Science and Technology Abstracts* (1985). **17**, 12G3.)

Gwozdz, E. (1978). In: *XX International Dairy Congress, Paris, Conferences Science and Technique* No. 51ST.

Gwozdz, E. (1983). *La Technique Laitiere*, **979**, 19.

Hambraeus, L. (1982). In: *Developments in Dairy Chemistry – 1. Proteins* (Ed. P. F. Fox), Applied Science Publishers, London and New York, pp. 289–313.

Hambraeus, L. (1985). In: *Milk Proteins '84. Proceedings of the International Congress on Milk Proteins, Luxembourg, 7–11 May 1984*, Centre for Agricultural Publishing and Documentation (Pudoc), Wageningen, Netherlands, pp. 63–79.

Hammonds, T. M. and Call, D. L. (1970). *Utilization of Protein Ingredients in the US Food Industry. Part I. The Current Market for Protein Ingredients. Part II. The Future Market for Protein Ingredients*, Cornell University Agricultural Experimental Station, New York State College of Agriculture, Ithaca, New York.

Hamosh, M., Hong, M. H. and Hamosh, P. (1989). In: *Textbook of Gastroenterology and Nutrition in Infancy* (Ed. E. Lebenthal), Raven Press, New York, pp. 143–50.

Hansen, P. M. T. (1963). *Australian Journal of Dairy Technology*, **18**, 79.

Hansen, P. M. T., Harper, W. J. and Sharma, K. K. (1970). *Journal of Food Science*, **35**, 598.

Hayes, J. F., Southby, P. M. and Muller, L. L. (1968). *Journal of Dairy Research*, **35**, 31.

Hayes, J. F., Dunkerley, J. and Muller, L. L. (1969). *Australian Journal of Dairy Technology*, **24**, 69.

Heap, H. A. and Lawrence, R. C. (1984). *New Zealand Journal of Dairy Science and Technology*, **19**, 119.

Hedrick, T. I. (1969). *Dairy Industries*, **34**, 127.

Henderson, J. O. and Buchanan, R. A. (1973). *Australian Journal of Dairy Technology*, **28**, 7.

Herner, E. (1988). *Dairy Field*, **171** (7), 32.

Hidalgo, J., Wenner, V. and Forni, F. (1981). US Patent 4 303 580.

Hooker, P. H., Munro, P. A. and O'Meara, G. M. (1982). *New Zealand Journal of Dairy Science and Technology*, **17**, 35.

Hugunin, A. G. and Nishikawa, R. K. (1977). *Milk Derived Ingredients for Confectionery Products*. Dairy Research Inc., Rosemont, Illinois, USA.

Humphries, M. A. and Roeper, J. (1974). In: *XIX International Dairy Congress, Brief Communications*, India. Volume 1E, pp. 778–9.

Jackson, D. L. C. and Backwell, A. R. A. (1955). *Australian Journal of Applied Science*, **6**, 244. (*Australian Journal of Dairy Technology* (1955). **10**, 142.)

Janssen, F. W., Voortman, G. and de Baaij, J. A. (1987). *Journal of Agricultural and Food Chemistry*, **35**, 563.

Japan Finance Ministry (1990–91). *Statistics for Imports*. Commodities 3501.10-000 (Casein) and 3501.90-000 (Caseinate) for 1989 and 1990 (pp. 0222–3), Customs and Tariff Bureau, Ministry of Finance of Japanese Government, Tokyo, Japan.

Japan Ministry of Agriculture, Forestry and Fisheries (1989–90). *The 64th Statistical Yearbook of Ministry of Agriculture, Forestry and Fisheries, Japan, 1987–88*, pp. 532–533 (Imports for 1984–1987), 1989; *The 65th Statistical Yearbook of Ministry of Agriculture, Forestry and Fisheries, Japan, 1988–89*, p. 513 (Imports for 1988), 1990, Statistics and Information Department, Ministry of Agriculture, Forestry and Fisheries, Tokyo, Japan.

Jenness, R. and Patton, S. (1959a). *Principles of Dairy Chemistry*, John Wiley & Sons, New York, pp. 101–57.

Jenness, R. and Patton, S. (1959b). *Principles of Dairy Chemistry*. John Wiley & Sons, New York, pp. 1–29.

Jordan, P. J. (1983). *New Zealand Journal of Dairy Science and Technology*, **18**, 27.

Jordan, P. J., Lay, K., Ngan, N. and Rodley, G. F. (1987). *New Zealand Journal of Dairy Science and Technology*, **22**, 247.

Jorpes, J. E., Magnusson, J. H. and Wretlind, A. (1946). *The Lancet*, **2**, 228.

Jost, R., Dannenberg, F. & Gumy, D. (1991). In: *Proceedings of the XXIII International Dairy Congress*, Montreal, 8–12 October 1990, Volume 2, International Dairy Federation, Brussels, pp. 1481–91.

Kas, J., Holakova, M., Korostenska, M., Fukel, L. and Paluska, E. (1984). *Prumysl Potravin*, **35**, 326.

Kastens, M. L. and Baldauski, F. A. (1952). *Industrial and Engineering Chemistry*, **44**, 1257.

Kelly, P. M., O'Keefe, A. M. and Phelan, J. A. (1983). In: *Proceedings of IDF Symposium on Physico-Chemical Aspects of Dehydrated Protein-Rich Milk Products*, 17–19 May 1983, Statens Forsogsmejeri, Hillerod, Denmark, pp. 260–75.

Kester, J. J. and Fennema, O. R. (1986). *Food Technology, Chicago*, **40** (12), 47.

King, D. W. (1970). *New Zealand Journal of Dairy Science and Technology*, **5**, 100.

King, D. W. and Jebson, R. S. (1970). In: *XVIII International Dairy Congress*, Volume IE, p. 420.

Kinsella, J. E. (1970). *Manufacturing Confectioner*, **50** (10), 45.

Kinsella, J. E. and Fox, P. F. (1986). *CRC Critical Reviews in Food Science and Nutrition*, **24** (2), 91.

Kinsella, J. E. and Whitehead, D. M. (1987). In: *Milk the Vital Force. Proceedings of the XXII International Dairy Congress*. The Hague, 29 September–3 October, 1986, D. Reidel Publishing Co., Dordrecht, Netherlands, pp. 791–804.

Kirk, D. J. (1973). *Bakers Digest*, **47** (5), 76.

Kitabatake, N. and Doi, E. (1982). *Journal of Food Science*, **47**, 1218.

Knightbridge, J. P. and Goldman, A. (1975). *New Zealand Journal of Dairy Science and Technology*, **10**, 152.

Knightly, W. H. (1968). *Food Technology, Chicago*, **22**, 731.

Knightly, W. H. (1969). *Food Technology, Chicago*, **23**, 171.

Knights, R. J. (1985). In: *Nutrition for Special Needs in Infancy. Protein Hydrolysates* (Ed. F. Lifshitz), Marcel Dekker, New York, pp. 105–15.

Koide, K. (1985). *New Food Industry*, **27** (6, 7), 60–74, 53–63. (*Dairy Science Abstracts* (1987). **49**, 5553.)

Korobkina, G. S., Danilova, E. N., Pokrovskii, A. A., Arbatskaya, N. I. and Agienko, K. S. (1974). In: *XIX International Dairy Congress, Brief Communications*, India. Volume 1E, p. 566.

Korolczuk, J. (1982). *New Zealand Journal of Dairy Science and Technology*, **17**, 135.

Krause, G. A. (1953). German Federal Republic Patent 896 449.

Kretchmer, N. (1991). *Médicine et Nutrition*, **27**, 140. (*Dairy Science Abstracts* (1992). **54**, 447.)

L. D. Schreiber Cheese Co. (1978). French Patent 2 381 474.

Lagua, R. T., Claudio, V. S. and Thiele, V. F. (1974). *Nutrition and Diet Therapy Reference Dictionary*, 2nd edn, C. V. Mosby Co., St Louis, Missouri, USA, pp. 314–17.

Laiteries Triballat (1979). French Patent Application 2 418 627.

Lakshminarayana, Y., Vijayakumar, M. T., Srinivasan, K. S. V. and Joseph, K. T. (1986). *European Polymer Journal*, **22**, 143.

Landmann, A. W. (1962). *Journal of the Society of Leather Trades' Chemists*, **46**, 97.

Langmaier, F. (1967). *Kozarstvi*, **17** (9), 273. (*Dairy Science Abstracts* (1968) **30**, 366.)

Lim, D. M. (1980). *Functional Properties of Milk Proteins with Particular Reference to Confectionery Products*, Scientific and Technical Survey No. RA 120, British Food Manufacturing Industries Research Association, Leatherhead, Surrey.

Lippe, F., Ottenhof, H. A. W. E. M. and de Boer, R. (1983). US Patent 4 407 747.
Little, L. L. (1966). US Patent 3 236 658.
Lodge, N. and Heatherbell, D. A. (1976). *New Zealand Journal of Dairy Science and Technology*, **11**, 263.
Lorient, D., Closs, B. and Courthaudon, J. L. (1991). *Le Lait*, **71**, 141.
Manson, W. (1980). *Bulletin, International Dairy Federation*, Document **125**, 60.
Markh, A. T., Feldman, A. L., Ponomarenko, S. F. and Strashnenko, E. S. (1982). In: *XXI International Dairy Congress, Brief Communications*, Moscow. Volume 1, Book 2, p. 85.
Maubois, J. L. and Leonil, J. (1989). *Le Lait*, **69**, 245.
McCarthy, J. (1990). *Bulletin of the International Dairy Federation*, No. 249, 45.
McCormick, R. D. (1973). *Food Product Development*, **7** (2), 16.
McDowall, F. H. (1971). *New Zealand Journal of Dairy Science and Technology*, **6**, 128.
McDowell, A. K. R. (1968). In: *Annual Report, New Zealand Dairy Research Institute*, New Zealand Dairy Research Institute, Palmerston North, New Zealand, p. 52.
McDowell, A. K. R., Southward, C. R. and Elston, P. D. (1976). *New Zealand Journal of Dairy Science and Technology*, **11**, 40.
McGilliard, A. D. (1972). *Journal of the American Oil Chemists' Society*, **49**, 57.
McIntire, J. M. (1971). *American Journal of Public Health*, **61**, 157.
McMeekin, T. L., Reid, T. S., Warner, R. C. and Jackson, R. W. (1945). *Industrial and Engineering Chemistry*, **37**, 685.
Meisel, H. (1991). In: *Proceedings of the XXIII International Dairy Congress*, Montreal, 8–12 October 1990, Volume 2, International Dairy Federation, Brussels, pp. 1534–49.
Meisel, H. and Frister, H. (1989). In: *Milk Proteins: Nutritional, Clinical, Functional and Technological Aspects* (Ed. C. A. Barth and E. Schlimme), Steinkopff Verlag, Darmstadt, Germany, pp. 150–4.
Melsheimer, L. A. and Hoback, W. H. (1953). *Industrial and Engineering Chemistry*, **45**, 717.
Migliore-Samour, D., Floc'h, F. and Jolles, P. (1989). *Journal of Dairy Research*, **56**, 357.
Milk Marketing Board (1986–1991). *EEC Dairy Facts and Figures* 1986, pp. 150–1; 1987, pp. 164, 165; 1988, pp. 137, 164, 165; 1989, pp. 134, 162, 163; 1990, pp. 166, 167; 1991, pp. 164, 165.
Miller, M. J. S., Witherly, S. A. and Clark, D. A. (1990). *Proceedings of the Society for Experimental Biology and Medicine*, **195**, 143.
Minami, J., Shigato, M. and Ishibashi, S. (1979). UK Patent Application 2 010 658A. (*Dairy Science Abstracts* (1980). **42**, 1951.)
Mitchell, J. R. (1986). In: *Developments in Food Proteins – 4* (Ed. B. J. F. Hudson), Elsevier Applied Science Publishers, London, pp. 291–338.
Modler, H. W., Larmond, M. E., Lin, C. S., Froehlich, D. and Emmons, D. B. (1983). *Journal of Dairy Science*, **66**, 422.
Morinaga Milk Industry Co. (1975). US Patent 3 901 979.
Morr, C. V. (1982). In: *Developments in Dairy Chemistry – 1. Proteins* (Ed. P. F. Fox), Applied Science Publishers, London and New York, pp. 375–99.

Morr, C. V. (1987). In: *Milk the Vital Force. Proceedings of the XXII International Dairy Congress*, The Hague, 29 September–3 October 1986. D. Reidel Publishing Co., Dordrecht, Netherlands, pp. 763–8.

Morr, C. V. (1989). In: *Developments in Dairy Chemistry – 4. Functional Milk Proteins* (Ed. P. F. Fox), Elsevier Applied Science, London, pp. 245–84.

Muir, D. D. (1988). *Dairy Industries International*, **53** (5), 25.

Muir, D. D. and Banks, W. (1984). In: *Milk Proteins '84. Proceedings of the International Congress on Milk Proteins*, Luxembourg, 7–11 May 1984, Centre for Agricultural Publishing and Documentation (Pudoc), Wageningen, Netherlands, pp. 120–8.

Muir, D. D. and Banks, W. (1987). *Hannah Research*, 1986, 83.

Muller, L. L. (1971). *Dairy Science Abstracts*, **33**, 659–674.

Muller, L. L. (1982a). In: *Developments in Dairy Chemistry – 1, Proteins* (ed. P. F. Fox), Applied Science Publishers, London and New York, pp. 315–37.

Muller, L. L. (1982b). In: *Food Proteins* (Ed. P. F. Fox and J. J. Condon), Applied Science Publishers, London, pp. 179–89.

Muller, L. L. and Hayes, J. F. (1963). *Australian Journal of Dairy Technology*, **18**, 184.

Muller, L. L., Hayes, J. F. and Snow, N. (1967). *Australian Journal of Dairy Technology*, **22**, 12.

Mulvihill, D. M. (1989). In: *Developments in Dairy Chemistry – 4. Functional Milk Proteins* (Ed. P. F. Fox), Elsevier Applied Science, London, pp. 97–130.

Mulvihill, D. M. (1991). *Food Research Quarterly*, **51**, 65.

Mulvihill, D. M. and Fox, P. F. (1989). In: *Developments in Dairy Chemistry – 4. Functional Milk Proteins* (Ed. P. F. Fox), Elsevier Applied Science, London, pp. 131–72.

Mulvihill, D. M. and Murphy, P. C. (1991). *International Dairy Journal*, **1**, 13.

Mummery, W. R. (1949). *New Zealand Journal of Science and Technology*, **B30**, 297.

Munro, P. A. and Vu, J. T. (1983). *New Zealand Journal of Dairy Science and Technology*, **18**, 93.

Munro, P. A., Southward, C. R. and Elston, P. D. (1980). *New Zealand Journal of Dairy Science and Technology*, **15**, 177.

Munro, P. A., Vu, J. T. and Mockett, R. B. (1983). *New Zealand Journal of Dairy Science and Technology*, **18**, 35.

National Casein Co. (1989). *Gluing Guide*, National Casein Company, Chicago, Illinois.

National Dairy Council (1968). *US Dairy Council Digest*, **39**, 7.

Nesmeyanov, A. N., Rogozhin, S. V., Slonimsky, G. L., Tolstoguzov, V. B. and Ershova, V. A. (1971). US Patent 3 589 910.

New Zealand Dairy Board (1988). *Annual Report for Year Ended 31 May 1988*, New Zealand Dairy Board, Wellington.

New Zealand Dairy Board (1991). *New Zealand Dairy Industry Statistical Bulletin*, October 1991, New Zealand Dairy Board, Wellington.

New Zealand Department of Statistics (1989–91). Exports 1985–86, 1986–87, 1987–88, *Key Statistics*, p. 91 (February 1989); Exports 1988–89, 1989–90, 1990–91, *Key Statistics*, p. 98 (December 1991). New Zealand Department of Statistics, Wellington.

Nichols, B. L., Jr., Klish, W. J. and Potts, V. E. (1983). US Patent 4 419 369.

Nienhaus, A. (1988). *Bulletin, International Dairy Federation*, Document 224, 27.

O'Neil, I. (1971). In: *The Art of the Painted Finish for Furniture and Decoration. Antiquing, Lacquering, Gilding and the Great Impersonators*, William Morrow & Co., New York, pp. 139–59, 171, 196–7, 343.

Owen, G. M. (1969). *American Journal of Clinical Nutrition*, **22**, 1150.

Pader, M. and Gershon, S. D. (1965). US Patent 3 224 883.

Patel, P. D., Stripp, A. M. and Fry, J. C. (1988). *International Journal of Food Science and Technology*, **23**, 57.

Paterson, L. O. (1955). US Patent 2 721 861.

Petka, T. E. (1976). *Food Product Development*, **10** (10), 26.

Pfaff, W. (1974). *Fleischwirtschaft*, **54**, 1740.

Phillips, L. G., German, J. B., O'Neill, T. E., Foegeding, E. A., Harwalker, V. R., Kilara, A., Lewis, B. A., Mangino, M. E., Morr, C. V., Regenstein, J. M., Smith, D. M. and Kinsella, J. E. (1990). *Journal of Food Science*, **55**, 1441.

Poarch, E. A. (1967). In: *International Dairy Federation Seminar on Caseins and Caseinates*, 31 May–2 June, 1967, Paris, France, PARIS-SEM Subject 6.

Powell, L. A. (1974). US Patent 3 843 805.

Ramshaw, E. H. and Dunstone, E. A. (1969a). *Journal of Dairy Research*, **36**, 203.

Ramshaw, E. H. and Dunstone, E. A. (1969b). *Journal of Dairy Research*, **36**, 215.

Rasic, J., Bosic, Z. and Vracar, L. (1978). In: *XX International Dairy Congress, Brief Communications*, Paris, pp. 987–8.

Reimerdes, E. H. and Lorenzen, P. C. (1983). In: *Proceedings of IDF Symposium on Physico-Chemical Aspects of Dehydrated Protein-Rich Milk Products*, 17–19 May 1983, Statens Forsogsmejeri, Hillerod, Denmark, pp. 70–93.

Reis, P. J. and Tunks, D. A. (1969). *Australian Journal of Agricultural Research*, **20**, 775.

Renner, E. and Eiselt-Lomb, U. (1985). *Milchwissenschaft*, **40**, 526.

Reynolds, E. C. and Black, C. L. (1987). *Caries Research*, **21**, 445 (*Dairy Science Abstracts* (1988). **50**, 5693).

Reynolds, E. C. & Black, C. L. (1989). *Caries Research*, **23**, 368 (*Dairy Science Abstracts* (1990). **52**, 4282).

Rhone-Electra (1990). New Zealand Patent 229 173.

Roberts, J. G. (1959). US Patent 2 878 126.

Robertson, H. R. and Smaill, D. W. (1947). *Canadian Medical Association Journal*, **56**, 59.

Robinson, G. H. (1964). Casein in the New Zealand Dairy Industry. Some Economic Implications and some International Marketing Aspects, MA Thesis, Victoria University of Wellington, New Zealand.

Roeper, J. (1974). *New Zealand Journal of Dairy Science and Technology*, **9**, 128.

Roeper, J. (1976). *New Zealand Journal of Dairy Science and Technology*, **11**, 62.

Roeper, J. (1977a). *New Zealand Journal of Dairy Science and Technology*, **12**, 182.

Roeper, J. (1977b). In: *Proceedings of Jubilee Conference on Dairy Science and Technology*, 15–17 March 1977, New Zealand Dairy Research Institute, Palmerston North, pp. 81–83.

Roeper, J. (1982). *Bulletin, International Dairy Federation*, Document 147, 21.

Roeper, J. and Winter, G. J. (1982). In: *XXI International Dairy Congress, Brief Communications*, Moscow, Volume 1, Book 2, pp. 97–8.

Roff, W. J. and Scott, J. R. (1971). *Fibres, Films, Plastics and Rubbers – A Handbook of Common Polymers*, Butterworth & Co., London, pp. 197–208.

Rusch, D. T. (1971). *Food Technology, Chicago*, **25**, 486.

Saito, Y. (1990). *Japan Fudo Saiensu*, **29**, 21. (*Chemical Abstracts* (1991). **113**, 170516s.)

Salmon, M. (1983). US Patent 4 423 081.

Salzberg, H. K. (1964). *American Paint Journal*, **48** (42), 104; (43), 110; (45), 90.

Salzberg, H. K. (1965). In: *Encyclopaedia of Polymer Science and Technology*, Volume 2 (Ed. H. F. Mark, N. H. Gaylord and N. M. Bikales), Interscience Publishers, New York, pp. 859–71.

Salzberg, H. K. (1967). *American Perfumer and Cosmetics*, **82**, (Nov), 41.

Salzberg, H. K. (1977). *Handbook of Adhesives* (Ed. I. Skeist), 2nd edn, van Nostrand Reinhold Company, New York, pp. 158–71.

Salzberg, H. K. and Marino, W. L. (1975). In: *Protein Binders in Paper and Paperboard Coating* (Ed. R. Strauss), Tappi Monograph Series No. 36, Technical Association of the Pulp and Paper Industry, Atlanta, Georgia, pp. 1–74.

Salzberg, H. K. and Simonds, M. R. (1965). US Patent 3 186 918.

Salzberg, H. K. Georgevits, L. E. and Karapetoff Cobb, R. M. (1961). In: *Synthetic and Protein Adhesives for Paper Coating* (Ed. L. H. Silvernail and M. W. Bain), Tappi Monograph Series No. 22, Technical Association of the Pulp and Paper Industry, New York, pp. 103–66.

Salzberg, H. K., Britton, R. K. and Bye, C. N. (1974). In: *Testing of Adhesives* (Ed. R. G. Meese), Tappi Monograph Series No. 35, Technical Association of the Pulp and Paper Industry, Atlanta, Georgia, pp. 30–51.

Schär, W., Flückiger, E., Rüegg, M. and Puhan, Z. (1987). *Schweizerische Milchwirtschaftliche Forschung*, **16** (4), 89. (*Dairy Science Abstracts* (1988). **50**, 2989.)

Schlimme, E., Meisel, H. and Frister, H. (1989). In: *Milk Proteins: Nutritional, Clinical, Functional and Technological Aspects* (Ed. C. A. Barth and E. Schlimme), Steinkopff Verlag, Darmstadt, Germany, pp. 143–9.

Scholz, H. A. (1953). *Industrial and Engineering Chemistry*, **45**, 70.

Schuette, H. A. (1939). In: *Casein and Its Industrial Applications*, (Ed. E. Sutermeister and F. L. Browne), 2nd edn, Reinhold Publishing Co., New York, pp. 366–90.

Scott, T. W., Cook, L. J., Fergusson, K. A., McDonald, I. W., Buchanan, R. A. and Loftus Hills, G. (1970). *Australian Journal of Science*, **32**, 291.

Segalen, P. (1982). In: *Proteines Animales. Extraits, Concentres et Isolats en Alimentation Humaine* (Ed. C.-M. Bourgeois and P. LeRoux), Technique et Documentation Lavoisier, Paris, pp. 155–71.

Segalen, P., Boullé, M. and Gwozdz, G. (1985). In: *Laits et Produits Laitiers. Vache-Brebis-Chèvre. Volume 2. Les Produits Laitiers. Transformation et Technologies* (Ed. F. M. Luquet), Technique et Documentation Lavoisier, Paris, pp. 395–442.

Sharp, J. L. (1987). *Florida Entomologist*, **70**, 225.

Shaw, M. (1984). *Journal of the Society of Dairy Technology*, **37**, 27.

Skurray, G. R. and Osborne, C. (1976). *Journal of the Science of Food and Agriculture*, **27**, 175.

Smith, D. R. and Snow, N. S. (1968). *Australian Journal of Dairy Technology*, **23**, 8.

Smithers, G. W. and Bradford, R. S. (1991). *Food Research Quarterly*, **51**, 92.

Snow, N. S., Townsend, F. R., Bready, P. J. and Shimmin, P. D. (1967). *Australian Journal of Dairy Technology*, **22**, 125.

Sobotka, J. J. (1971). *Current Therapeutic Research*, **13**, 636.

Societe des Produits Nestlé (1990). New Zealand Patent 223 400.

Solms-Baruth, H. Graf zu (1972). *Deutsche Milchwirtschaft, Hildesheim*, **23** (48), 2057.

Southward, C. R. (1974). *Food Technology in New Zealand*, **9** (8), 11.

Southward, C. R. (1986). In: *Modern Dairy Technology, Volume 1. Advances in Milk Processing* (Ed. R. K. Robinson), Elsevier Applied Science Publishers, London, pp. 317–68.

Southward, C. R. (1989). In: *Developments in Dairy Chemistry – 4. Functional Milk Proteins* (Ed. P. F. Fox), Elsevier Applied Science, London, pp. 173–244.

Southward, C. R. and Aird, R. M. (1978). *New Zealand Journal of Dairy Science and Technology*, **13**, 77.

Southward, C. R. and Elston, P. D. (1976). *New Zealand Journal of Dairy Science and Technology*, **11**, 144.

Southward, C. R. and Goldman, A. (1975). *New Zealand Journal of Dairy Science and Technology*, **10**, 101.

Southward, C. R. and Goldman, A. (1978). *New Zealand Journal of Dairy Science and Technology*, **13**, 97.

Southward, C. R. and Walker, N. J. (1980). *New Zealand Journal of Dairy Science and Technology*, **15**, 201.

Southward, C. R. and Walker, N. J. (1982). In: *CRC Handbook of Processing and Utilization in Agriculture*, Volume 1 (Ed. I. A. Wolff), Chemical Rubber Company Press, Boca Raton, Florida, pp. 445–552.

Southward, C. R., Humphries, M. A. and Creamer, L. K. (1978). In: *XX International Dairy Congress, Brief Communications*, Paris, p. 910.

Spellacy, J. R. (1953). *Casein, Dried and Condensed Whey*, Lithotype Process Co., San Francisco, California.

Stavrova, E. R. and Mochalova, K. V. (1978). In: *XX International Dairy Congress, Brief Communications*, Paris, pp. 981–2.

Stichting Nederlands Instituut voor Zuivelonderzoek (1982). Netherlands Patent Application 82.04923.

Suchkov, V. V., Grinberg, V. Ya, Bikbov, T. M., Muschiolik, G., Schmandke, H. and Tolstoguzov, V. B. (1988). *Nahrung*, **32**, 669. (*Food Science and Technology Abstracts* (1989). **21**, 2G26.)

Swartz, M., Walker, N., Creamer, L. and Southward, R. (1991). In: *Encyclopaedia of Food Science and Technology* (Ed. Y. H. Hui), John Wiley & Son Inc., New York, pp. 310–18.

Szpendowski, J., Smietana, Z. and Zuraw, J. (1983). *Milchwissenschaft*, **38**, 577.

Thomas, T. D. and Lowrie, R. J. (1975a). *Journal of Milk and Food Technology*, **38**, 269.

Thomas, T. D. and Lowrie, R. J. (1975b). *Journal of Milk and Food Technology*, **38**, 275.

Tolstogusow, W. B., Tschimirow, Ju. I., Braudo, E. E., Wajnermann, E. S. and Kosmina, E. P. (1975). *Nahrung*, **19**, 33.

Tolstoguzov, V. B., Grinberg, V. Ya and Gurov, A. N. (1985). *Journal of Agricultural and Food Chemistry*, **33**, 151.

Tornberg, E. (1980). *Journal of Food Science*, **45**, 1662.

Towler, C. (1974). *New Zealand Journal of Dairy Science and Technology*, **9**, 155.

Towler, C. (1976a). *New Zealand Journal of Dairy Science and Technology*, **11**, 24.

Towler, C. (1976b). *New Zealand Journal of Dairy Science and Technology*, **11**, 285.

Towler, C. (1976c). *New Zealand Journal of Dairy Science and Technology*, **11**, 140.

Towler, C. (1977). In: *Proceedings of Jubilee Conference on Dairy Science and Technology*, 15–17 March 1977, New Zealand Dairy Research Institute, Palmerston North, pp. 83–85.

Towler, C. (1978). *New Zealand Journal of Dairy Science and Technology*, **13**, 71.

Towler, C., Creamer, L. K. and Southward, C. R. (1981). *New Zealand Journal of Dairy Science and Technology*, **16**, 155.

Townsend, F. R. and Buchanan, R. A. (1967). *Australian Journal of Dairy Technology*, **24**, 113.

Tuohy, J. J., Downey, G. and Burgess, K. J. (1983). *Progress in Food Engineering* (Ed. C. Contarelli and C. Peri), Forster-Verlag AG, Küsnacht, Switzerland, pp. 703–13.

United States Department of Agriculture (1989a–90). Foreign Agriculture Circular *FD2-89* (1989), pp. 28, 29; *FD1-90* (1990), p. 28. Foreign Agricultural Service, US Department of Agriculture, Washington, D.C.

United States Department of Agriculture (1989b). *Agricultural Statistics, 1989*. United States Department of Agriculture, US Government Printing Office, Washington D.C., p. 345.

United States International Trade Commission (1979). *Casein and Its Impact on the Domestic Dairy Industry*. USITC Publication 1025. United States International Trade Commission, Washington, D.C.

University of Melbourne and Victorian Dairy Industry Authority (1987). New Zealand Patent 211 745.

Vakaleris, D. G. (1980). In: *Proceedings from the First Biennial Marschall International Cheese Conference, September 10–14, 1979*. Marschall Dairy Ingredients Division, Miles Laboratories, Inc., Madison, Wisconsin, pp. 261–71.

van Kreveld, A. (1969). *Voeding*, **30**, 231.

Vernon, H. R. (1972). *Food Product Development*, **6** (5), 22.

Visser, F. M. W. (1982). *Bulletin, International Dairy Federation*, Document 147, 42.

Visser, F. M. W. (1985). In: *Milk Proteins '84. Proceedings of the International Congress on Milk Proteins*, Luxembourg, 7–11 May 1984, Centre for Agricultural Publishing and Documentation (Pudoc), Wageningen, Netherlands, pp. 206–16.

Voropalva, V. S. and Ionkina, A. A. (1979). *Molochnaya Promyshlennost'*. No. 10, 37. (*Dairy Science Abstracts* (1980). **42**, 3319.)

Wade, N. (1977). *Science, USA*, **197** (4306), 844.

Waite, R. (1972). *Journal of the Society of Dairy Technology*, **25**, 92.

Wakodo Co. (1969). New Zealand Patent 150 402.

Walker, N. J. (1972). *Journal of Dairy Research*, **39**, 231.

Walker, N. J. (1973). *Journal of Dairy Research,* **40**, 29.

Walker, N. J. and Manning, D. J. (1976). *New Zealand Journal of Dairy Science and Technology,* **11**, 1.

Watts, D. A., Ough, C. S. and Brown, W. D. (1981). *Journal of Food Science,* **46**, 681.

Weal, B. C. and Southward, C. R. (1974). *New Zealand Journal of Dairy Science and Technology,* **9**, 2.

Webb, B. H. (1970). In: *By Products from Milk* (Ed. B. H. Webb and E. O. Whittier), 2nd edn, AVI Publishing Co., Westport, Connecticut, pp. 285–330.

Wiedmann, W. and Millauer, C. (1986). US Patent 4 605 444.

Wilson, G. F. (1970). *Proceedings of the New Zealand Society of Animal Production,* **30**, 123.

Wong, N. P. and Parks, O. W. (1970). *Journal of Dairy Science,* **53**, 978.

Wormell, R. L. (1954). *New Fibres from Proteins,* Butterworths Publications, London.

Young, V. R. and Pellett, P. L. (1991). *Journal of Nutrition,* **121**, 145.

Zalewska, S. and Swiderski, F. (1984). *Acta Alimentaria Polonica,* **10** (3/4), 255 (*Dairy Science Abstracts* (1986). **48**, 183).

Zavagli, S. B. and Kasik, R. L. (1978). US Patent 4 115 376.

Zittle, C. A., Dellamonica, E. S. and Custer, J. H. (1956). *Journal of Dairy Science,* **39**, 1651.

Chapter 9

Automation in the Dairy

W. M. Kirkland

APV Baker Ltd., PO Box 4, Gatwick Road, Crawley, West Sussex, UK

Automation of the manufacturing processes, the collection and the transportation of raw milk and finished products is an essential feature of today's dairy industry. The consequences have been that many processes operated by skilled staff or tasks that required large numbers of manual or semi-skilled labour to pack products or clean pipelines and equipment, have been replaced by electronic controls. Plants are operated remotely, often from a central control room, daily processing many hundreds of thousands of litres of raw product with minimum manual intervention. Such plants could not be operated manually without risk to product quality or excessive wastage of product or cleaning chemicals.

Growth Through the Fifties and Sixties

The aim of automation has been to reduce the amount of manual effort required to execute any given operation or task. As the size of unit operations in the dairy industry grew, the difficulties of cleaning plant and maintaining production came to the fore. Out of this need cleaning-in-place (CIP) technology developed. This involved close attention to pipework design, valve configurations, spraying devices and chemical systems, but most of all the need for automatic operation.

The basic requirement of these systems was to provide interlocking to avoid risk of contamination and accurate timing for each operation in the cleaning programme. A typical cleaning programme is made up of a pre-rinse with water recovered from an earlier clean, detergent clean, inter-rinse with fresh water, sterilisation with chemicals or hot water and a final rinse. The automation would also be set up to recover water and chemicals and to respond if a component failure or an interlock violation occurred. The control systems of the 1950s and 60s that formed the basis of this automation relied on electromechanical devices such as electrical relays, rotating drums or cams for sequencing and timing, and mechanical or optical card readers for providing variable cleaning programmes or recipes.

Use of Programme Logic Systems – 1970 Onwards

The 1970s and 80s saw the electromechanical devices largely replaced by solid state electronics and, in particular, microprocessor and program-mable logic systems. A characteristic of these systems is that a highly structured approach can be adopted to their design and implementation. All equipment and plant is electrically wired to the controller (this is sometimes called field wiring) in a predefined manner completely independ-ent of the way in which the plant is required to operate. The operation of the plant is then described in a programme (control software) converted by a programming module into data which the control system's microprocessor can execute and download into the controller memory in a few seconds. Thus it was possible to change operations very quickly by reprogramming the control software without changing the wiring to the plant. This was a tremendous advantage over the earlier electromechanical systems where the equivalent of the programme or control software was embedded in the electrical wiring in the control panel and making changes often resulted in adding additional electrical components and rewiring the panel. Such tasks were time-consuming, needed the electrical power to be removed and were often difficult to check out, thus reducing plant availability and introducing risk. The control panels built to house these designs were also very large, being made up of many individual components linked together by electrical wiring with consideration given to regular access for possible changes to the operation of the control logic. The microprocessor or programmable control system is much simpler to configure, more compact because it requires less access and can therefore use high density electronics. It is also less costly to produce and much easier to maintain and document. This has led to rapid growth in the application of automation in the dairy industry.

BENEFITS OF AUTOMATIC CONTROL

Size, cost and ease of application are important criteria when deciding how to invest in a new process or mechanical handling plant, but there are other fundamental benefits to be obtained from automatic control systems.

(1) Increased productivity and more efficient use of labour, particularly where a plant may be subject to seasonal variations in raw materials.
(2) Elimination of the dependence upon the human factor, thereby reducing risk of error and variations to operating parameters, giving higher standards of hygiene, improving product quality and consistency and reducing waste and process cleaning costs.
(3) Improvements in plant utilisation due to reduced shutdown times through the scheduling of operations, integration of plant cleaning and the minimisation of manual intervention.
(4) Improvements in plant operation through analysis of data from the process which is readily available within the control system.
(5) Mobility of operating staff by programming the control system to contain many of the operating procedures as well as the control logic. This allows operator skill levels and the training periods to be reduced dramatically.

CONTROL SYSTEM DESIGN AND ARCHITECTURE

Early automatic control systems were little more than a remote control facility bringing together all the control information (plant feedbacks) and the control actuation (command signals) to one central point, but relying upon human intervention to make all the logical decisions to operate the plant. Under full automatic control, all process items are in communication with the control equipment which itself determines, from its programmed logic, exactly how to drive the process equipment.

Process Plant Components

From a control systems viewpoint, all components associated with the process plant and connected electrically to the control equipment fall into the following categories.

Digital outputs
Commands or output signals used to operate equipment such as valves or pumps or signal lamps that have only two conditions – on/off, open/close, running/stop.

Digital inputs
Feedback or input signals from valves or motors that allow the control system to monitor for correct operation, or signals from push-buttons initiating actions within the programmed logic.

Analogue outputs
Output signals that can be varied by the control system over a predetermined range to drive equipment such as variable speed pumps or modulating control valves.

Analogue inputs
Feedback signals continuously variable over a predefined range that allow the control system to monitor process variables such as temperature, pressure, flow, pH, weight and density.

Electrical Signals

There are many standards for ensuring electrical compatibility between the plant components and the control system. These are the most common.

> Digital input/outputs: 24 V DC, 24 V AC, 48 V DC, 110 V AC
>
> Analogue input/outputs: 4–2 mA, 0–10 V DC

Some control systems also cater for the direct connection of electrical transducers such as resistance thermometers, thermocouples or conductivity cells. Alternatively, suppliers of these sensors include electronic circuitry to provide compatibility with the common analogue standards of 4–20 mA.

In the European dairy industry, 24 V DC is the commonly accepted standard for digital signals as this reduces risk to personal injury. In the USA, it is the practice to use 110 V AC. However in recent years there has been a trend towards the lower 24 V DC signal in North America.

Dairy Plant Input/Outputs

Table I provides an indication of the mix of control signals for typical dairy processes and process plants. For most applications it is usual to find a predominance of digital inputs and digital outputs, reflecting that

TABLE I
Dairy process plant – typical input/output configurations

Plant	Digital inputs	Digital outputs	Analogue inputs	Analogue outputs	PID loops
City milk 1000 k l	2600	1500	90	10	30
Milk pasteuriser	24	15	5	2	5
Bottling plant 250 k l	550	220	30	–	5
Yoghurt plant 100 k l	1000	480	64	–	5
CIP unit (single pump)	35	15	5	–	1
Cheese vat	5	10	4	1	1
UHT milk processing	40	30	12	–	2

most factories require product to be distributed through pipework systems to different buffer storage areas.

Cleaning systems also increase the number of process valves in a plant by 15–20 per cent. The control equipment is connected to electrical devices whether they are sensors or actuators. A good example is the process valve (Fig. 1), a device operated by high pressure air switched by

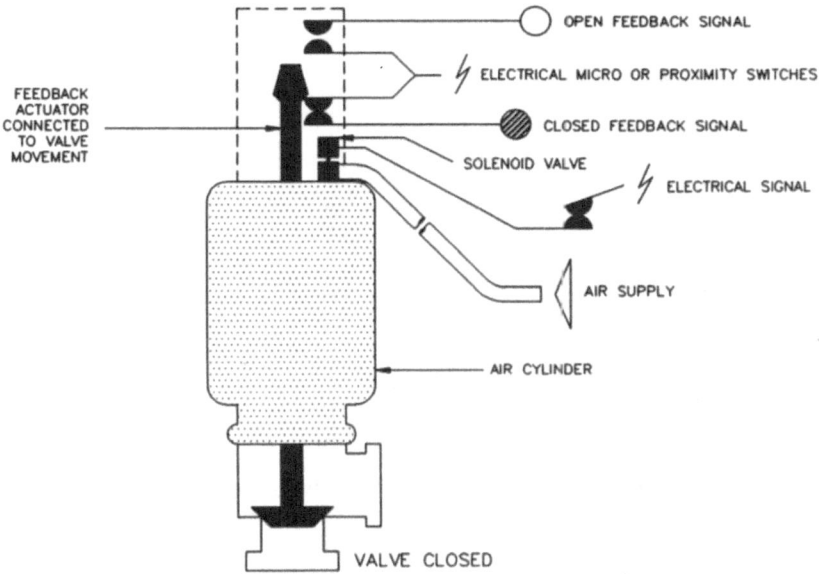

Fig. 1. Hygienic process valve with the solenoid and feedback switches incorporated into valve head.

an electrical solenoid valve driven from a digital output within the control system. The valves are fitted with a mechanical arm attached to the valve movement. This arm is used to operate microswitches or proximity sensors that provide electrical feedback signals to digital inputs within the control system.

Control System Construction

Input/output modules

Control systems are built up of modules to cater for each of the different types of signal. Each module has circuits for between 8 and 32 separate signals. Table II shows a typical product range of modules (often called input/output modules or I/O) that would satisfy the requirements of dairy industry applications. The I/O modules have terminals to connect the electrical field wiring from the plant to the control system. The I/O modules also provide a barrier or electrical isolation between the control wiring to the plant equipment and the low voltage signals used by the electronic circuits of the microprocessor. This isolation barrier protects

TABLE II
Typical range of input/output modules available with control systems

Module type	Number of channels	Electrical spcification
Digital output	16	24V DC
Digital output	32	24V DC
Digital input	16	24V DC
Digital input	32	24V DC
Digital output	16	120/240V AC
Digital input	16	120V AC
Analogue output	8	± 10V DC
Analogue output	4	0–10V DC
Analogue output	6	4–20mA DC
Analogue input	8	4–20mA DC
Thermocouple input	12	J–K
Resistance thermometer	6	100 ohm

the electronics from damage due to faults in the plant wiring, and also prevents malfunctions from electrically noisy signals which could interfere with the low voltage signals of the microprocessor. A device called an optical isolator is the most common component used to achieve this barrier, although other components such as electromechanical relays or transformers are also used.

Racks
The I/O modules are housed in a rack which provides mechanical support but, more importantly, includes a collection of electronic connections (called a bus) through which the I/O modules communicate with the central processing unit (CPU).

Central processing unit (CPU)
The CPU is the heart of the control system and houses the microprocessor and the memory which stores the control software logic. In addition, the memory holds software which organises the operation of the system and provides what is called a 'real-time multi-tasking' environment, essential to handle the many different simultaneous actions required from the control system. Other electronic circuits on the CPU deal with monitoring the power source to provide safe shutdown in the event of the loss of the electrical supply, battery back-up for the control software, timers to generate time of day clock and calendar information and communications signals which comply with the electrical protocols used with printers and visual displays. The protocols are usually in the form of a string of pulses (serial) encoded to present the characters commonly used in English text. The codes that represent each character have been laid down in a standard known as the ASCII standard (American Standard Codes for Information Interchange).

The electrical signals which represent the codes have been defined by the Electrical Industries of America (EIA) and are known as RS232. The signal level usually has a 24 V swing; the complete specification is RS232 V24 and is normally called a serial interface circuit. Figure 2 shows the organisation of a typical control system and Fig. 3 an example of a commercial control system.

Distributed Input/Output

The system described above involves routing all the electrical signals from the plant components to the centrally located control equipment. The signals are grouped in multicore cables to reduce costs, and this is often a significant element in the total cost of control system installation (30–50 per cent). Commercial control systems are equipped with circuitry that effectively extends the CPU bus and allows the I/O modules to be distributed around the process area and close to the plant components. Therefore, the majority of the signal cables are relatively short and longer cables, which communicate to the control centre, have only a small number

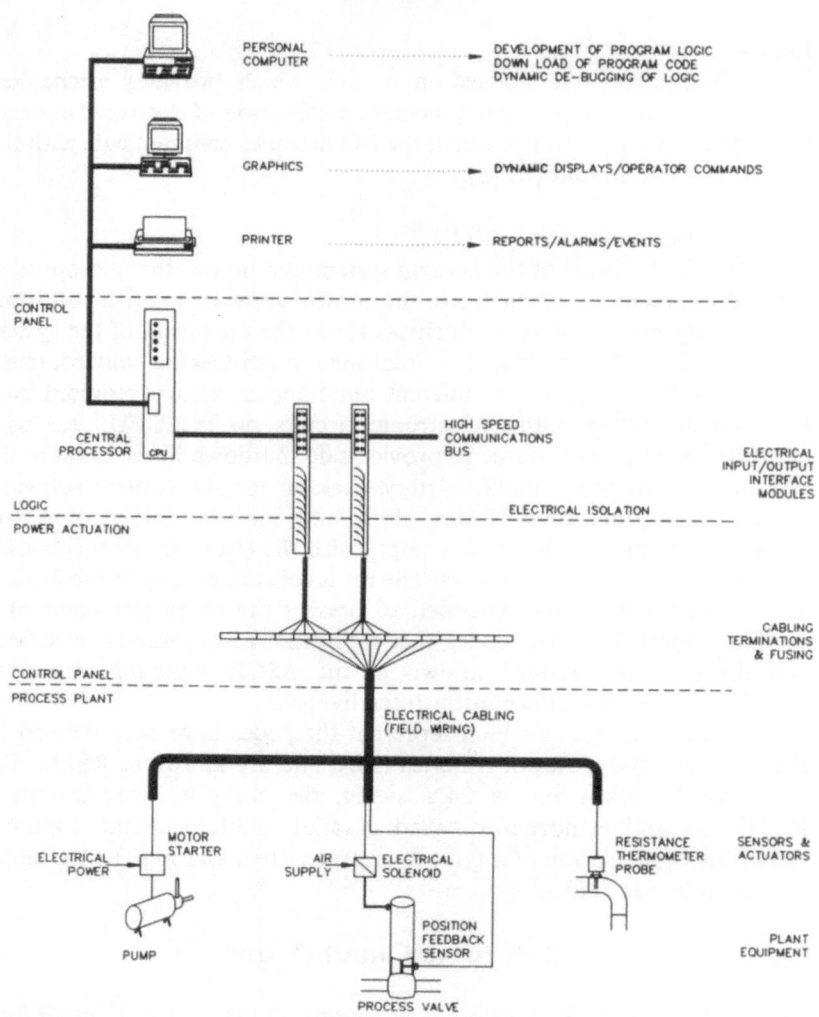

Fig. 2. Control system configuration.

of signal wires. In applications where long distances or electrically noisy environments are encountered, low loss coaxial cables or optical fibres are used. Such a system is called distributed I/O or sometimes remote I/O. Figure 4 shows how distributed I/Os are applied to a dairy process plant.

Fig. 3. Commercial microprocessor based control system.

Control System for a Dairy Plant

Figure 5 shows a typical dairy plant receiving milk by road tanker and converting it into a range of products in a variety of packages. There are four cleaning units, two handling the processes associated with raw milk, the other dealing with finished products. In such a plant there are several factors to be taken into account when building a control system to automate the process: plant size and complexity, method of operation and the number of operators required, accessibility of equipment and electrical cabling distances, the installation programme (which may be phased over a long period) and future expansion plans.

Using this information, the control engineer configures a 'best value' solution. This would involve using more than one controller (or CPU) if the system had a large number of I/Os or if there were a requirement to store or retain a large number of process operating variables or very complex logic operations. On a medium size dairy, processing 250,000 to 500,000 l per day, it would not be unusual to have two CPU systems each controlling a different process area (raw milk and milk products).

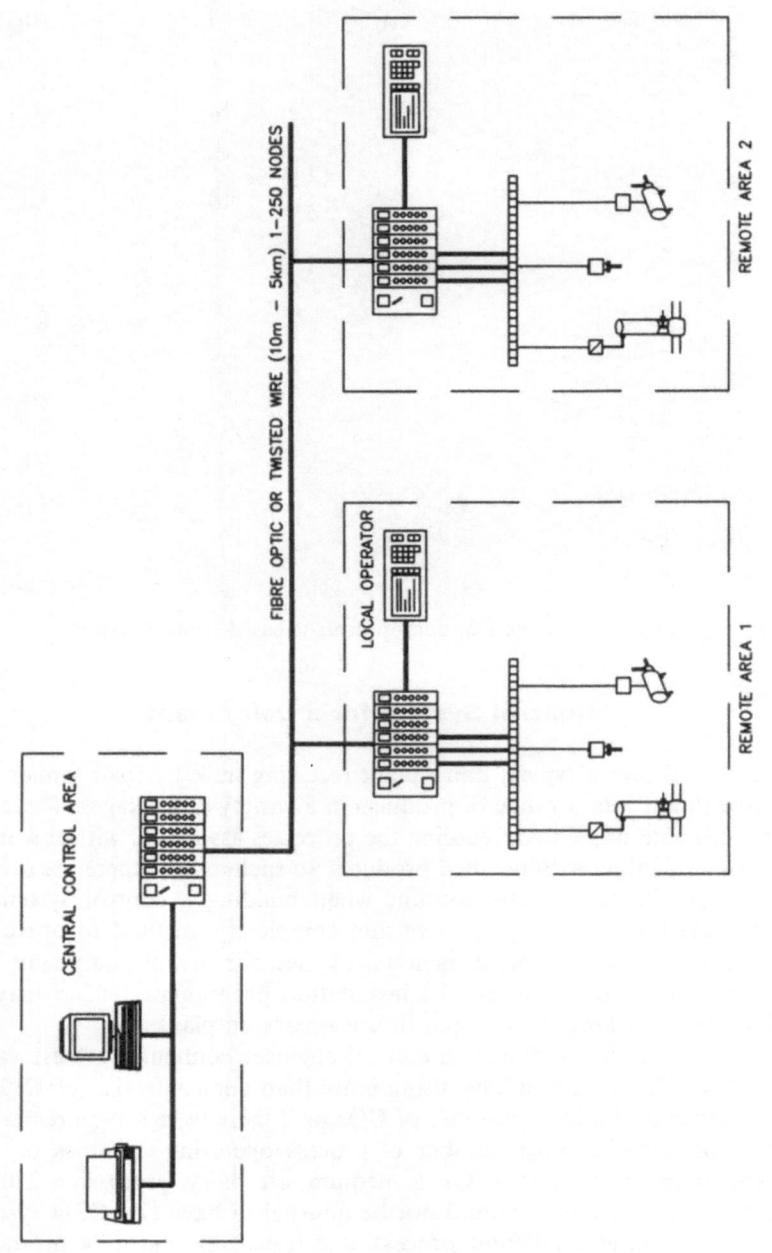

Fig. 4. Distributed input/output system.

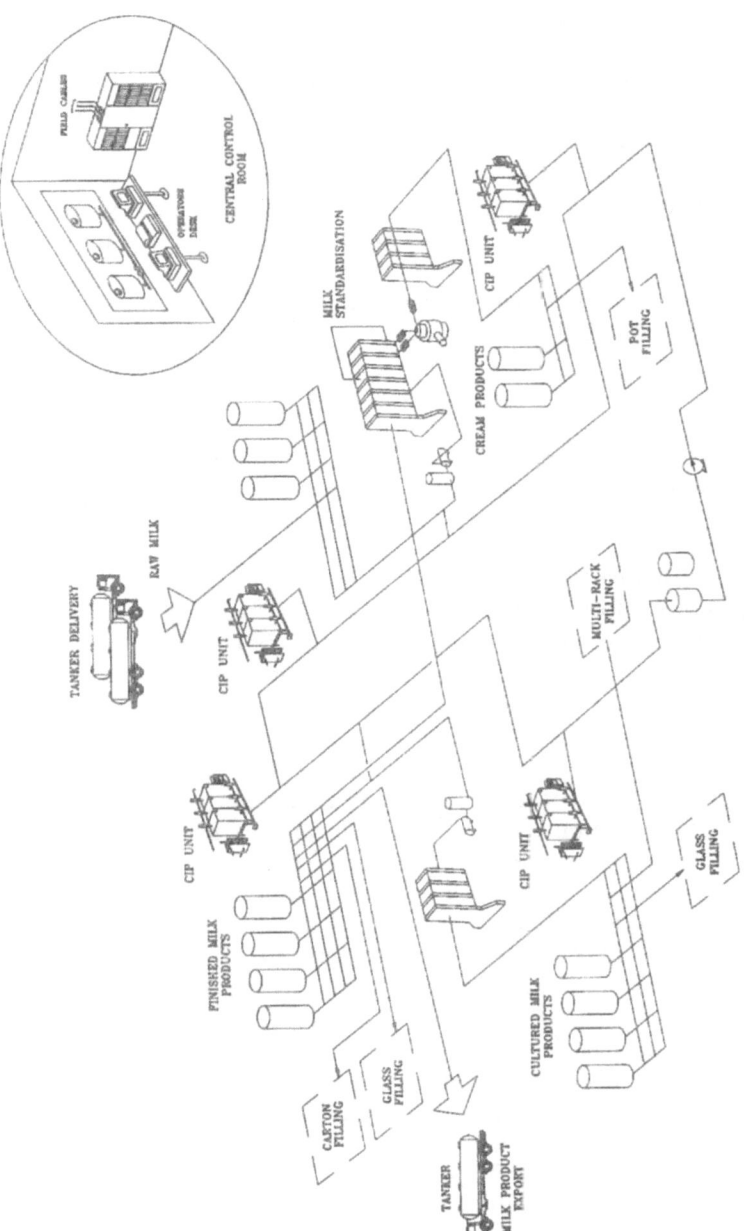

Fig. 5. Typical dairy process plant.

Where these needed to interlock, they would be connected together by a high performance electrical network operating at 300,000 to 2,000,000 bits/second. This is called distributed control and the concept is regularly used to 'share' the electronic data processing load and to add an element of redundancy into the configuration, such that if one system or CPU fails the whole processing plant is not totally affected.

Distributed I/O would be employed in an existing factory, as it can be disruptive and costly to feed all electrical connections for plant items to a central control room. And where large distances are involved, the cable cost can be reduced by employing distributed I/O. If the plant is to be run by one operator, it is advantageous to collect all operating parameters at a single source.

Figure 6 shows a configuration for the control system for the dairy shown in Fig. 5 which would be used for a centrally located control system, and Fig. 7 shows an alternative solution which offers both central and local operational capabilities. This may be particularly attractive if a night-shift operator is required to manage plant-wide CIP from a central location.

The distributed I/O system employed in Fig. 7 uses the CPU, housed in the central control room, to drive remote I/O and remote CPUs via a signalling system called a field bus. The field bus concept is used by many manufacturers all operating on different communication systems, and therefore it is not possible to mix equipment purchased from different sources of supply. However, new field bus standards are emerging and these are covered by IEEE standards whose specifications are being developed jointly by makers, users and academics.

A similar situation also exists with local area networks (LANs) and again it is most common to find supplier based networks installed. However, with the widespread use of Ethernet standards for PC office networks, the cost of implementation has fallen to levels where it is feasible to use this network for process control.

General Motors have attempted to face up to the problem of mixed supplier-based systems and how to address their configuration into a computer integrated factory environment by specifying a set of hardware and software known as manufacturing automation protocol (MAP) for which a number of suppliers are now producing products. MAP operates at very high speed (10 Mbits/second) and uses a principle called token passing. Although the electronics are now produced in silicon large-scale integrated circuits, the cabling system and the connection of each node to it makes the system too costly for most applications within the dairy

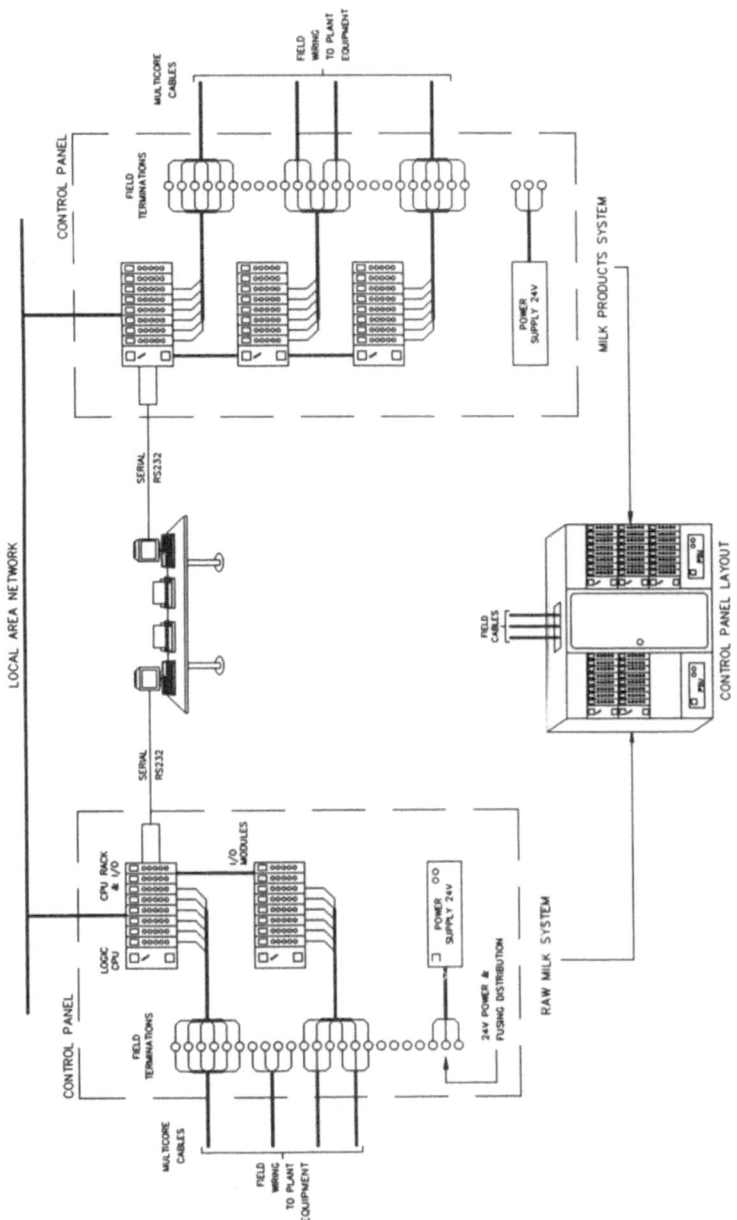

Fig. 6. Central control system concept.

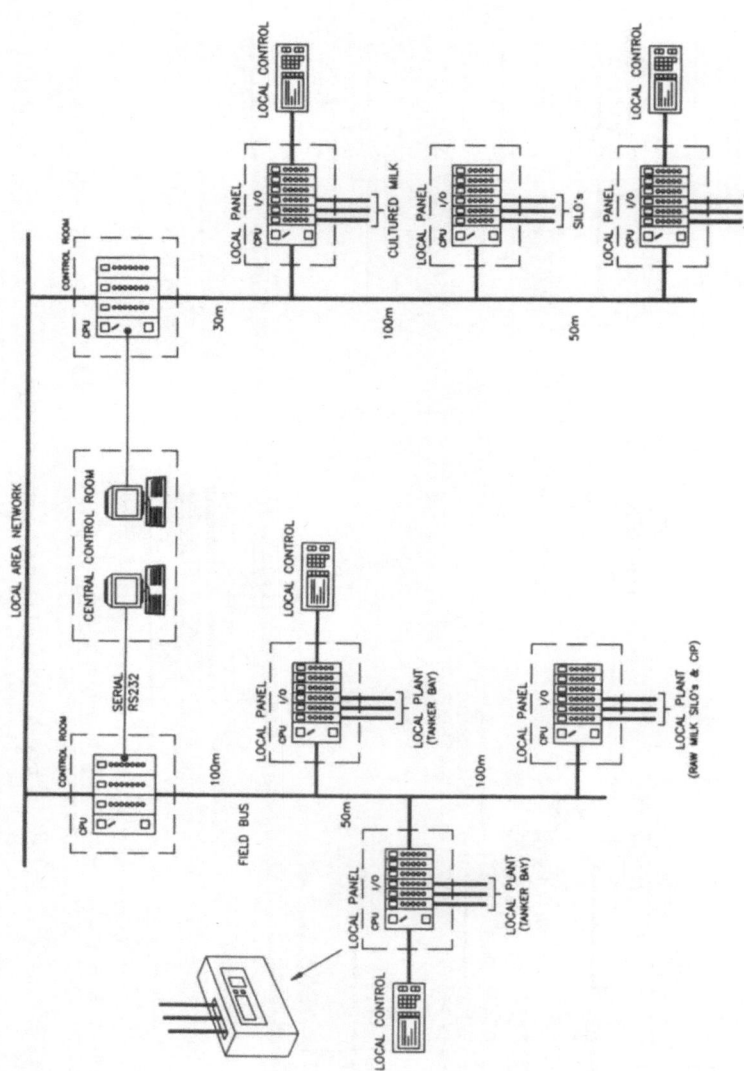

Fig. 7. Local/central control utilising distributed I/O.

industry. Low loss coaxial cables are used for high speed networks; twisted wire pairs for lower fequencies. (Twisted wires cancel out electrical interference that may be induced into the cable.) Standards are under development for fibre optic LANs which offer the benefits of high speed and immunity to electromagnetic interference.

Production Control and Management Information

Effective process control is vital to the operations of a dairy. It also offers plant management a communication and data collection facility to measure the performance and efficiency of each process or plant component. When used properly, this can provide a focus for improvements in operation and utilisation of equipment and a precise record of all automation and manual events and alarms on the plant. Figure 8 shows the addition of a production control system to the process control configuration used for the dairy example in Figs. 6 and 7.

As quality standards and procedures, e.g. ISO 9000 or BS 5750, become the norm in the dairy industry, these records will become an essential part of all operations. Supermarkets and other large buyers of dairy products have precise requirements for the maunfacture and handling of their products and require 'audit trails' for each production batch. The audit trail provides a complete record of the raw material sources, additive sources and a detail of the batch's processing history. It identifies alarms or deviations from specified operating parameters and links them to cleaning history records.

If problems are found with a particular production batch, audit records can be used to identify all other batches with a similar processing history. These may also be suspect. If there is a problem with a plant component, all batches which used that component can quickly be identified.

The data processing and storage requirements of these systems are usually very demanding and there is often a need for access from a wide variety of users, i.e. production management, plant maintenance, quality control, planning and plant operations – a true multi-user requirement. The technology, both hardware and software, for this application would therefore be significantly different from that of the process controller. Typical systems utilise a relational database, Unix operating software and Winchester disk technology for storing data and programmes.

Table III shows a simple comparison between memory, storage, CPU performance and serial I/O for production control and process control

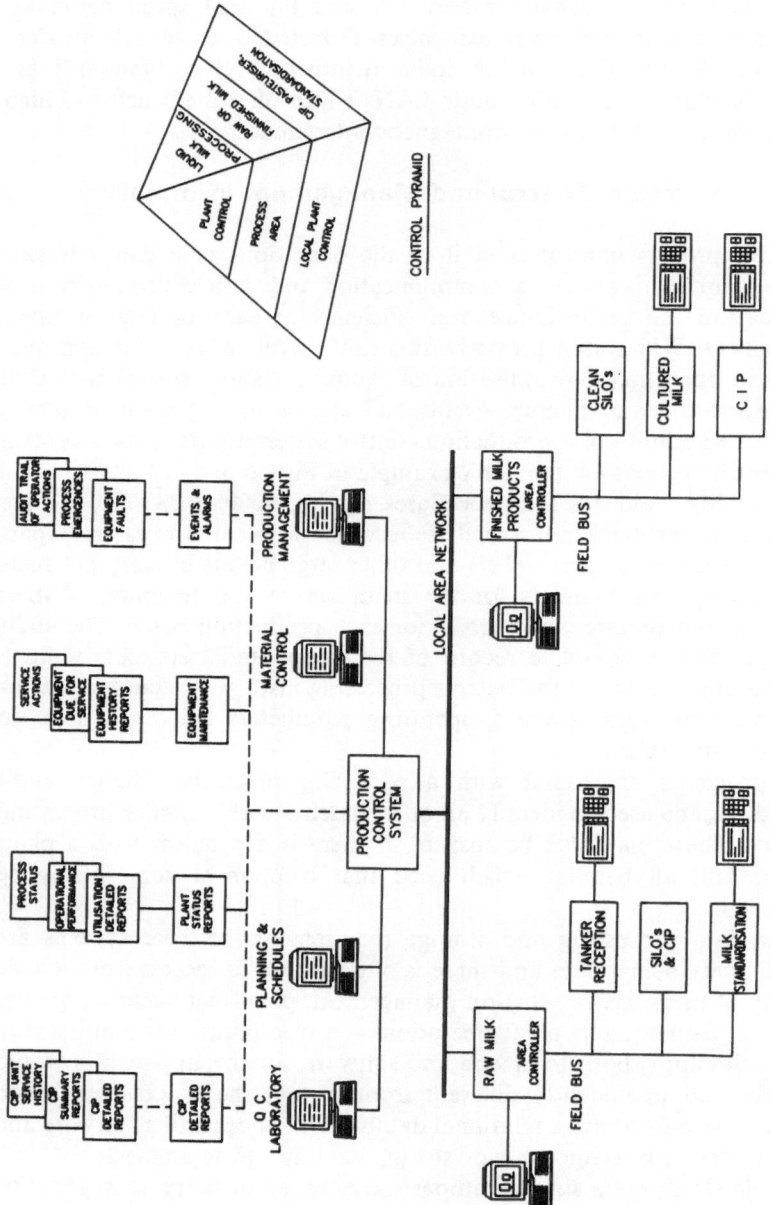

Fig. 8. Production control for the dairy industry.

TABLE III
Comparison of performance specification between production and
process control systems

System resource	Production control system	Process control system
CPU speed	33 MHz	8–16 MHz
RAM memory	8 Mbyte	256 Kbyte
Disk memory	300 Mbyte	0
Serial I/O channels	10–24	4–12
Network speeds	10 Mbits s^{-1}	50 k–1 Mbits s^{-1}
EPROM memory	0	512 Kbyte

systems. Figure 9 shows examples of production, CIP and maintenance reports generated from data collected from the process system.

Computer Integrated Manufacture (CIM)

The production control system described above serves as an ideal bridge between factory floor processes and computers employed to manage the business and commercial operations of the dairy.

Developments in the standards for computer system network architecture, the protocols for exchange of data and the operating programmes for their management and execution have paved the way for a number of effective applications in the dairy and related industries of ice cream products. Typical examples are production scheduling and plant utilisation, production costing, materials usage, energy usage and plant efficiency. Figure 10 shows a typical example of a CIM configuration that incorporates a collection of data from filling machines as well as processing lines.

Plant Utilisation and Production Scheduling

In most dairies, rough cut planning from sales orders is carried out to determine how raw milk is to be allocated to the product mix. The precise utilisation of the factory equipment is managed by the operator from experience. This can invariably lead to unscheduled overruns in production time or loss of availability of key process plant due to untimely cleaning operations leaving manned filling machines idle or no product for a key account customer. The dairy industry has a particularly difficult series of interrelated factors to resolve and many manufacturers are

Evaporator Efficiency Report

Report Start Date   12-03-90   Report End Date   24-03-92

	Evap 1	Evap 2
Time on Production	94.8%	95.2%
Time going forward	80.2%	81.8%
Time on CIP	7.8%	8.1%
Time on Water	6.8%	4.9%
Kg Steam / Kg Milk intake	0.13	0.12
Kg Steam / Kg Concentrate	0.56	0.55
Kg Steam / Kg Water Evaporation	0.13	0.15
Steam first effect	68250 Kg	68540 Kg
Steam pre-heater	91008 Kg	91392 Kg

CIP Detail Report

Report Start Date: 10-02-92   Report End Date: 10-02-92
CIP ID Number: 170   Description: Raw Milk Silo 1
Bulk Unit: 3   CIP Station: A
Start Time: 14.32   Complete: YES Alarmed :   YES

			Alarm	Actual min	Target min	Actual litres	Target litres
1	Pre - Rinse	14.32	1	22	20	5204	5200
2	Caustic Charge	14.54		7	6	1700	1690
3	Caustic Circulation	15.01		6	6	N/A	
4	Post Caustic Rinse	15.07		7	7	1982	1960
5	Acid Charge	15.14		2	2	562	570
6	Acid Circulation	15.16		5	4	N/A	
7	Post Acid Rinse	15.21		3	3	871	850
8	Final Rinse	15.24		7	6	1705	1720
9	Drain	15.31		5	5	N/A	

Plant Item Maintenance Report   **Centrifugal Pump A1.01**

Unit description	Room 1 Area 3	Time since last replacement	28250 hrs
Unit type / part number	Puma Pump 2-7-11	Time on since last replacement	17450 hrs
Interface type / location	Card 3 Rack 2	Number of operations since replaced	2920
Replacement Due	2-4-1992	Time since last service	6150 hrs
Service Due	2-4-1991	Time on since last service	3422 hrs
Last Replacement	20-4-1988	Number of operations since service	600
Last Service	20-2-1990	Replacement period	35000 hrs
Last Failure	10-8-1989	Max time on before replace	25000 hrs
Number Replacements	1	Max operations before replace	4000
Number Services	2	Service period	9000 hrs
Number Failures	2	Max time on before service	6000 hrs
Time Failed	0.3 hours	Max operations before service	1000

Item History Report   Fault (F) , Service (S), Installation / Replacement (R)
2-4-1984 (R) Pump installed
10-2-1988 (R) Pump replaced - damaged by forklift
10-8-1989 (F) Shaft seal replaced part 512-302
20-2-1990 (S) Routine Service no problems found

Fig. 9. Production reports using data collected from the process control system.

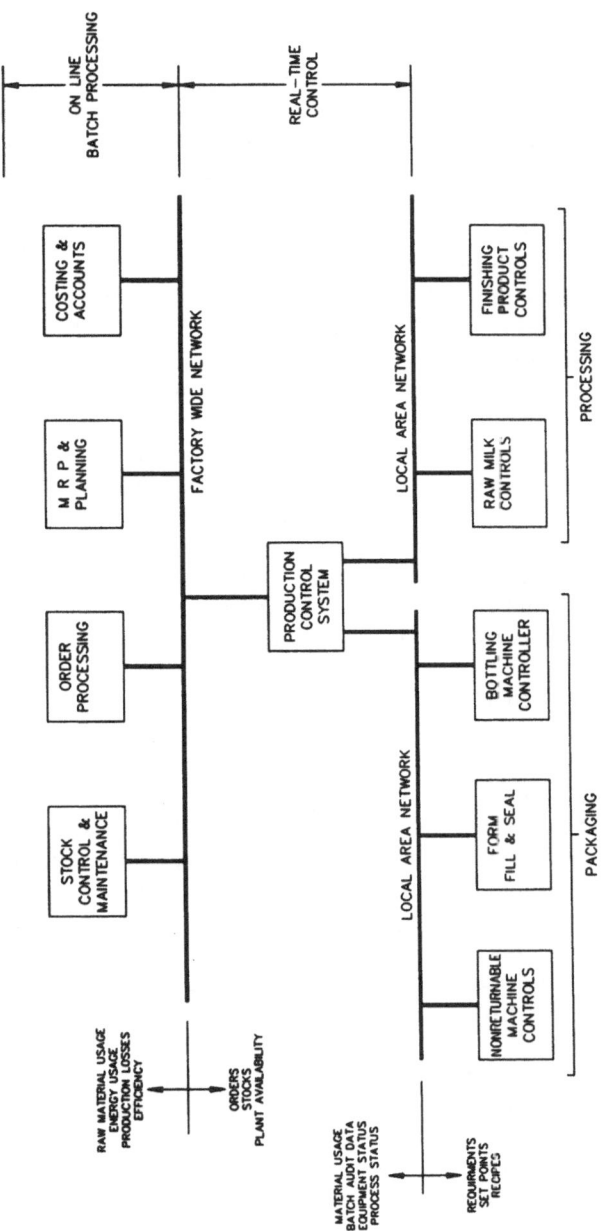

Fig. 10. Computer integrated manufacture in the dairy industry.

turning to computer assisted modelling aids to provide detailed plans for production and plant utilisation.

Figure 11 shows an example of a model based planning and scheduling system, where the planner defines the plant configuration, product processing methods and production requirements. This data is processed by a scheduling algorithm which computes how each production batch is to be processed, identifying bottlenecks, idle time and overdue batches. The planner adjusts his requirement to achieve a best fit on his production by a 'what if' approach.

Figure 12 shows a schedule produced by a computer based model which identifies manning levels and utilisation and details all processing operations. The planner can experiment with different plant configurations if he is aware that equipment will be unavailable at certain times of the day, giving the production management the ability to determine with accuracy what the impact of planned maintenance or plant downtime will have on overall production throughput. The technology of the production computer can accommodate the planning and scheduling programmes, which allows the system to be enhanced to monitor the planned performance against actual output with rescheduling carried out automatically, thus providing an early warning system of production delays or early completion of planned operations.

Control of Unit Processes and Machinery

Unlike the systems described above, many dairies operate on a mixture of manually operated and automatically controlled processes. Bottling plants producing relatively few variations in product mix are typical examples and would have manual control of milk reception and finished product process, automated control of pasteurisers, standardisation control, CIP, filling and packaging machines. This type of dairy operation requires a complete package to combine processing equipment and controls with a predefined specification. Where dairies are subject to a phased refurbishment programme, this is often the most cost-effective solution.

Figure 13 shows examples of unit process equipment combining the control equipment onto a single skid-mounted assembly. The control technology for these units has advanced to meet the growing demand for performance and diagnostic data and the operator displays shown in Fig. 14 illustrate the amount of data available from these packages. As a dairy develops, there may be benefits to be gained by integrating these unit

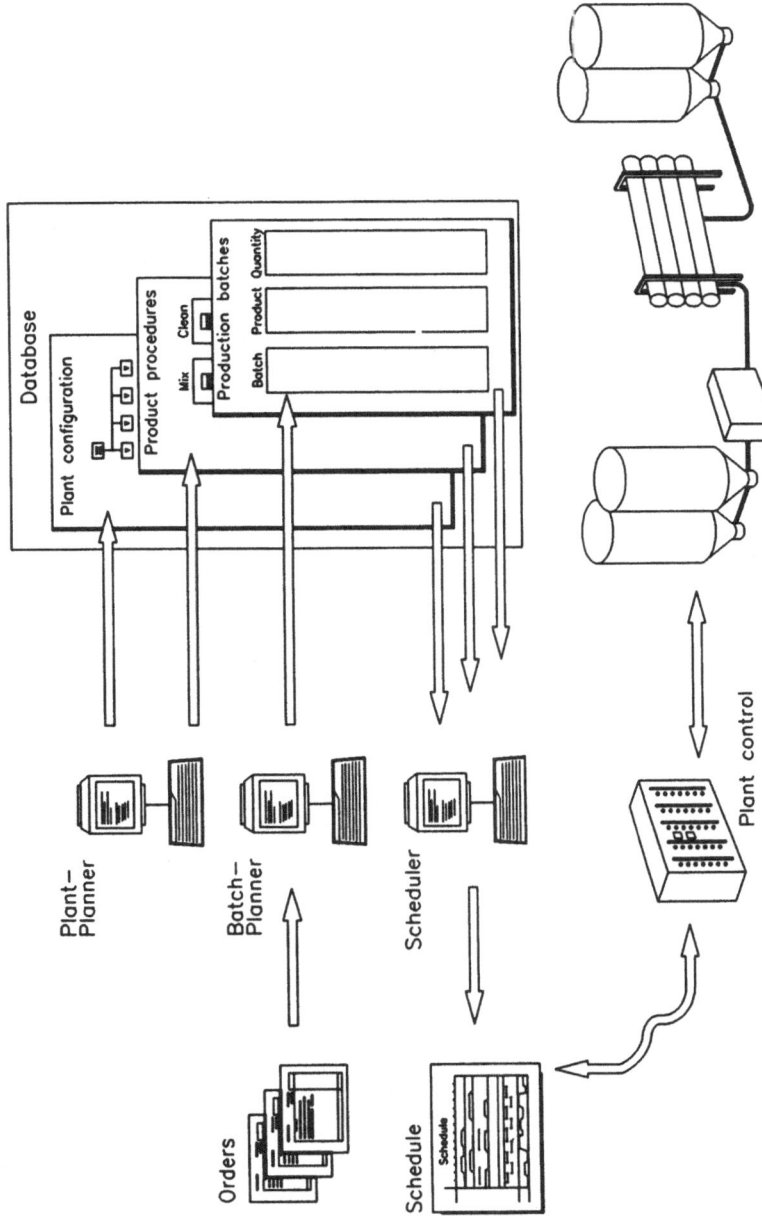

Fig. 11. Model based planning and scheduling system for the dairy industry.

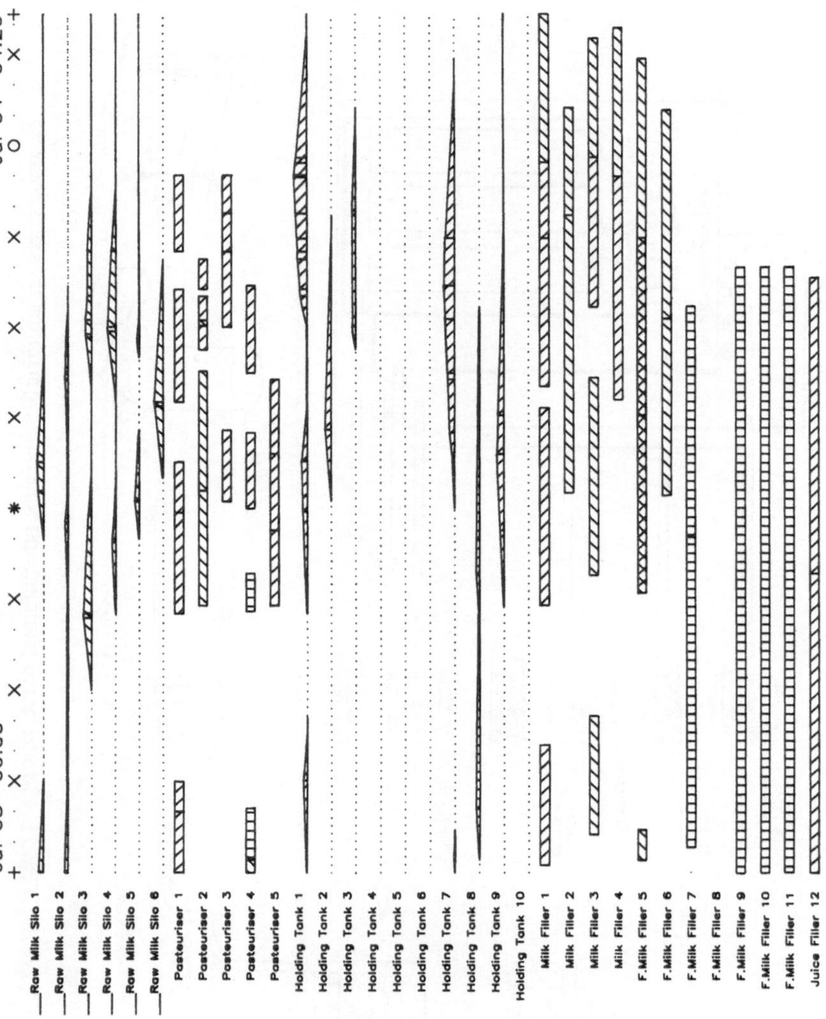

Fig. 12. Production utilisation produced by a computer model of a dairy process.

Fig. 13. Unit process combining controls and process equipment.

processes into a centrally operated automation scheme. The unit process controllers can either be connected to the peer to peer network or to the distributed I/O systems to provide full time integration.

CONTROL SOFTWARE

Control software organises the operation of the control system equipment or hardware and comprises a suite of computer programmes stored in memory and executed by the system CPU. Some control systems (multiprocessor systems) have more than one CPU and in this instance, the control software is usually split into specific functions such as I/O scanning, serial I/O communications, network management and control logic execution. Each software module is operated by its own CPU.

Control software falls into two categories: the operating system and the application programmes (Fig. 15). The operating system organises all the internal functions. It provides a real time execution of control logic, manages peripherals such as printers, visual displays and graphics, carries out housekeeping to keep track of the time, handles low power and

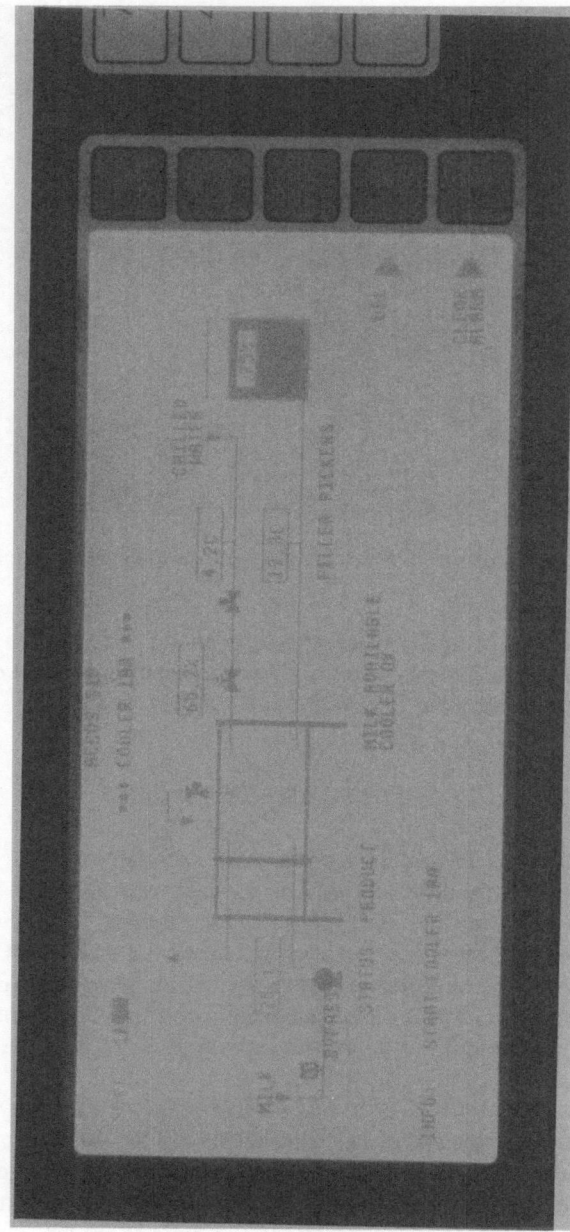

Fig. 14. Data available to process operator from control system.

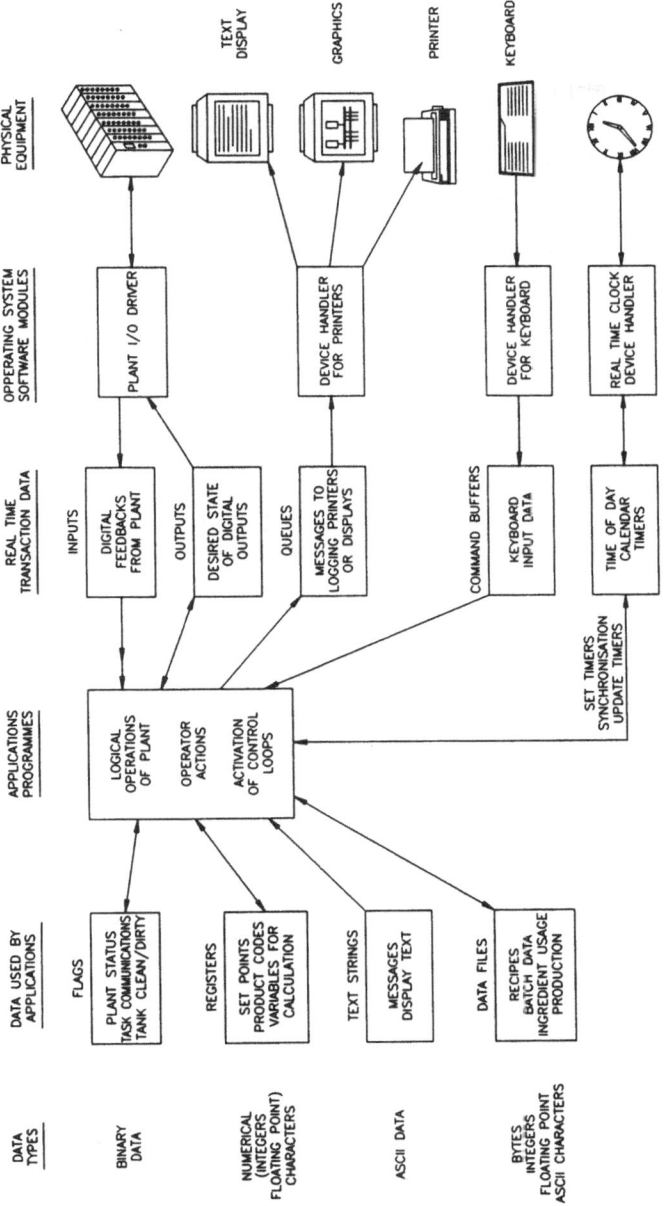

Fig. 15. Software organisation for process control.

executes start-up from power-up. These programmes reside in read only memory and are not subject to changes by the end-user. For this reason they are often called firmware. The application programmes are effectively all the logical rules and requirements of running the process plant connected to the control system. They are designed by an engineer who understands the process requirements rather than a computer programming specialist. Manufacturers of control systems provide applications programme languages to facilitate the translation of a design specification into a computer programme.

Process Languages

In all applications, the control system must be capable of managing and executing many concurrent tasks and in real time responding in *fractions* of a second to any stimulus so as to avoid overfilling, contamination or excessive losses. It is therefore important that the applications programming language has a structure that reflects these requirements.

Figure 16 shows an example of a process language that breaks down each task into one or more sequences. As far as the design engineer is concerned, each sequence operates independently, but sequences related to a specific task can be chained together in a 'tree'. Through this structure, the operating software links process equipment failures to sequences and sequence trees which allow automatic safety and start-up routines to be executed. Each sequence is made up of a number of steps and each step contains a command directed at the process equipment internal data registers or peripheral devices. The instructions are abbreviated into mnemonic codes.

ENGE VALVE 1	Energise valve 1
DENG VALVE 1	De-energise valve 1
WAIT 10	Wait for 10 s on this step
GOTO 50	Go to step 50
MESS 20	Output message number 20

A typical medium size dairy would have 600–700 sequences describing the process, CIP and other operational tasks. Figure 17 shows an example of a sequence that empties a storage vessel to a road tanker with a following air purge to clear the line.

The applications control software is prepared on a personal computer and downloaded into the control system usually via a serial I/O module. This process language is readily understood by all disciplines of engineer

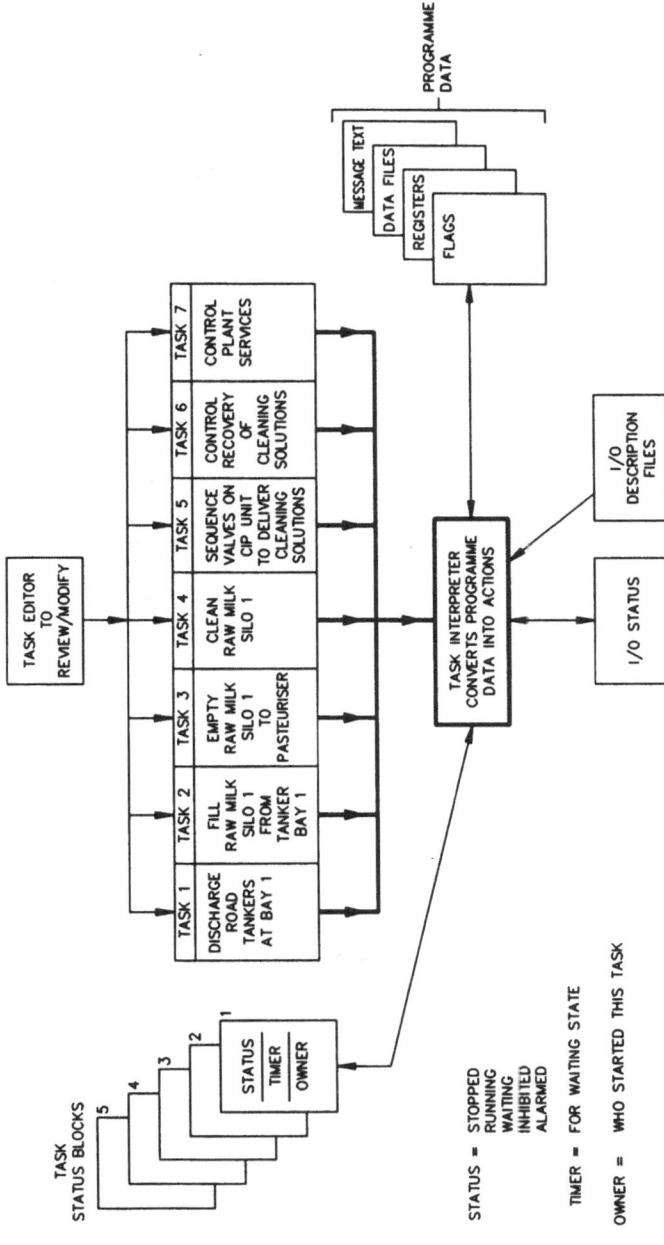

Fig. 16. A multitask control software.

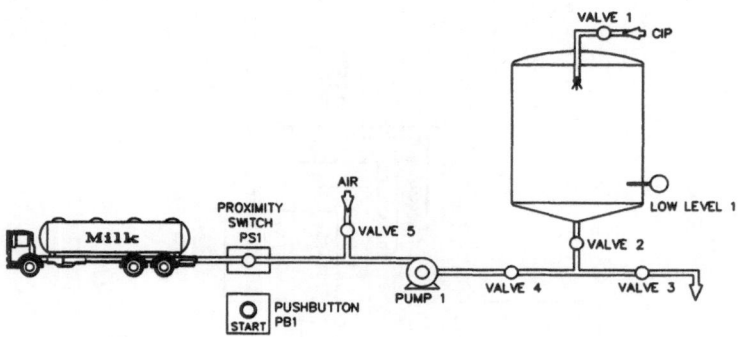

SEQUENCE STEP	INSTRUCTION		DESCRIPTION
1.000	WUEN	PB1	Wait for Pushbutton to be pressed (or energised)
1.001	CHECK	PS10	Check flowplate proximity switch is in correct position
1.002	WUDE	VALVE 1	Wait until CIP Valve 1 is De-energised
1.003	WUDE	VALVE 3	Wait until Drain Valve 3 is De-energised
1.004	ENGE	VALVE 2	Energise Tank Outlet Valve 2
1.005	ENGE	VALVE 4	Energise Product Routeing Valve 4
1.006	WAIT	5	Wait for 5 seconds for Product to prime pump
1.007	ENGE	PUMP1	Energise Pump 1
1.008	WUEN	LOW LEVEL 1	Wait until Low Level is energised (Tank is empty)
1.009	WAIT	10	Allow tank to drain out for 10 seconds.
1.010	DENG	VALVE 2	De-energise Tank Outlet Valve 2
1.011	DENG	VALVE 4	De-energise Routeing Valve 4
1.012	DENG	PUMP 1	De-energise Feed Pump 1
1.013	ENGE	VALVE 5	Energise Air Purge Valve to clear line
1.014	WAIT	12	Purge out for 12 seconds
1.015	STOP		

Fig. 17. A process control sequence.

or technician. They easily learn how to change programs to improve operation or to accommodate process plant modifications. In some systems where there are very demanding production requirements, these changes must be made whilst the plant is operating on-line.

Ladder Logic

An alternative to the process language is to describe the control logic in the form of an electrical relay circuit diagram similar to those produced for control logic in the 1950s and 60s. In Fig. 18 each relay coil is typically associated with an output, such as a valve solenoid or a lamp, and each feedback or push-button is represented by a contact. The circuit diagram takes the form of a number of individual circuits comprising a set of logical conditions (the contacts) driving an output (the coil) which resembles a ladder made up of a number of rungs – hence ladder logic. This type of design tends to be favoured by electrical engineers and whilst

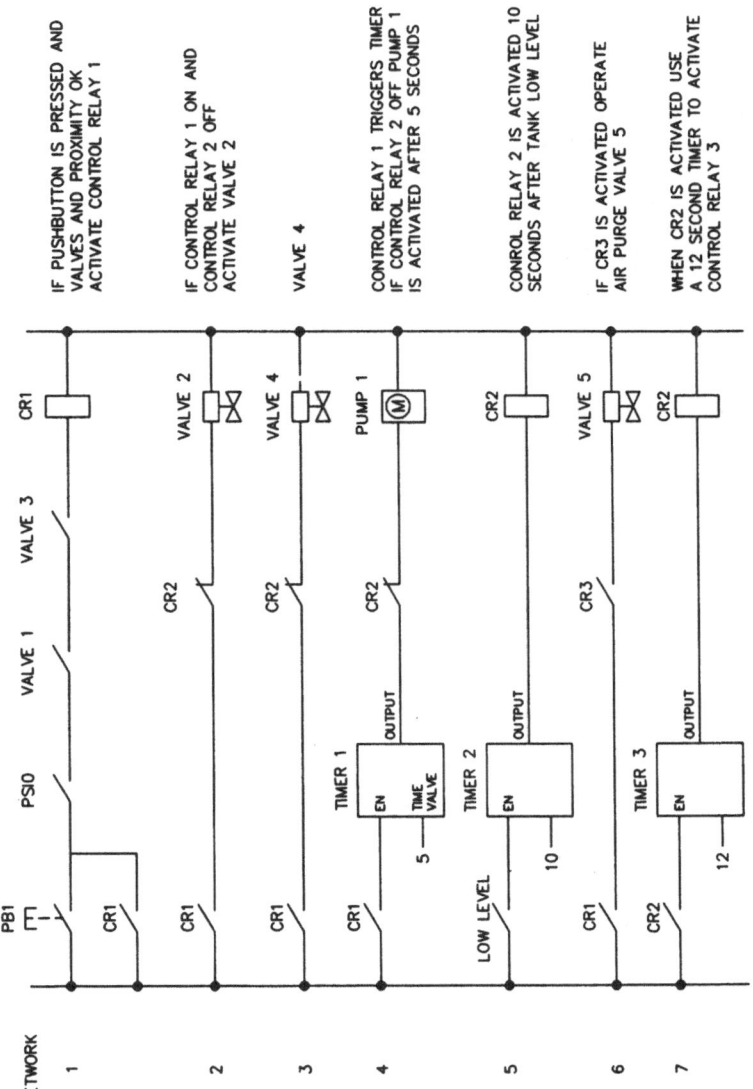

Fig. 18. A ladder logic sequence for emptying a silo.

there are many function blocks for timers, PID algorithms and mathematical operations, it is not ideal for large I/O systems where there are many time dependent sequences. Control system suppliers all implement their ladder logic differently. This has prompted the IEC to define standards for the implementation of ladder logic controllers in an attempt to improve the 'portability' of applications from one system to another, and to ease the learning curve for engineers picking up a system for the

Fig. 19. Plant simulators in use to validate system software.

first time. In on-line operations the programming device is connected to the controller to give animated operation of the ladder logic diagram for debugging problems.

Grafset

Grafset is a design tool to overcome many of the difficulties of configuring ladder logic applications. The application programme is designed graphically by specifying a number of operations or states and defining the logic which causes a transition from one state to another. Grafset standards are also defined by the IEC and like the ladder logic diagrams can be animated by connection to the on-line system.

Software Test and Simulation

The design, configuration and testing of the control hardware is relatively simple as it is a highly structured operation. Occasionally difficulties arise through poor design of field cabling but these are rarely more than frustrations. The applications software is crucial to the operation of the control, and it is often this element of the design which is the last area to be specified as it depends very much on the details of the process. Any pressures to release control software onto an operating plant before it has been thoroughly tested, introduce unnecessary risk and should be avoided at all costs. A full simulation is essential, preferably using a simulator for the process. It can also be used to train plant operators and to fine tune some of the design detail, those aspects unclear at the time of specification. Fig. 19 shows plant simulators in use to test out control software.

INSTRUMENTATION AND CONTROLS

Control Loops – Regulation

Most control or regulation in the dairy industry, with the possible exception of some aspects of spray dryers and evaporators, uses an industry standard known as 'three term control' or proportional, integral and derivative action (PID). Figure 20 shows the principle of PID control where the controller measures the variable to be controlled (e.g. product temperature) compares this to the desired value or set-point (to produce

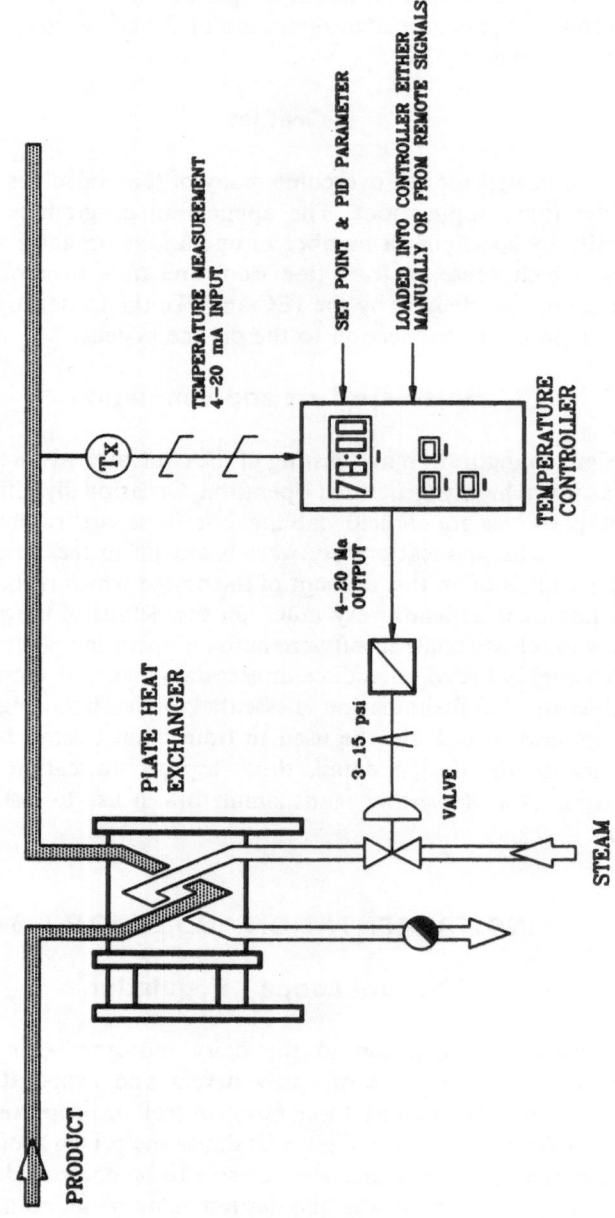

Fig. 20. Principle of PID control (temperature).

an error value) and then adjusts the output variable (e.g. steam flow) by raising or lowering the modulating valve to bring the process variable into specification, i.e. to reduce the error to zero. The rules for making this adjustment, the control algorithm, in a PID controller take into account the value of the error (proportional action), the rate of change of the error (derivative action) and its long-term deviation from the set-point (integral action). Each factor is given a weighting (the gain) and then combined to give the required output setting.

The control engineer tunes the controller to match the dynamics of the process so as to obtain a fast response to any deviations from the set-point. The tuning is carried out by adjusting the gain factor of each of the PID terms. This requires experience and a good knowledge of the process to obtain best results.

Self-Tuning Intelligent Controllers

Since the 1970s, when the early pneumatic controllers were largely superseded by electronic and later microprocessor based technologies, much work has been done on improving the sensitivity of the three-term control algorithm and its set-up in the field. This work has included the development of 'self-tuning' or intelligent controllers which automatically adjust the three terms in the algorithm to give best results from direct process measurements. Manufacturers have experimented with different techniques for developing these self-tuning features, some utilising stochastic techniques, others – probably the most successful – emulating the actions of the experienced control engineer. For most dairy processes, the use of derivative action is too severe and produces a responsive, but very sensitive, control and therefore two-term proportional and integral control is most commonly used.

Direct Digital Control

As an alternative to external or individual controllers, many applications embed the control algorithms into the stored programmes of their control systems. The measurement signals are connected to analogue inputs, the modulating valve is connected to an analogue output, and the set-points and tuning constants are entered through the operator's workstation. Most control system suppliers provide a standard function within their system to compute the output required from a PID algorithm. This solution has the advantage that it is easy to switch the loop on and off

TABLE IV

Commonly used sensors in the dairy industry

Variable	Sensor used	Application
Temperature	Platinum resistance thermometer	General use vessels and pipelines
Level	Conductivity switches Vibronic switches	Vessels (full, empty, specific points)
	Pressure { resistive/capacitive/ Piezoresistive/DP capacitance }	Vessels (contents of)
Pressure	resistive/capacitive Piezoresistive	General use vessels and pipelines
Flow (rate or volume)	Electromagnetic (analogue/pulse)	General use product and CIP/water
Weight	Load cells	Vessels, filling containers
Mass/density	Mass meter	Product mass flow Product density
Conductivity	Electrode Inductive }	Product interfaces CIP interfaces/strength
Turbidity	Turbidity	Product interface CIP interface
pH	pH	Product quality/QA Incubation

and to implement direct control of the output which may be necessary when a process has complex start-up and shutdown requirements. It is often called direct digital control (DDC).

Sensors

Although control technology has made advances, there have been few major developments in the use of on-line sensors for process measurement. Table IV lists the most commonly used sensors in the dairy process.

PLANT OPERATION

Most dairy plants require a mix of local and central operation.

Local Control Station

Where a manual function is required on the factory floor, it will usually be necessary to provide communications with the automatic control system. This is achieved through a local signal panel electrically wired to the I/O system of the control system. In its simplest form, it comprises push-buttons which the operator presses to select a function or acknowledge an action, and indicator lamps driven by the control system to inform the operator of plant status or to act as a prompt for his action. For instance, milk reception may require the road tanker driver to connect flexible hoses and then press a start button to activate the logic within the control system. The control system will respond by illuminating an 'OK' lamp or, if the connection is not properly made, by flashing an alarm lamp if there is an interlock preventing operation. In more complex operations where audit trails are in operation, the tanker driver may be required to enter a security code or identify the source of the raw milk. The local panel will therefore require an information display and data keypad.

Certain unit processes make considerable use of the display/keypad operator station as this provides a better quality of information at a much lower cost than conventional push-button and lamps. The operator functions are usually laid out in the form of a menu with prompts and help information in full descriptive form rather than encoded into a flashing lamp. Figure 21 shows an example of local control panels.

Fig. 21. Local panel operated by menu selections.

Centrally Located Control Rooms

Most modern dairies have a control centre, usually in a clean room overlooking the processing area. Where the control centre also houses the electronics, it is common practice for environmental controls to be installed. In the 1970s these control rooms were dominated by a process mimic diagram, essentially a copy of the process and instrumentation diagram for the plant. Each active component is represented by a lamp (or light emitting diode) for digital items or an indicator for analogue items. Electronic graphics systems are more flexible, smaller, easier to modify and have a much lower cost. They have largely replaced the mimic diagram. Figure 22 shows a central control room.

Fig. 22. Operator control room for a large dairy.

Process Graphics

Many dairies prefer to use the graphic station to provide a mirror image of the process plant configuration. This requires the operator to interpret the data presented. Other systems use the intelligence of the graphics to provide the operator with a high level overview of plant operations backed up by detailed plant graphics if required. Figure 23 shows examples of process graphics for dairy processes that use a windowing technique to generate displays of each area of the plant from a master menu.

Plant Maintenance and Service

Supervisory systems are often configured to provide plant maintenance staff with detailed historical reports of all plant equipment. These reports highlight operational times, faults and service actions. Systems with large data storage facilities can extend these functions to predict when preventative action is required, electronically record on alarms and events for retrieval and analysis, and provide in-line documentation to avoid the need for carrying operating instruction manuals.

Control systems can easily be linked into the public or factory telephone system through a device called a modem. This control system can therefore contact a maintenance or service engineer by automatically dialling the appropriate number. Using either a terminal or a personal computer, the control system can be remotely accessed to investigate

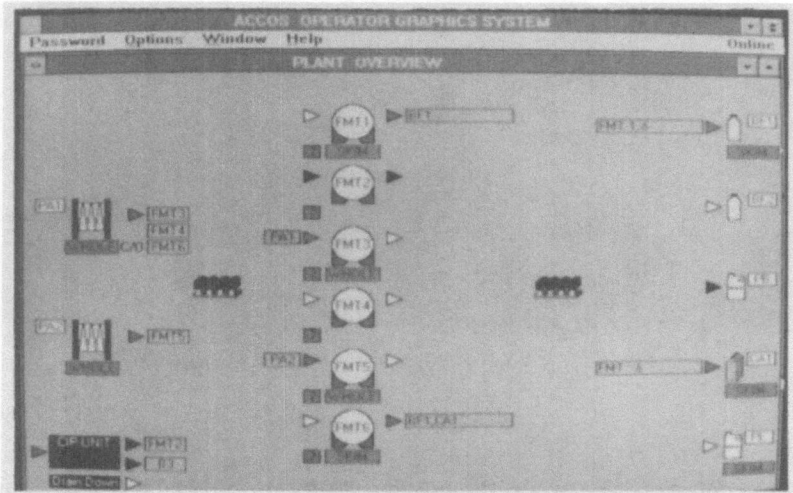

Fig. 23. Process graphics for monitoring and operation of dairy plant.

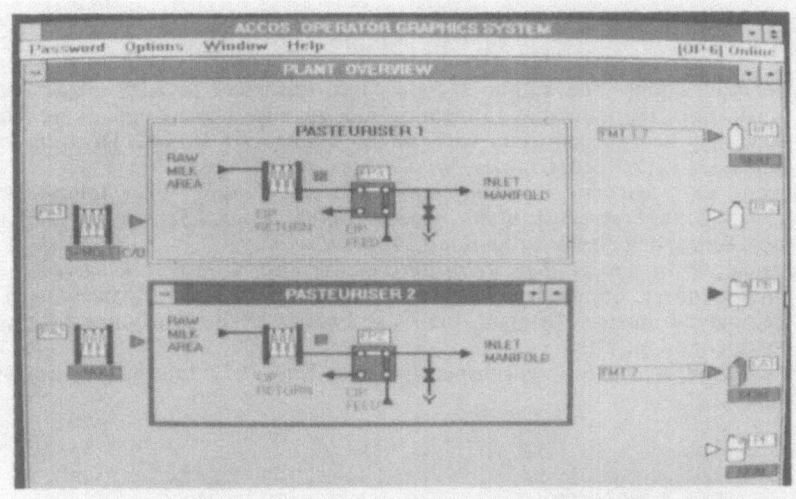

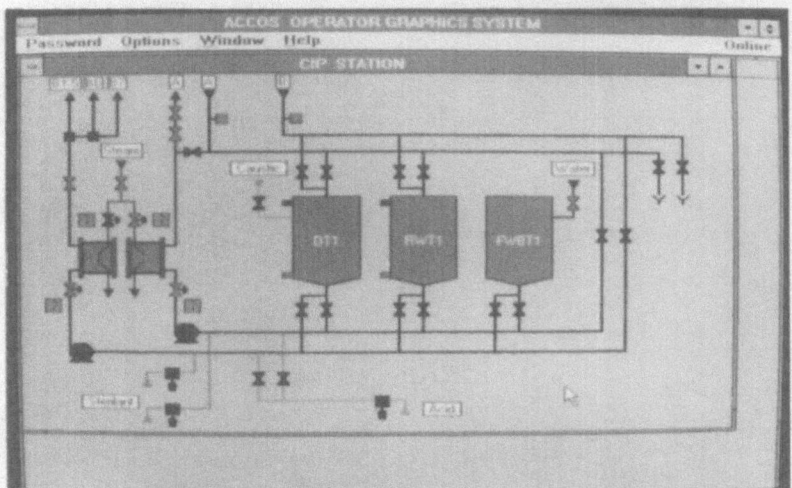

Fig. 23 *Contd.*

the fault reported. This is particularly useful where overnight CIP requirements are preprogrammed and left to run with an unmanned control room. The plant operator can be paged if a fault which holds up CIP is detected.

FURTHER READING

APV Publications. The author acknowledges references to APV papers and publications too numerous to mention individually. These company publications may be obtained from: APV Baker Automation Process Division, PO Box 4, Gatwick Road, Crawley, West Sussex, RH10 2QB.

Bibbero, R. J. *Microprocessors in Instruments and Control*, Wiley Interscience.

Jakeman, C. In: *Proceedings of 8th World Congress of Food Science and Technology*, Scheduling Needs of the Food Process Industry.

Jakeman, C. In: *Proceedings of Institute of Chemical Engineers Conference*, St John's College, Cambridge, 30 March–1 April, 1992. Food Engineering in a Computer Climate – Bridging the Gap. Process Plant Scheduling for both Design and Operation.

Kirkland, W. *et al. Food Processing Magazine*, May 1992, Information Revolution.

Index

N-acetyl-β-glucosaminidase 14
Acid casein 377
  manufacture 379–81
  uses of 404
Acidity 227–9
Activation energy 4–5, 39
Adhesives, casein 397–401
Aerosol cream 97
After-crystallisation process 252
Age thickening 43–5
Agglomeration process 235–9, 243
Air cooling system 187
Air distributor 195
Air filtration 182–3
Air heating system 186–93
Air inlet temperature 231–2
Air outlet temperature 232–3
Alcohol production from whey
  streams 347–9
Alcohol stability 53
Aldehydes 52
Alfablend process 142
Alkaline phosphatase 14
Alumina membranes 280
American Dry Milk Institute (ADMI)
  220–1, 225–7
Amino-acids
  biological availability 395
  chemical analysis 395
  loss of 13–14

Ammonium caseinate 386
Ammonium lactate 349–50
Analogue inputs 436
Analogue outputs 436
Anhydrous milk fat (AMF) 107, 133
  preparation 116
  production 76
Animal feeds, casein in 404
Arrhenius equation 4
Arthrobacter 15
Aseptic packaging 54–5, 95
Aspergillus niger 351, 355, 356, 362
Aspergillus oryzae 355
Atomisation degree 233
Atomisation speed 237
Atomiser 184–6, 195, 249
Automatic control 434–5
  see also Control systems
Automation 433–72
  see also Control systems

Baby food powder 239–43
Bacillus 15
Bacillus cereus 9, 15
Bacillus stearothermophilus 24, 31,
  356
Bacterial proteinases 15
Bacteriological contamination 199,
  242
Bacteriological estimate 226

473

Bacteriological quality
  infant feeds 242
  skim-milk 307
Bactofuges 81
Bag filter, fire-fighting in 260
Bag-in-the-box systems 55, 99
Baked products, casein in 408
Bifidobacterium 363
Bifidus factor 363
Binders 364
Biogas, production of 349
Biological oxygen demand (BOD)
    245, 295, 348
Biological value (BV) 395
Bi-pro 338–9
Blood serum albumin 41
Bregott 140, 142
Bulk density 230–1, 234, 236
  determination 226–7
Burners, low NO$_x$ 191, 193
Butanol production 351
Butter
  bulk repackaging 130–1
  comminution 131
  handling and packing 127–30
  hardness 137
  oxidative rancidity 130
  plastic packaging 130
  reblending 131
  recombined 132–7
  spreadable 137–9
  structure of 107
  see also Buttermaking
Butter cream 94
Butter mixer/blenders 131
Butter oil 238
Butter silos 127–8
Buttergrains 116, 119, 121
Buttermaking 107–58
  alternative processes for ripened
      butter 124–6
  batch 107, 115–16, 122, 127, 131
  continuous 107, 108, 116, 123, 127,
      131
  control of 126–7
  conversion of cream to butter
      115–22

microbiological aspects 127
preparation of ripened cream
    111–14
preparation of sweet cream 108–11
principles of 108–32
process parameters 126–7
salted butter 122–4
transfer of cream to buttermaker
    114–15
see also Butter; Cream
Buttermilk 116, 119, 121, 122, 127,
    153, 243–4

Calcium 242
Calcium caseinate 386, 387, 391
Calcium ion concentration 53
Calcium phosphate 53
Campylobacter 9
Candida curvata 350
Carbery process 347–9
Carbohydrate 242
Carbohydrate addition 101
Carbon disulphide 50
Carbonyl sulphide 50
Casein 375–432
  adhesives 397–401
  amino-acid analysis 394–5
  composition of 388–9
  co-precipitates 381–2
    composition of 390–1
    manufacture of 388
    properties of 391–5
  edible products 406–17
  exports and imports 397
  extraction from milk 376–82
  fibre production 402
  flavour of casein products 394
  heavy metal derivatives 387
  in animal feeds 404
  in baked products 408
  in cereals 408
  in cheesemaking 415–16
  in coffee whiteners 410–11
  in confectionery 414–15
  in creamers 410–11
  in cultured milk products, soups
      and gravies 412

in frozen desserts 409–10
in high-fat powders, shortenings
and spreads 413
in ice cream 409–10
in imitation milk beverages 412
in infant foods 413–14
in instant breakfast formulations
411
in leather industry 403
in meat products 407–8
in paints 402–3
in pasta products and snack foods
414
in pet foods 404
in pharmaceutical products 416–17
in whipped toppings 408–9
industrial applications 419
industrial uses 397–405
manufacturing processes 378
new developments 419–20
nutritional properties 394–5, 412
precipitation 217
production statistics 395–7
properties of 389
stability 216
texturisation applications 393–4
whipping and foaming applications
393
yield from skim-milk 381
see also Acid casein; Rennet casein
α_{s1}-Casein 44, 53
β-Casein 44, 53
Casein-based aqueous phases 153
κ-Casein 44
Casein fraction of milk 2
Casein micelles 43, 44
Caseinates 153
composition of 389
functional properties 391–2
gelation 392–3
manufacture of 382–7
properties of 391–5
solubility in water 391
viscosity 392–3
water absorption 392
Cellulose acetate membranes 279,
298, 304

Central processing unit (CPU) 439
Centrifugal atomiser 260
Centrifugal separators 65–7
disc configuration 75
Centripetal pumps 67–9
Ceramic membranes 342–3
Cereals, casein in 408
Cheese whey solids 245
Cheesemaking
casein in 415–16
reverse osmosis 305
ultrafiltration 293–5
whey disposal in 244
Chemical parameter (C^*) 36
Chemical reactions, reaction kinetic
parameters 7
Chlorine 242
Citric acid production 350–1
Cladosporium 151
Clarifiers 81
Clarifixators 81
Cleaning-in-place (CIP) 79, 210–12,
433, 471
Clostridium botulinum 4, 23
Clostridium perfringens 9
Clotted cream 93
Clotting properties 45
Clover 142, 144
Coagulation 43–5
Coffee cream 90, 95
Coffee whiteners, casein in 410–11
Colour of milk 42–3
Computer integrated manufacture
(CIM) 449
Computer programmes 455–8
Concentrate heating 217–19, 234
Concentrated milk 162
Concentration polarisation 276
Condensate 172
Condensers 169
Confectionery, casein in 414–15
Confectionery cream 94
Control loops 463–5
Control software 455–8
Control systems
centrally located control rooms
469

Control systems—*contd.*
configuration 440, 444
construction 438–9
dairy plant 441–7
design and architecture 435
distributed input/output 439–40, 444
electrical signals 436
input/output configurations 436–8
input/output modules 438
local control station 467
plant maintenance and service 470–1
plant operation 467–71
process graphics 469–70
process plant components 435–6
unit processes and machinery 452–5
Cooked flavour in milk 45–52
Corning process 356
Counterions 308
Cream
bulk standardisation 82
coagulation 85
consumer products 90–4
conversion to butter 115–22
cryogenic freezing of 100–1
density 84
drying 101–2
fat globules 62
freezing 98–101
heat treatment 85
homogenisation of 92
microbiological aspects 127
milkfat content 84
pasteurisation 85–90, 94, 108–9, 111
physico-chemical properties 61
powder 102
preservation and packing 94–103
rapid freezing 99–101
recombined 102
reconstituted 102
separation and processing developments 61–105
solids-not-fat (SNF) content 90
souring 85
standards 61, 81–4

sterilisation 94–8
summer 112, 113
transfer to buttermaker 114–15
treatment modifications 112
UHT 95–8
whipping properties 91–2
winter 112, 113
Cream cheese separators 81
Cream concentrators 80
Cream liqueurs 103
Cream processing tank 110
Creamers, casein in 410–11
Creaming process 63
Critical self-ignition temperature 257–8
Cryogenic freezing of cream 100–1
Cryogenic freezing tunnel 100
Crystallisation process 247–8, 251
Crystallisation tank 248
Cultured cream 90, 412
Cultured dairy foods 413
Cyclone separators 197–8
losses from 198
L-cysteine 51
L-cystine 51

Dairy based spreads 139–56
batch process 140, 147
continuous techniques 140, 147
definition 107
low-fat 147–53
reduced-fat 144–5
very-low-fat 155–6
Data processing and storage requirements 447
Delight Extra Low 155
Demineralisation of whey 310, 330–2
Denaturation of whey protein 12–13, 40–1, 219–20, 222, 325
Densitometer 84
Desludging type separators 81
Diacetyl 111, 112
Diafiltration 278
Differential scanning calorimetry (DSC) 41, 113
Digital inputs 436
Digital outputs 435
Dimethyl sulphide 50

Direct digital control (DDC) 465–7
Dispersibility 235
Disulphide groups 48–9
Dornic degrees 228–9
Double cream 93
Drying chamber 193–7
    active volume of 268
    conical bottom 193–4
    fire-fighting in 258–60
    flat bottom 193–4
    inlet temperature 258
    insulation 195
    volume 195
Drying process
    cream 101–2
    lactose-hydrolysed products 359
    milk and milk products 159–254
        advantage 159
        disadvantage 159
        history and development 159–61
        matching evaporator and dryer
            173–5
        roller drying 179–81
        spray drying 181–2
        whey based products 328–9
Ducts, explosion suppression in 270
Dust collection 200
Dust explosion conditions 255–6

Earthing of plant components 258
Economiser 187–8
Electrical energy conversion to heat
    56
Electrodialysis 273–4, 307–10
    application 310
    concentration polarisation 309
    membrane configurations 309
    principles of 307–9
    whey 330
Energy saving, reverse osmosis 306
Enzymes 14, 44
    reaction kinetic parameters 7
Escherichia coli 9
Ethanol production 348
Evaporated milk 159, 162
Evaporation
    whey 245
    see also Vacuum evaporation

Evaporators
    time-temperature conditions 223
    see also Vacuum evaporators
Explosion doors 264–7
Explosion pressure, dimensioning for
    261
Explosion pressure relief device
    262–9
Explosion protection in spray dryers
    255–72
Explosion reaction force 267
Explosion suppression 261–2
    in duct 270

F_0 value 24
Fans, vibration control 261
Fat content
    measurement of 82–4
    monitoring 84
Fat emulsification 393
Fat globules
    in cream 62
    in milk 62
    size distribution 78
    theoretical velocities of 63
    velocity of 75
Feed pump 184
Fermentation
    lactose utilisation by 322
    of whey 346–52
Fermented products, milk
    concentrates in 295
Filter bags 201, 203
Filtermat dryer 101, 215–16
Finisher 171
Fire-fighting
    fluid bed 260
    in bag filter 260
    in drying chamber 258–60
    in external fluid bed 260
    instructions and training 270–1
    personnel 270–1
Fire protection in spray dryers
    255–72
Flame arresters 270
Flash pasteurisation 11
Flat sheet membrane modules 282–3
Flavour changes of milk 45–52

Flavour changes of milk—*contd.*
  primary phase 46
  secondary phase 46–8
Flavour improvement in UHT milks
    50–2
Flavour of milk 45–9
Flavour taints 86
Flavour volatiles in UHT milk 50
Fluid bed, fire-fighting 260
Fouling of heat exchanger surface 52
Fractionation processes 273–4
Freezers 99
Freezing of cream 98–101
Frozen desserts, casein in 409–10
α-Fucosidase 14
Fuel-bound nitrogen 190
Fuel gas, production 349
Furosine 39

Gamma irradiation 55
Gas-fired air heaters 188–92
Gelation 43–5
Gluconic acid 364
Gluconobacter oxydans 364
γ-glutamyl transpeptidase 14
Goat's milk 54
Gold Lowest 155
Grafset 463
Granular sodium caseinate 385
Gravies 413

Half-cream 90
Hazard Analysis Critical Control
    Points (HACCP) 55–6
Heat classification 219, 229
Heat exchangers 203–4, 285
  fouling 52–4
Heat recovery 203–4
Heat treatment of milk 1–60
  batch process 16–20
  changes occurring in 2
  chemical changes in 39
  continuous processing 16–20
  effects of raw milk quality on heat
    stability 8–9
  effects on colour 42–3
  effects on micro-organisms 3–4
  process development 56

purpose of 1–2
  reactions taking place in 39
  variation of heat resistance with
    temperature 4–9
Hermetic separators 69–71, 80
High fat creams 93–4
High pressure liquid
    chromotography (HPLC) 41
High temperature-short time (HTST)
    process 6, 14–17, 19, 108
  plant layout 18
  typical results 22
Homogenisation 14, 42
Homogenising effect 234
Human breast milk 240–2
Hydrogen peroxide 55
Hydrogen sulphide 50, 51
Hydroxymethyl furfural (HMF) 25, 39

Ice cream 305
  casein in 409–10
IDF Standard 87 238–9
Ignition sources 256
Ignition temperature 256–8
  for dust cloud 256
  in powder layer 256
Imitation milk beverages, casein in
    412
Infant foods, casein in 413–14
Inorganic membranes 279–80
Instant breakfast formulations, casein
    in 411
Instant milk powder 235–9
Instrumentation 463–7
Ion exchange, whey 330, 332
Ion selective membranes 307
Iron deficiency 242

Just-Hatmaker process 179–80

Kjeldahl method 221
Kluyveromyces marxianus var. *lactis*
    346
Kluyveromyces marxianus var.
    marxianus 346–8, 362

Lactalbumin 219, 249, 333–4
α-Lactalbumin 41, 50, 334, 336

Lactic acid 85, 90, 112, 228, 249, 250, 349–50
Lactic acid casein 377, 380–1
Lactitol 362
Lactobacillus delbrueckii subspecies *bulgaricus* 350
Lactobacillus helveticus 124
Lactobionic acid 363
Lactococcus diactylactis 111
Lactococcus lactis 124, 228
Lactoglobulin 219, 249
β-Lactoglobulin 40, 41, 44, 50, 53, 334, 336
Lactoperoxidase 14
Lactose 247, 251
α-Lactose 247
β-Lactose 247
Lactose hydrolysis 352–60
  acid hydrolysis/cation exchange 353
  determination of degree 358
  drying of hydrolysed products 359
  enzymatic hydrolysis 353–4
  immobilised enzyme systems 355
    sanitation 357
  regulatory aspects 360
  single-use enzymes 354
  storage and properties of hydrolysed syrups 359–60
  sweetness of hydrolysed syrups 358–9
  ultrafiltration 354–5
Lactose production 360–1
Lactose utilisation
  and chemical derivatives 361–5
  by fermentation 322
Lactosyl urea 361–2
Lactotococcus lactis subspecies *cremoris* 111
Lactulose 362–3
Ladder logic 460–3
Latt och Lagom (L & L) 147
Leather industry, casein in 403
Lecithin 238
Lethality 8, 24
Lethality profile 8, 24
Leuconostoc mesenteroides subspecies *cremoris* 111, 124

Liquid phase heating 187
Local area networks (LANs) 444
Lysino-alanie 39

Magnesium 242
Magnesium caseinate 387
Management information 447–9
α-Mannosidase 14
Manufacturing automation protocol (MAP) 444
Margarines 139
Mastitis 9
Meat products, casein in 407–8
Membrane cleaning 292, 304
Membrane deposits 292, 304
Membrane flux decline 341
Membrane fouling 341, 342
Membrane geometry
  electrodialysis 309
  reverse osmosis 299
  ultrafiltration 281
Membrane modules
  reverse osmosis 299
  ultrafiltration 281–4
Membrane processing 273–311
  physical properties 274
  properties of milk relevant to 274–5
  whey 339–46
Membranes
  reverse osmosis 298–9
  ultrafiltration 279–84
Methanethiol 50
Methyl ketones 51
Microbacterium 15
Microbial growth during ultrafiltration 292
Microbial quality 55
Microbiological parameter (B^*) 36
Microfiltration 273–4, 306–7
  fat separation 307
  skim-milk 307
Micro-organisms
  effect of heat 3–4
  inactivation of 4
  reaction kinetic parameters 7
Mikromaster 2000S 146

Milk
  casein fraction of 2
  drying. *See* Drying process
  fat globules in 62
  heat treatment. *See* Heat treatment
    of milk
  nutritional value 2
  protein coagulation in 2
  sensory characteristics 2
Milk-based drinks, definition 11
Milk Can 54
Milk components, molecular sizes of
  274
Milk concentrates
  in fermented products 295
  reverse osmosis 304–5
  ultrafiltration 293–5
Milk powder 162, 164, 186, 189
  as powder layer/dust cloud 255
  factors affecting properties 227–31
  fat content 194
  free fat content 227, 237
  particle density 237
  porous agglomerates 237
  process conditions affecting
    properties 231–5
  quality 225–52
  reverse osmosis 305
Milk products, drying. *See* Drying
  process
Milk solids, recovery from dairy rinse
  water 305
Milk solids not fat (MSNF) 133, 138
Milkoscans 82
Milkotesters 82, 84
Mineral acid casein 379
Mineral content 242
Mineral fractions 41
Mock cream 94
Moisture content 229, 234, 238
Molecular sizes of milk components
  274
Mucor pusillus protease 340
Multiple-effect evaporator 172
MVR evaporators 165, 175–8
Mycobacterium tuberculosis 12, 14

Nanofiltration 273

Natural circulation evaporator 163–4
Net protein utilisation (NPU) 395
Nitrates 189, 192
Nitrites 189, 192
Nitrogen oxides 189–91
Nitrosamines 189, 192
NIZO process 124–6
Nozzle atomiser 184–6, 195
Nozzle-bowl centrifuges 81
Nuclear magnetic resonance (NMR)
  113
Nutritional value 41–2

Ohmic heating 56
Oil-fired air heaters 188
Oil production 350
Oligosaccharide formation 357–8

P* value 20
Paints, casein in 402–3
Paper coating 400
Paring discs 67–9
Pasta products 414
Pasteurisation 9–23
  cream 85–90, 94, 108–9, 111
  definition 10
  design, operation and control
    19–20
  microbiological standards 12
  plant layout 18
  processing conditions 11–16
  statutory regulations 12
  time-temperature combinations
    11–12
Pasteurisation unit (PU) 20
Pearson's square 81
Penicillium 151
Permeate flux 270–1, 275
Pet foods, casein in 404
Pharmaceutical products, casein in
  416–17
Phosphorus 242
Pillsbury dryer 215–16
Plant maintenance 470–1
Plant operation, control systems
  467–71
Plant utilisation 449–52
Plate evaporator 163

Plate heat exchanger 163
Plug flow 20
Pneumatic powder conveying system 203
Polyamide membranes 279
Polysaccharides, production of 351
Pop-out panels 264
Post-processing contamination (PPC) 15
Potassium 242
Potassium caseinate 386
Potassium iodate 50
Powder manufacture to certain heat classification 219–25
Powder recovery system 197–203
Preheaters 169
Pressure relief area, calculation of 267
Primary explosion 264
Process graphics 469–70
Process isolation 269–70
Process languages 458–60
Production control 447–9
Production scheduling 449–52
Programme logic systems 434
Proportional, integral and derivative action (PID) control 463–5
Protein coagulation in milk 2
Protein efficiency ratio (PER) 395
Protein recovery and modification 333–9
Pseudomonas fluorescens 9, 44
PSM buttermaker 140
Pumps 172
Purity test 226
Pyridosine 39

Q_{10} value 6–8
Quality standards and procedures 447

Radionucleotides in whey 314
Raw milk
  bacteriological quality 227–8
  effect of quality on heat stability 8–9
  effect on separation efficiency 78
  quality of 9–10
Regeneration efficiency (RE) 17–18

Rennet casein 377, 378, 392
  manufacture 381
  uses of 405–6
Residence time 20–3
  bacteriologically effective 23
  measuring technique 22
Reverse osmosis 273–4
  cheesemaking 305
  composition of the product 303–4
  energy saving 306
  engineering equipment 300–1
  flow velocity 302–3
  membrane geometry 299
  membrane modules 299
  membranes 298–9
  milk 301–4
  milk concentrates 304–5
  milk powder 305
  operating pressure 302–3
  plant design 299–301
  principles of 296–8
  temperature effect 302–3
  whey 321, 331, 339–46
Reynolds number 21
Ripened butter, alternative processes for 124–6
Ripened cream, preparation of 111–14
Roller drying 179–81
  milk 240
  sodium caseinate 384–5
Rotary-drum freezers 99
Rotary sluice valves 270
Rotating atomiser 184–6, 195
Rupture discs 264

Saccharomyces cerevisiae 347
Saccharomyces lactis 357
St Ivel Gold 151–2
Salmonella 9
Salted butter 122–4
Scopulariopsis 355
Scorched particle test 226
Scorched particles, causes of 230
Scraped-surface heat exchangers (SSGH) 133–6, 139, 149–50, 155
Secondary explosion 264

Sediment tests 203, 226
Seeding 249
Self-desludging separators 71–3
Self-ignition 257–8
Self-tuning intelligent controllers 465
Sensors 467
Separation process 63
    automation and control of 79–80
    bowl speed effect 75
    factors affecting 73–8
    flow rate effects 76–8
    raw milk effects 78
    temperature effects 74–5
Separators 65–73
    advances in 79
    bowl speed 75
    commercial types 79
    disc-bowl 81
    disc configuration 75–6
    modified 81
    non-hermetic 77
    placement of channels in discs 76
    semi-open 79
    Soft-Stream system 77
Simon-Freres Contimab machines
    122
Single-cell protein 346–7
Single cream 90
Single-use enzymes 354
Sizing 401
Skim-milk 42, 73–6, 153, 173, 175
    bacterial quality of 307
    casein yield from 381
    fat contents 78
    microfiltration 307
    ultrafiltration 291
Skim-milk powder 133, 189–90, 194,
    219, 247
    ignition temperature 256
    self-ignition 257–8
Slide valve 269
Snack foods 414
Sodium 242
Sodium caseinate 101, 153, 189, 391,
    392, 404, 408–10, 412, 413
    extruded 386
    fat emulsification 393
    granular 385

preparation of concentrated
    solutions 385–6
roller dried 384–5
spray dried 382–4
Sodium citrate 153
Sodium hydroxide 53, 228
Sodium phosphate 153
Software test and simulation 463
Solubility 230, 234
Solubility index 226, 230
Solubility test 230
Soups 413
Sour cream 30, 412
Sour cream buttermilk 243
Soxhlet Henkel degrees 228
Spherosil 338
Spirally wound membrane modules
    283–4
Spoilage rates 55
Spray balls 210
Spray bed dryer 212–15
Spray dryers 192
    components of 182–205
    protection against fire and
        explosion 255–72
    single-stage 182
    thermal efficiency 204–5
Spray drying 181–2, 192
    butter milk 243
    cream 101
    principle of 184
    sodium caseinate 382–4
    whey 245
Spreadable butters 137–9
Spreads. *See* Dairy based spreads
Standard plate count 226
Staphylococcus 9
Starter cultures 111–12, 124, 126
Sterilisation 55
    cream 94–8
    milk 23–8
        UK heat treatment regulations
            27–8
    ultra-high temperature (UHT) 95
Sticking process 250
Sticking temperature of whey powder
    250
Stokes' Law 63

Straight-through process 236
Streamline flow 20, 21
Streptococcus salivarius subspecies
 thermophilus 15
Sulphur components in milk 50–2
Sulphydryl compounds 225
Sulphydryl disulphide 50
Sulphydryl groups 48–9
Sulphydryl interactions 344
Sulphydryl oxidase 51
Sumitomo system 356
Sweet cream
 preparation of 108–11
 thermal processes 114
Sweet curdling 15
Sweetened condensed milk 159, 161,
 162

Tall form spray dryers 212
Temperature-time profiles 8, 24, 34–6
Tetra Pak system 97, 98
Textile filters 200–1
Texture of milk 43
Thermal efficiency 204–5
Thermisation 10
Thermocompressors 174
Thermoplasticity 250
Three-stage drying 207–9
Three-term control 463
Time-temperature profiles 8, 24, 34–6
Titratable acidity 228
Trace metals 242
Trichosporon cutaneum 350
Tubular falling film evaporator
 163–5, 175
Tubular membrane modules 281–2
Turbidity test 27
Turbulent flow 20, 22
TVR evaporators 165–9, 173
Two-stage drying 205–9

Udder disease 9
UHT cream 95–8
 bulk packaging 98
UHT milk 23
 colour of 42
 texture of 43
 thickening and coagulation 43–5

UK regulations 28–30
UHT process 6, 18, 25–7
 comparison of direct and indirect
  processes 34–9
 direct methods 32–4
 first generation 37
 plant 30–2, 55
 second generation 37
 third generation 38
UHT whipping cream 96
UK heat treatment regulations for
 sterilised milk 27–8
UK Milk Special Designation
 Regulations (1989) 11
Ultra-high temperature (UHT)
 process. *See* UHT
Ultrafiltration 245, 273–4
 batch operation 287
 cheesemaking 293–5
 composition of product 290–1
 continuous operation 286–7
 engineering equipment 287–9
 factors determining rate of 276
 flow velocity 289–90
 future developments 310
 growth of bacteria during 292
 lactose hydrolysis 354–5
 membranes for 279–84
 milk concentrates 293–5
 operating pressure 289–90
 plant design 284–9
 plant performance on milk 289–92
 principles of 275–8
 protein concentrate 133
 retention coefficients 290–1
 skim-milk 291
 temperature effects 289–90
 use of permeate 295–6
 whey 339–46

Vacreation 86–90
 disadvantage of 88
 steam quality 88
 taint removal 88
Vacreator 86–90
 new versions 89
Vacuum evaporation 161–78
 history and development 161–2

Vacuum evaporation—*contd.*
  principles of 162–3
Vacuum evaporators
  history and development 162
  matching to dryer 173–5
  principal types 163–78
Vacuum producing equipment 171
Valio system 356
VDI 3673 (1979) 268
Vent ducts 264
Vent ratio method 268
Ventex valve
  detector controlled 269
  explosion pressure controlled 269
Venturi gas-air mixer 191
Viscosity 230
Vitamin A 242
Vitamin C 14, 242, 243
Vitamin D 242
Vitamins 14, 41–2

Wall sweep arrangement 207
Weaving flow 87
Wet cleaning 210
Wet collector 198–9
Wet scrubber 198–9
Wetting agent 239
Wetting rate 172–5
Whey 244–52
  acid 313, 314
  analysis of 323
  changes occurring during storage
    329–30
  characteristics 323–7
  chemical properties 323–4
  clarification 327
  classification 313
  composition 314–17
  concentration 328
  definition 313
  dehydration 328–9
  demineralisation 310, 330–2
  disposal in cheesemaking 244
  effect of heat 324–6
  evaporation 328
  fermentation 346–52
  flavour 324

  lactose in 247
  manufacturing procedures 329
  membrane processes 339–46
  nutritional aspects 326–7
  pH values 313
  physical properties 323–4
  polluting effect 245
  process and product options 321
  processing 327–30
  production 313
  production economics 320–3
  production statistics 317–20
  radionucleotides in 314
  reverse osmosis 321, 331, 339–46
  seasonal variations 314–17
  storage 327
  sweet 313, 314
  ultrafiltration 339–46
  utilisation of 313–73
Whey concentrate, drying 251, 252
Whey powder
  quality 249
  sticking temperature of 250
Whey protein 219
  adsorption onto insoluble supports
    338
  co-precipitation with soluble
    polymers or phosphates 334–9
  denaturation 12–13, 40–1, 219–20,
    222, 325
  functionality 343–6
  isolates by selective precipitation
    334
  modification 339
  recovery and modification 333–9
Whey protein concentrates (WPC)
    242, 245, 343–6
  foaming and emulsifying properties
    345
  gelation strength 344
  manufacture of 346
Whey protein nitrogen (WPN) index
    220–4
Whey solids
  constituents of 245
  utilisation of 245
Whipped toppings, casein in 408–9

Whipping cream 91–3
  factors of importance for 91
  UHT 96
Wingea robertsii 347
Wood glues 397–400

Xanthine oxidase 14

Yield of components 277–8
Yoghurt 305, 413

Z value 5, 20, 36
Zirconium oxide membranes 280, 342